THE COLLECTED PAPERS
OF GERHARD GENTZEN

STUDIES IN LOGIC

AND

THE FOUNDATIONS OF MATHEMATICS

Editors

A. HEYTING, *Amsterdam*
A. MOSTOWSKI, *Warszawa*
A. ROBINSON, *New Haven*
P. SUPPES, *Stanford*

Advisory Editorial Board

Y. BAR-HILLEL, *Jerusalem*
K. L. DE BOUVÈRE, *Santa Clara*
H. HERMES, *Freiburg i. Br.*
J. HINTIKKA, *Helsinki*
J. C. SHEPHERDSON, *Bristol*
E. P. SPECKER, *Zürich*

NORTH-HOLLAND PUBLISHING COMPANY
AMSTERDAM · LONDON

THE COLLECTED PAPERS OF GERHARD GENTZEN

Edited by

M. E. SZABO

Sir George Williams University
Montreal

1969

NORTH-HOLLAND PUBLISHING COMPANY
AMSTERDAM · LONDON

© NORTH-HOLLAND PUBLISHING COMPANY - AMSTERDAM - 1969

All rights reserved. No part of this publication may be reproduced, stored in a retrieval system, or transmitted, in any form or by any means, electronic, mechanical, photocopying, recording or otherwise, without the prior permission of the copyright owner.

Library of Congress Catalog Card Number: 71-97201

Standard Book Number: 7204 2254 X

PUBLISHERS:

NORTH-HOLLAND PUBLISHING COMPANY - AMSTERDAM
NORTH-HOLLAND PUBLISHING COMPANY, LTD. - LONDON

PRINTED IN THE NETHERLANDS

ACKNOWLEDGMENTS

The editor would like to thank Michael Dummett for his stimulating tutorials in 1958 and the initial interest which he created in the work of Gerhard Gentzen; Paul Bernays for contributing towards giving the present volume its final shape by providing unlimited information and counsel, and for suggesting several valuable improvements of the Introduction; John Denton for spending a much appreciated week in the editor's office helping to compare the final manuscript with Gentzen's original texts; C.A.S.A. of Sir George Williams University for making secretarial assistance possible; Mr. M. D. Frank of the North-Holland Publishing Co. for bringing the Gentzen project to its fruition, and H. J. Stomps and Mrs. O. Troelstra-Bakker for their technical expertise in seeing the manuscript through the press; finally, Lothar Collatz, Gentzen's closest friend, for making available what appears to be the only existing photograph of Gerhard Gentzen.

BIBLIOGRAPHY

This volume is based on the following papers of Gerhard Gentzen:

#1 Über die Existenz unabhängiger Axiomensysteme zu unendlichen Satzsystemen, Mathematische Annalen **107** (1932) 329–350 29

#2 Über das Verhältnis zwischen intuitionistischer und klassischer Arithmetik, Galley proof, Mathematische Annalen (1933) received on 15th March 1933. (Kindly made available by Prof. Paul Bernays.) 53

#3 Untersuchungen über das logische Schliessen, Mathematische Zeitschrift **39** (1935) 176–210, 405–431. (Accepted as Inaugural Dissertation by the faculty of mathematics and natural science of the university of Göttingen.) 68

#4 Die Widerspruchsfreiheit der reinen Zahlentheorie, Mathematische Annalen, **112** (1936) 493–565. Appendix: Galley proof of sections IV and V, Mathematische Annalen (1935) received on 11th August 1935. (Galley proof kindly made available by Prof. Paul Bernays.). 132

#5 Die Widerspruchsfreiheit der Stufenlogik, Mathematische Zeitschrift **41**, No. 3, (1936) 357–366. 214

#6 Der Unendlichkeitsbegriff in der Mathematik, Semester-Berichte, Münster in/W., 9th Semester, Winter 1936–37, 65–80. 223

#7 Die gegenwärtige Lage in der mathematischen Grundlagenforschung, Forschungen zur Logik und zur Grundlegung der exakten Wissenschaften, New Series, No. 4, Leipzig (Hirzel), (1938) 5–18; also: Deutsche Mathematik **3** (1939) 255–268. 234

#8 Neue Fassung des Widerspruchsfreiheitsbeweises für die reine Zahlentheorie, Forschungen zur Logik und zur Grundlegung der exakten Wissenschaften, New Series, No. 4, Leipzig (Hirzel), (1938) 19–44. 252

#9 Beweisbarkeit und Unbeweisbarkeit von Anfangsfällen der transfiniten Induktion in der reinen Zahlentheorie, Mathematische Annalen **119**, No. 1, (1943) 140–161. (Submitted as D. Phil. Habil. thesis to the faculty of mathematics and natural science of the university of Göttingen.). 287

#10 Zusammenfassung von mehreren vollständigen Induktionen zu einer einzigen, published posthumously in the Archiv für mathematische Logik und Grundlagenforschung **2**, No. 1, (1954) 1–3. (Dedicated to H. Scholz on his 60th birthday, 17th December 1944). 309

BIOGRAPHICAL SKETCH

Gerhard Gentzen was born in Greifswald, Pomerania, on November 24th, 1909. He spent his childhood in Bergen on the Isle of Rügen in the Baltic Sea, where his father was practicing law. There he attended elementary school and the local *Realgymnasium*. After his father's death in the First World War, his mother decided in 1920 to move to Stralsund, where Gentzen completed his secondary education at the *Humanistische Gymnasium*. On February 29th, 1928, he was granted the *Abitur* with distinction, having attained the highest academic standing in his school and, on the recommendation of his headmaster, he received a university scholarship from the *Deutsche Studentenwerk* enabling him to continue his higher education.

Even as a young boy, Gentzen is said to have displayed exceptional mathematical ability and had declared categorically that the only subject which he would ever be able to study was mathematics. He enrolled at the University of Greifswald for two semesters and there earned Hans Kneser's respect as 'a particularly gifted student'. From Greifswald Gentzen went to Göttingen, where he matriculated for the first time on April 22nd, 1929. After two semesters he went to Munich, studied there for one semester, and after a further semester at Berlin, he finally returned to Göttingen and worked under Hermann Weyl. Five semesters later, in the summer of 1933, Gentzen sat his *Staatsexamen* and, at the age of twenty-three, was granted a doctorate in mathematics. The great mental strain which his studies had involved and his delicate constitution forced him to interrupt his academic career and to return home for an extended period of rest.

The major turning point in Gentzen's academic life came undoubtedly with his appointment in 1934 as Hilbert's assistant in Göttingen, where he continued to work even after Hilbert's retirement. During these years Gentzen published some of his most important papers and was also given the responsible task of reviewing numerous works of eminent researchers from many countries for the *Zentralblatt für Mathematik*. These reviews attest his extraordinary range of interest and the great extent of his involvement in the international community of scholars. In 1937, he was invited to the Philosophical Congress in Paris and delivered an address on the 'Concept of Infinity and the Consistency of Mathematics'.

At the outbreak of the Second World War, Gentzen was conscripted into the armed forces and was given an assignment in Telecommunications in Braunschweig. Within two years he became seriously ill and spent three months in a military hospital. Upon his release, he was freed from military service for reasons of ill health. After a period of rest, he rejoined the University of Göttingen, where in 1942 he attained the Dr. phil. habil. degree for his papers on the 'Provability and Nonprovability of Restricted Transfinite Induction in Elementary Number Theory'.

Upon the request of the director of the Mathematical Institute of the German University of Prague, Gentzen was subsequently appointed *Dozent* at that University in the autumn of 1943. He taught there until, on May 5th, 1945, he and all other professors at the University were taken into custody by the new local authorities. On August 4th, 1945 amid the turmoil and confusion that must have marked that period, Gentzen died tragically in his cell of malnutrition after several months of extreme physical hardship. One of his friends writes: "I can still see him lying on his wooden bunk thinking all day about the ((mathematical)) problems which preoccupied him. He once confided in me that he was really quite contented since now he had at last time to think about a consistency proof for analysis. He was in fact fully convinced that he would succeed in carrying out such a proof. He also concerned himself with other questions such as that of an artificial language, etc. Now and then he would give a short talk . . . . We were continually reassured that the formalities of our release would take only a few days longer . . . . He was hoping to be able to return to Göttingen and devote himself fully to the study of mathematical logic and the foundations of mathematics. He was dreaming of an Institute for this purpose, perhaps together with H. Scholz . . . ."

M.E.S.

TABLE OF CONTENTS

ACKNOWLEDGMENTS V
BIBLIOGRAPHY VI
BIOGRAPHICAL SKETCH VII
TABLE OF CONTENTS IX
INTRODUCTION 1
   NOTES TO THE INTRODUCTION 24

#1. ON THE EXISTENCE OF INDEPENDENT AXIOM SYSTEMS FOR INFINITE SENTENCE SYSTEMS. 29

   SECTION I. Notations and some lemmas 29
      § 1. The 'sentences'. 29
      § 2. The 'proof' of a 'sentence' from other 'sentences'. 30
      § 3. The equivalence of our concept of provability with that of P. Hertz 32
      § 4. The correctness and completeness of the forms of inference, the normal proof 33
      § 5. Systems of sentences 38
      § 6. The independence of axiom systems 39
      § 7. Maximal nets 41

   SECTION II. An infinite closed sentence system possessing no independent axiom system. 42
      § 1. Construction of the paradigm $\overline{\mathfrak{A}}$ 42
      § 2. Some lemmas about the system $\overline{\mathfrak{A}}$. 43
      § 3. Proof of the nonexistence of an independent axiom system for $\overline{\mathfrak{A}}$ 45

   SECTION III. Construction of an independent axiom system for a given denumerably infinite closed linear sentence system. 46
      § 1. Outline of procedure 46
      § 2. Construction of the axiom system $\mathfrak{T}$ 46
      § 3. The system $\mathfrak{T}$ is an axiom system of $\overline{\mathfrak{S}}$ 48
      § 4. The system $\mathfrak{T}$ is independent. 50

#2. ON THE RELATION BETWEEN INTUITIONIST AND CLASSICAL ARITHMETIC. 53
   Introduction 53

§ 1. Terminology and notations 53
§ 2. The formal structure of arithmetic 55
§ 3. The intuitionist validity of the law of double negation. 58
§ 4. Transformation of proofs of classical arithmetic into proofs of intuitionist arithmetic 60
§ 5. Consequences 65
§ 6. The consistency of arithmetic – the redundancy of negation in intuitionist arithmetic 66

#3. INVESTIGATIONS INTO LOGICAL DEDUCTION

SYNOPSIS 68

SECTION I. Terminology and notations 69

SECTION II. The calculus of natural deduction 74
§ 1. Examples of natural deduction 74
§ 2. Construction of the calculus *NJ* 75
§ 3. Informal sense of *NJ*-inference figures 78
§ 4. The three examples of § 1 written as *NJ*-derivations 79
§ 5. Some remarks concerning the calculus *NJ*. The calculus *NK* . . . 80

SECTION III. The deductive calculi *LJ*, *LK* and the *Hauptsatz* 81
§ 1. The calculi *LJ* and *LK* 81
§ 2. Some remarks concerning the calculi *LJ* and *LK* – the *Hauptsatz*. . 85
§ 3. Proof of the *Hauptsatz* 88

SECTION IV. Some applications of the *Hauptsatz* 103
§ 1. Applications of the *Hauptsatz* in propositional logic 103
§ 2. A sharpened form of the *Hauptsatz* for classical predicate logic. . . 106
§ 3. Application of the *sharpened Hauptsatz* to a new consistency proof for arithmetic without complete induction 110

SECTION V. The equivalence of the new calculi *NJ*, *NK*, and *LJ*, *LK* with a calculus modelled on the formalism of Hilbert 115
§ 1. The concept of equivalence 115
§ 2. A logistic calculus according to Hilbert and Glivenko 116
§ 3. Transformation of an *LHJ*-derivation into an equivalent *NJ*-derivation 117
§ 4. Transformation of an *NJ*-derivation into an equivalent *LJ*-derivation 120
§ 5. Transformation of an *LJ*-derivation into an equivalent *LHJ*-derivation. 123
§ 6. The equivalence of the calculi *LHK*, *NK*, and *LK* 128

#4. THE CONSISTENCY OF ELEMENTARY NUMBER THEORY . . . 132

SECTION I. Reflections on the purpose and possibility of consistency proofs 132
§ 1. The antinomies of set theory and their significance for mathematics as a whole 132
§ 2. How are consistency proofs possible? 136

SECTION II. The formalization of elementary number theory 139
§ 3. The formalization of the propositions occurring in elementary number theory 139

§ 4. Example of a proof from elementary number theory. 143
§ 5. The formalization of the forms of inference occurring in elementary number theory. 149
§ 6. Derived concepts and axioms in elementary number theory 155

SECTION III. Disputable and indisputable forms of inference in elementary number theory 158
§ 7. Mathematics over finite domains of objects 158
§ 8. Decidable concepts and propositions over an infinite domain of objects 160
§ 9. The 'actualist' interpretation of transfinite propositions 161
§ 10. Finitist interpretation of the connectives $\forall$, &, $\exists$ and $\vee$ in transfinite propositions. 163
§ 11. The connectives $\supset$ and $\neg$ in transfinite propositions: the intuitionist view 167

SECTION IV. The consistency proof 170
§ 12. The elimination of the symbols $\vee$, $\exists$ and $\supset$ from a given derivation 171
§ 13. The reduction of sequents 173
§ 14. Reduction steps on derivations. 179
§ 15. Ordinal numbers and proof of finiteness. 186

SECTION V. Reflections on the consistency proof 193
§ 16. The forms of inference used in the consistency proof 193
§ 17. Consequences of the consistency proof 198

APPENDIX TO #4 201

#5. THE CONSISTENCY OF THE SIMPLE THEORY OF TYPES 214

§ 1. The formal structure of the simple theory of types 214
§ 2. The consistency proof. 217

#6. THE CONCEPT OF INFINITY IN MATHEMATICS 223

#7. THE PRESENT STATE OF RESEARCH INTO THE FOUNDATIONS OF MATHEMATICS. 234

§ 1. The different points of view concerning the question of the antinomies and the concept of infinity. 234
§ 2. Exact foundational research in mathematics: axiomatics, metalogic, metamathematics. The theorems of Gödel and Skolem 238
§ 3. The continuum. 243
§ 4. The possibility of reconciling the different points of view 247

#8. NEW VERSION OF THE CONSISTENCY PROOF FOR ELEMENTARY NUMBER THEORY. 252

§ 1. New formalization of number-theoretical proofs 253
§ 2. Outline of the consistency proof 260
§ 3. A reduction step on a contradictive derivation 261
§ 4. The ordinal numbers – concluding remarks 277

#9. PROVABILITY AND NONPROVABILITY OF RESTRICTED TRANS-FINITE INDUCTION IN ELEMENTARY NUMBER THEORY 287
  § 1. TJ-derivations 287
  § 2. Characterizations of TJ-derivations 292
  § 3. Demonstrations of nonprovability 297

#10. FUSION OF SEVERAL COMPLETE INDUCTIONS 309

NOTES 312

GLOSSARY 318

INDEX OF SYMBOLS 321

INDEX OF AUTHORS 322

INDEX OF SUBJECTS 325

Excerpt from a handwritten draft copy of Gentzen's dissertation. Kindly made available by Prof. Paul Bernays.

GERHARD GENTZEN (1909–1945)

INTRODUCTION

The papers in this collection were written during a period of approximately ten years between 1932 and 1942, and comprise Gerhard Gentzen's extant contributions to logic and metamathematics. The editor first became interested in collecting and translating these papers during his undergraduate days at Oxford, when this task seemed far less arduous and delicate than it later turned out to be. Had it not been for an encouraging telegramme from Prof. Paul Bernays, the project would in fact never have been undertaken and completed. This debt of gratitude is here gratefully acknowledged.

The present introduction is intended to place Gentzen's work in its historical context and to trace some of the germane subsequent developments back to their origin in the papers below. These papers are not necessarily discussed in the chronological order in which they appear. An attempt has been made to provide ample cross-references between the various papers in order to motivate the reader to proceed from the nontechnical to the technical levels of Gentzen's writings. In this way the study of some lengthy and detailed arguments is postponed and the striking originality and freshness of Gentzen's ideas is not obscured and lost. Numerous references to other writers have been included in order to place Gentzen's achievements in their proper perspective and to facilitate the locating of relevant sources. Whenever possible, these references have been chosen from publications that have appeared in English.

Gentzen's public career began on the 2nd February 1932, when, at the age of twenty-two, he submitted to the *Mathematische Annalen* a paper *'On the Existence of Independent Axiom Systems for Infinite Sentence Systems'*. Today this paper is relatively unknown and it has received less attention since its publication than it deserves. Gentzen here presented a study of the theory of sentence systems as developed by Paul Hertz and at once displayed his great mathematical ability by giving a complete answer to a difficult open question concerning the existence of 'independent' axiom systems for arbitrary sentence systems. He showed by the construction of a counterexample that not all sentence systems possess independent axiom systems, whereas all 'linear' sentence systems do possess an independent axiomatization. In the course of his proof, Gentzen brought about a simplification of Hertz's rules of inference and noticed in particular that Hertz's 'syllogism' could be transformed into a special form which he called a 'cut'. In view of the special role which the 'cut' later played in Gentzen's celebrated discovery in predicate logic, the importance of Hertz's ideas for a totally new approach to logistic enquiries becomes apparent. Gentzen himself acknowledges this fact by using not only some of Hertz's methodology, but also some of Hertz's terminology in his 'Investigations' and in most of his other writings. In generalizing Hertz's 'sentences' to 'sequents', for example, Gentzen speaks of the component parts of a sequent as the 'antecedent' and the 'succedent' and in doing so follows Hertz's terminology. The notion of 'logical consequence' as used by Gentzen is also largely inspired by Hertz's work. In this connection Bernays observed that Hertz's theory of sentence systems constitutes an area of research in axiomatics and logic in which the possibilities of enquiry and discovery have by no means been exhausted[0]. It will be interesting to observe to what extent the recent use of Gentzen-type deductive systems in the solution of open problems in 'categorical algebra' by Joachim Lambek[1], for example, will give new importance to Hertz's work and also perhaps add yet another dimension to the remarkable influence and normative character of Gentzen's mathematical thinking.

It is historically significant that the semantic notion of *logical consequence* which Tarski introduced explicitly in 1936[2], was already developed by Gentzen to a considerable degree in #1. When Beth writes that 'the notion of *logical consequence* was introduced by Tarski (1936) as a semantical counterpart to the syntactical notion of *derivability*'[3], this should be compared with Gentzen's penetrating discussion of these concepts in § 4 of #1. In a somewhat modified form the notion of *consequence* is also implicit in Gentzen's calculus of *natural deduction* developed in #3 and published in 1934. Recently K. Schröter has in fact shown that Gentzen's calculus of natural deduction yields an exact formalization of the notion of *consequence*, as long as this notion is understood in the sense in which it was first used by Bolzano[4]. Yet even Tarski's notion of *logical conse-*

quence is analogous to a far-reaching extent to that of Bolzano, presented over a hundred years earlier[5].

Gentzen's interest in *axiom systems* and, more generally, in *axiomatic methods* stems in part from his concern with the aims of Hilbert's programme of setting classical mathematics on a secure foundation in order to eliminate the crisis by which mathematics had been beset because of the appearance of the antinomies of set theory. As early as 1904, one year after the publication of Russell's antinomy, Hilbert had outlined his views on the foundations of logic and arithmetic[6] and in a succession of papers he drew up what Gentzen refers to as 'Hilbert's programme'[7]. Hilbert's papers provide the appropriate historical background and motivation for Gentzen's discussion in #4, section I, and #6 and #7. In #6, for example, Gentzen expounds Hilbert's ideas and classifies mathematics into three distinct levels according to the degree to which the concept of infinity is used in the various branches of mathematics. The classification actually goes back to Weyl[8]. The first level is called *elementary number theory*, i.e., the theory of the natural numbers, the second level *analysis*, i.e., the theory of the real numbers, and the third level *general set theory*, i.e., the entire theory of Cantor's cardinal and ordinal numbers.

Gentzen attributes the occurrence of the antinomies of set theory to an excessively liberal use of the concept of infinity in general set theory and on this point finds himself in agreement with such prominent mathematicians as Poincaré, Brouwer, and Weyl, the foremost proponents of 'intuitionism'[9]. Yet since the intuitionists also reject some rather fundamental principles of classical logic such as the validity of the *law of the excluded middle* in situations involving infinity, a careful comparison of Brouwer's and Hilbert's methods of reconstructing mathematics became vital and in #2 Gentzen establishes that at least at the first level, intuitionist and classical mathematics are proof-theoretically coextensive.

#2 constitutes a natural link in the development of Gentzen's ideas and methods, and the editor has decided to include it in the present volume in spite of the fact that, in 1933, Gentzen himself had withdrawn the corrected galley proof from publication when Gödel's comparable discovery became known[10]. It should be noted, incidentally, that as early as 1925, five years before the publication of Heyting's formalization of intuitionist logic, Kolmogorov had the idea of reinterpreting classical arithmetic in terms of intuitionist arithmetic and in this way seems to have anticipated the intuitionist consistency of classical arithmetic[11]. The essential significance of #2, on the other hand, lies in the fact that here Gentzen begins to develop his *formalist* approach to foundational problems and discusses the connection between intuitionist and classical logic which is at the heart of his methods in #3. Also mentioned here for the first time is the question of the *consistency*[12], i.e., the freedom from contradiction, of the various levels of mathematics as seen from Hilbert's *finitist* point of view, a criterion of reliability in mathematics that became the central theme of all of Gentzen's further investigations. At this time, however, a consistency proof of the different levels of mathematics seemed singularly out of reach and Gentzen ends #2 on a pessimistic note concerning the realizability of Hilbert's programme in view of the obstacles presented to such an undertaking by Gödel's proof of the impossibility of 'internal' consistency proofs[13]. In order to preserve its historical flavour, no attempt has been made to bring #2 in line with Gentzen's further work by standardizing the notations employed. In #2, Gentzen still uses Hilbert's quantification symbols, for example, whereas from #3 onwards, a variety of new symbols is consistently used, including the $\exists$ of the *Principia Mathematica* for the existential quantifier and, by analogy, the symbol $\forall$[14] for the universal quantifier, a notation that is prevalent in mathematics textbooks today.

#3 is perhaps the most widely read and universally acclaimed source of Gentzen's influence. Here Gentzen breaks away from the traditional formulations of predicate logic as they were developed by Frege, Russell, and Hilbert[15] and presents two basically different versions of predicate logic now customarily referred to as the *N-systems*[16] and the *L-systems*[17]. The *N*-systems are the outcome of Gentzen's attempt to find a more 'natural' approach to formal reasoning[18]. Similar independent attempts were made by Jaśkowski in Poland, incidentally, and in 1934, the same year in which Gentzen published his 'Investigations', Jaśkowski published a different version of 'natural deduction' based on ideas first put forward by Łukasiewicz in seminars as early as 1926[19]. In view of Gentzen's efforts to find more 'natural' methods in mathematical logic, it is not surprising that his first consistency proof for elementary number theory (#4) is formalized in terms of an *N*-type calculus (*NK*), where simplicity and elegance of procedure are sacrificed to the demands of 'naturalness'. In #8, Gentzen reverses his methods and uses an *L*-type calculus (*LK*) in order to simplify his consistency proof, but, in doing so, jeopardizes some of the naturalness in procedure[20]. Gentzen elicits the properties which an

N-system should have from a detailed analysis of Euclid's classical proof of the nonexistence of a largest prime number. This analysis reveals very clearly and simply how Gentzen is led to a calculus based on assumption formulae instead of the usual axiom formulae, and how the 'natural' separation of the rules for the logical operators into 'introduction' rules and into 'elimination' rules evolved. In addition, Gentzen achieves a very natural formulation of intuitionist logic, since classical logic follows from it by the mere inclusion of the single basic formula $\mathfrak{A} \vee \neg \mathfrak{A}$. In this connection, Kneale justly points out[21] that 'Gentzen's success in making intuitionistic logic look like something simpler and more basic than classical logic depends ... on the special forms of the rules he uses and that certain rules yield the classical logic more 'naturally' than the intuitionistic'. This fortuitous state of affairs is of course in harmony with Gentzen's intention of using intuitionist logic and intuitionist arguments in his metamathematical investigations, feeling as he does that the intuitionists have drawn the most radical conclusions from the crisis that had emerged in the foundations of mathematics[22].

Several detailed studies of natural deduction have recently appeared in the literature. Leblanc, for example, has written a lucid account of the relationship between Gentzen's various N-systems developed in #3 and #4[23]. In 1965, Dag Prawitz published perhaps the most comprehensive survey of the developments that have taken place in natural deduction up to 1965[24]. Prawitz generalizes and extends to N-systems the results which Gentzen has established for L-systems in #3. This monograph covers a wide variety of topics including the application of natural deduction to modal logic.

When Gentzen examined the specific properties of the 'natural calculus', he was led to the conjecture that it should be possible to bring purely logical proofs into a certain 'normal form' in which all concepts required for the proof would in some sense appear in the conclusion of the proof[25]. In order to enunciate and prove his theorem which he called the *Hauptsatz* (literally rendered as the *main theorem* and also known in English as the *elimination theorem*) for both intuitionist and classical predicate logic, however, Gentzen had to abandon his natural 'assumption' calculi and formulate logistic 'sequent' calculi in which the logical rules can be divided into 'structural' and 'operational' rules[26]. The *Hauptsatz* then refers to the fact that one of the 'structural' inference figures of the calculus, the 'cut', can can be eliminated from *purely logical* proofs; as a corollary Gentzen obtains the *subformula property*, which says that in a cut-free proof all formulae occurring in the proof are compounded into the 'endsequent', i.e., the formula to be proved. In section 4 of #3, Gentzen gives several applications of the *Hauptsatz*: a consistency proof for classical and intuitionist predicate logic; the solution of the decision problem for intuitionist propositional logic; and a new proof of the nonderivability of the law of the excluded middle in intuitionist logic. In the *L*-systems the difference between intuitionist and classical logic is again characterized very simply by restricting the number of 'succedent formulae' in intuitionist sequents to a single formula[27].

The *Hauptsatz* can be brought into a sharper form in the case of classical logic. Here a proof can be broken up into two parts: one part belonging exclusively to propositional logic, and the other essentially consisting only of application instances of the rules of quantification. Gentzen calls his generalized *Hauptsatz* the '*verschärfter Hauptsatz*' (literally rendered as the '*sharpened Hauptsatz*' and also known in English as the '*midsequent theorem*', the '*normal form theorem*', the '*strengthened Hauptsatz*', and the '*extended Hauptsatz*')[28].

A comparable theorem for the usual logical calculus was obtained by Herbrand in his '*Investigations in proof theory*'[29]. Thanks to the work of B. Dreben, J. Denton and others, Herbrand's result is now better understood, and it has become possible to clarify the connection between Herbrand's theorem and Gentzen's *sharpened Hauptsatz*. Gentzen himself considered Herbrand's theorem as a special case of his own result[30]. He presumably had in mind the less general version of Herbrand's theorem popularly known at that time in the form in which it later also appeared in volume II of *Hilbert and Bernays*. Herbrand's full theorem, however, applies not only to formulae in prenex form, as Gentzen had supposed, but to any formula, and with his '*domains*' Herbrand actually supplies more information about the 'midsequent' than Gentzen's *sharpened Hauptsatz* is able to provide. However, the *sharpened Hauptsatz* can and has been extended to intuitionist logic and to various modal calculi, whereas this has not been done so far with Herbrand's theorem[31]. The possibility of extending the *sharpened Hauptsatz* to modal logic was first announced by Curry in 1950 in his *A theory of formal deducibility*, which also contains a new proof of the *Hauptsatz*[32], and his result subsequently appeared in print in 1952[33]. Curry's solution applies essentially to Lewis's calculus S4[34]. Since then the extension has also been carried out for the calculi S2, S5, and M by

Ohnishi and Matsumoto[35], who have given a decision procedure for these calculi as well as for S4 by means of Gentzen-type arguments.

An important achievement in another direction is Kanger's first model proof of the *Hauptsatz*[36]. This result, which includes a completeness theorem for a modified Gentzen formalism, was simplified further by Rasiowa and Sikorski in 1960[37].

Since Gentzen's *sharpened Hauptsatz* and Herbrand's theorem are analogues in certain quantificational situations, it has become customary to refer to both theorems together as the '*Herbrand-Gentzen theorem*', and an interesting modification of this theorem is presented by Craig in his *Linear reasoning*[38], where three substantial uses of the theorem are made, affording a valuable connection between model theory and proof theory. The model and proof-theoretic approaches, incidentally, are exploited fruitfully by Kleene in his *Mathematical logic*[39] in which the advantages of Gentzen's L-systems are displayed for the first time at an introductory level.

Gentzen's analysis of the predicate calculus in terms of 'natural deduction' has had a marked influence on the development of logic in another direction: Beth reformulated a system of natural deduction by means of '*semantic tableaux*'[40], a method for the systematic investigation of the notion of 'logical consequence'. This method is based on the construction of a counter-example in cases in which a given formula is *not* a logical consequence of a given list of formulae. Beth's system satisfies Gentzen's *subformula property* and also brings within easy reach the theorems of Herbrand and Löwenheim-Skolem-Gödel. Beth proves the completeness of his system of natural deduction but not of course its decidability since it cannot be estimated in advance how many steps in the construction will result in a formal derivation or in a counterexample[41]. The counterexample method is also discussed and extended to a Gentzen-type L-system by Kleene in his *Mathematical logic*[42]. Beth himself discovered another important application of Gentzen's methods of #3 when he proved his well-known theorem on definability[43] by using a modified version of the *sharpened Hauptsatz*. Craig's first use of the Herbrand-Gentzen theorem in the paper mentioned earlier, incidentally, consists of a generalization of Beth's results on definability from 'primitive predicate symbols to arbitrary formulas and terms'[44]. Also mentioned at this point should be Ketonen's proof of the completeness of the predicate calculus by a method closely akin to procedures involved in the proof of the *Hauptsatz*[45] in which a strengthening of Gödel's result is achieved[46]. Ketonen later formulated necessary and sufficient conditions for provability for a new version of Gentzen's calculus LK[47]. This is not of course a new discovery for predicate logic as such, and in his review of Ketonen's paper, Bernays rightly noted that Ketonen's result 'corresponds to a condition for provability in the usual calculus, which is one of the forms of stating the theorem of Herbrand'[48]. Ketonen's paper represents nevertheless a desirable refinement of Gentzen's methods and contains also several interesting elementary applications of these methods to plane geometry. Gentzen himself had applied the *sharpened Hauptsatz* to a new consistency proof for arithmetic without complete induction[49]. However, he himself has remarked that 'arithmetic without complete induction is of little practical significance since complete induction is constantly required in number theory. Yet the consistency of arithmetic with complete induction has not been conclusively proved to date'[50]. In #4, Gentzen explains why it is precisely the inference of complete induction which causes the difficulty in a consistency proof[51].

The year 1936 represents a significant turning point for the realizability of the aims of metamathematics. Since one of the main objectives of Hilbert's programme is the vindication of classical mathematics at its various levels, and since this requires the development of explicit consistency proofs for these levels, Gentzen's discovery of a way out of the impasse presented to the feasibility of formal consistency proofs by Gödel's theorem must be regarded as a major achievement. In order to enable the reader to follow the precise historical and conceptual evolution of Gentzen's ideas, excerpts from a galley proof are included in an appendix to #4, which give a good indication of how Gentzen had attacked the problem initially. The crucial part of this galley proof consists of the lemma in section 14.6 of the Appendix. Gentzen himself had withdrawn the galley proof from publication[52] when objections were raised against an alleged implicit use of the *fan theorem*[53] in the proof of the lemma. Gentzen achieved his major breakthrough when he discovered that a restricted form of transfinite induction could be used instead of his original procedure as a method which is not formalizable in elementary number theory, but which is nevertheless in harmony with acceptable principles of proof in this context[54]. As Gentzen puts it: 'A consistency proof is once again a mathematical proof in which certain inferences and derived concepts must be used; their reliability (especially their freedom from contradiction) must already be presupposed. There can be no *absolute consistency proof*. A consistency proof can merely reduce the correctness of certain forms of inference to the correctness of other forms of inference. It is therefore clear that in a consistency proof we can use only forms of inference that count as considerably more secure than the forms of inference of the theory whose consistency is to be proved'[55]. The following passage from Hilbert and Bernays, *Grundlagen der Mathematik*, should serve as an illustration of the kind of criterion for reliability which Gentzen adopted with Hilbert's *finitist* point of view[56]:

> 'Our examination of the beginnings of number theory and algebra has served the purpose of elucidating in its application and use the direct, informal reasoning, free from axiomatic assumptions, as it is carried out in mental experiments in terms of intuitively conceived objects. This kind of reasoning we shall, in order to have a concise expression for it, call *finitist* reasoning and shall also designate as the *finitist* attitude or the *finitist* point of view the methodological attitude underlying this reasoning. In the same sense we shall speak of finitisticly specified concepts and assertions by expressing with the word 'finitist' in each case the fact that the deliberation, assertion or definition proceeds within the bounds of the conceivability in principle of objects as well as the realizability in principle of processes and thus takes place within the realm of concrete considerations.'

This excerpt is not intended to convey the idea that a formal 'definition' of the finitist point of view has here been given or that such a definition can be given[57]. Gentzen points out that it cannot be 'proved' that his techniques of proof are 'finitist' since this concept is not unequivocably formally defined and cannot in fact be delimited in this way. He emphasizes that all that can be achieved in this direction is that individual inferences are examined from the 'finitist' point of view and that it must be assessed separately in each case whether the inferences concerned are in harmony with finitist intentions.

In this connection, Kleene remarks that

> 'to what extent the Gentzen proof can be accepted as securing classical number theory in the sense of that problem formulation is in the present state of affairs a matter of individual judgement, depending[58] on how ready one is to accept induction up to ε_0 as a finitary method.'

Tarski, for example, states that

'Gentzen's proof of the consistency of arithmetic is undoubtedly a very interesting metamathematical result, which may prove very stimulating and fruitful. I cannot say, however, that the consistency of arithmetic is now much more evident to me (at any rate, perhaps, to use the terminology of the differential calculus more evident than by an epsilon) than it was before the proof was given.'[59]

Black, on the other hand, ponders:

'Is it then permissible to regard Gentzen's leading idea as 'finitist'? I believe that anybody who tries to replace the specific pattern, constructed in the way described above, by a series of steadily preceding patterns, will receive a vivid impression of the elementary and intuitively convincing character of the type of argument involved.... It seems incredible that such types of argument should involve the mathematical paradoxes. It seems, then, that Bernays is justified in blaming the previous lack of success of the formalist programme upon its excessive ((überspannte)) restrictions on permissible modes of research (*Grundlagen der Mathematik*, vol. II, vii).'[60]

Black's account of Gentzen's innovation in proof theory is extremely illuminating and shows convincingly that Gentzen's *restricted transfinite induction* is entirely elementary (even though it may no longer be 'strictly' finitist) and independent of Cantor's theory of the second number class. Gentzen himself presents a careful analysis of the finitist point of view[61] and discussed in particular the intuitionist delimitation between permissible and nonpermissible forms of inference in number theory. It turns out that in certain respects the *finitist* methods are actually more restrictive than the *intuitionist* methods, in spite of the fact that 'the intuitionists (Brouwer) even object to forms of inference customary in number theory, not only because these inferences might possibly lead to contradictions, but because the theorems to which they lead have no actual sense and are therefore worthless'[62]. The intuitionist objection against certain classical procedures because of their alleged lack of 'sense' must not be construed to mean that intuitionist mathematics is 'obviously' securer than classical mathematics. One reason why Gentzen was unhappy about securing the consistency of classical arithmetic by establishing its intuitionist consistency, as was done in #2, is precisely that from the finitist point of view objections can be raised against the use of the *'implication'* in intuitionist reasoning[63]. In his address to the Mathematical Congress in Paris in 1937[64], Gentzen made the further point that Brouwer's intuitionism involves such liberal uses of complicated forms of the concept of infinity that consistency proofs should if at all possible be carried out with more restricted techniques. Here Gentzen may have had in mind techniques of proof of the kind found in Brouwer's paper '*On the domain of definition of functions*'[65] some of which are presumably no longer compatible with the narrower finitist point of view outlined above.

In addition to the obstacles presented to metamathematical procedures from within by the *finitist condition* on the methods of proof, Gödel's 'incompleteness theorem' imposes seemingly serious limitations on the significance of consistency proofs from without. Gentzen assesses the effect of Gödel's theorem on his formalization of elementary number theory by noting that 'whenever the present framework is transgressed, an extension of the consistency proof to the newly incorporated techniques is required. The consistency proof is already designed in such a way that this is to a very large degree possible without difficulty'[66]. He states further that 'the concept of the *reduction rule* has in fact been kept general enough so that it is not tied to a definite logical formalism but corresponds rather to the general concept of 'truth', certainly to the extent to which that concept has any clear meaning at all'.

The consistency proof centres around the concept of a 'reduction rule' for sequents and derivations: Gentzen proves the consistency of elementary number theory by characterizing the concept of the informal 'truth' of the provable formulae, in this case the sequents, of number theory in terms of a reduction procedure for the derivations of these formulae and shows that every derivable sequent can be brought into '*reduced form*'[67]. The consistency follows from the reducibility of the derivable sequents to 'reduced form'. If the sequent in question is a 'derived' sequent, a reduction step is carried out on the entire derivation of the sequent and this in some sense 'simplifies' the derivation. In order to show that a finite succession of reduction steps leads to the *reduced form* of the endsequent, i.e., the simplest possible derivation consisting of only a single formula whose truth can be decided by elementary calculation, Gentzen correlates with the derivations finite decimal fractions order-isomorphic with a well-ordered segment of Cantor's transfinite ordinal numbers up to the first ε-number. These decimal fractions serve as measures for the complexity of the derivations. The crucial step in the consistency proof consists in showing that the application of a reduction step diminishes the decimal fractions and thus achieves a 'simplification' of derivations and sequents involved.

In § 16, Gentzen explains why the inclusion of complete induction in the formalism makes it necessary to go beyond the natural numbers for measures of the complexity of a derivation. This complexity is surprisingly enough not primarily determined by the *number* of complete inductions occurring in a derivation, but rather by it together with the complexity of each induction proposition. This is proved in ≠10, where Gentzen shows that any finite number of complete inductions in a proof can be fused into a single complete induction.

In order to establish that a finite number of reduction steps on a derivation leads to the reduced form of the endsequent, Gentzen carries out a '*restricted transfinite induction*'[68] on the measure of complexity of the derivation. In his address to the Mathematical Congress in Paris[69], Gentzen explains the connection between his consistency proof and 'restricted transfinite inductions' as follows:

'In my proof the number-theoretical 'proofs' whose consistency is to be proved are arranged in a sequence in such a way that the consistency of a 'proof' in the sequence follows from the consistency of the preceding 'proofs'. This sequence may at once be put into one-to-one correspondence with the transfinite ordinal numbers up to the number ε_0. It is for this reason that the consistency of all 'proofs' follows from a transfinite induction up to the number ε_0.'

In ≠6, Gentzen illustrates directly the fundamentally constructive nature of the transfinite ordinal numbers up to the number ε_0, and explains in detail why he considers the use of 'restricted transfinite induction' to be in harmony with finitist intentions[70]. He furthermore cherishes the hope that an extension of transfinite induction over a larger segment of Cantor's second number class will eventually also yield a consistency proof for analysis[71].

Unfortunately Gentzen was prevented from pursuing his investigations of the consistency of analysis by his untimely and tragic death in 1945. The foundations had nevertheless been laid for the work of a large number of mathematicians who began to concern themselves with the question of the consistency of analysis, among them Ackermann, Fitch, Lorenzen, Schütte and Takeuti. The first major consistency result for analysis dates back to 1938, when Fitch proved nonconstructively the consistency of the ramified *Principia*[72], i.e., a modified version of the Russell-Whitehead system of analysis which *includes* the axiom of infinity, but *excludes* the axiom of reducibility. The first major *constructive* result was published by Lorenzen in 1951, and furnishes a proof of the consistency of the ramified theory of types[73]. An excellent discussion in English of Lorenzen's paper by Hao Wang has appeared in the Journal of Symbolic Logic[74] in which the principal steps of Lorenzen's proof are summarized and compared with Fitch's earlier result. Wang points out that 'the consistency proof for the ramified theory of types may be considered as an extension of Gentzen's dissertation[75] which strengthens and refines results of Herbrand'[76]. Lorenzen himself writes that[77]:

> 'Since the lattice-theoretical ideas are used only implicitly, the consistency proof appears as an extension of an earlier attempt by Gentzen in his dissertation to prove the consistency of arithmetic without complete induction. This is so since the consistency follows as an immediate consequence of the fact that every theorem of the calculus can be derived 'without detours'. The proof here given goes beyond Gentzen's proof since the calculus whose consistency is prove contains as an integral part arithmetic including complete induction. The calculus is equivalent with that used by Russell and Whitehead in the *Principia Mathematica*, without the axiom of reducibility. Since this axiom has not been included, our calculus does not encompass classical analysis, although the analytic forms of inference in this calculus may be represented within the limits imposed by the ramified theory of types. The extension of Gentzen's scheme to a so much richer calculus is effected without the inclusion of new techniques. Only the notion of the 'detourless derivability' is extended by admitting certain induction rules in which a conclusion with infinitely many premisses is deduced. The progress vis-a-vis Fitch's work lies in the constructive character of all occurring inferences. This is the crucial point which brings our proof within the scope imposed on consistency proofs since Hilbert.'

One of the most thorough and comprehensive exploitations of Gentzen's methods and ideas began to take shape with Takeuti's publication in 1953 of a generalization of Gentzen's classical calculus of sequents (LK) to a calculus (GLC) for simple type logic[78]. In GLC, derivations are built up, as in LK, from sequents by means of 'basic sequents' and rules of inference, but with the quantificational rules extended to apply to a hierarchy of 'types'. The consistency of the calculus GLC, which does not yet include the axiom of infinity, follows from Gentzen's consistency proof for the simple theory of types carried out in #5 of the present volume. Gentzen here presents a finitist consistency proof of the simple theory of types by extending the method by which Hilbert and Ackermann have proved the consistency of the restricted predicate calculus[79]. In his review of #5, P. Bachmann assesses the significance of Gentzen's proof by observing that[80]:

> 'The question of the consistency of the pure simple theory of types must be distinguished from the question of the consistency of the applied simple theory of types, or, more precisely, the formalisms which result if the axioms and rules of the simple theory of types are extended by the inclusion of axioms characterizing a definite domain of objects. The pure simple theory of types must be consistent if there exists a domain of objects so that the theory of types remains consistent when applied to this domain. Gentzen proves therefore the consistency of pure types theory by showing that the simple theory of types is consistent if it is specialized to the simplest nonempty domain of objects, i.e., the domain consisting of a single element .... A modification of this proof easily yields the consistency of the simple theory of types when it is specialized to an arbitrary finite domain of objects.'

Similar remarks apply to the consistency proof for the restricted predicate calculus[81]. H. Arnold Schmidt made the comment in his review of #5[82] that Gentzen had requested him to report that after the publication of his consistency proof for the simple theory of types, he became aware of the existence of a similar proof for type theory by Tarski[83]. The presentation of the theory of types and its proof of consistency in Tarski's paper forms however part of a more general context.

The consistency of the simple theory of types represents an important result for the applicability of formal techniques to the consistency problem of ramified analysis as tackled by Takeuti and Schütte. Takeuti has shown that various mathematical theories are representable by means of his calculus *GLC* if certain closed formulae are adjoined as mathematical axioms and he has proved that if Gentzen's *Hauptsatz* extends to the calculus *GLC*, then the consistency of the theory of the real numbers, i.e., classical analysis, follows at once. Schütte comments that he does not think that Takeuti's results can be extended to theorems about 'absolute' consistency, since the formal systems referred to in the hypotheses of Takeuti's key theorems have infinitely many basic objects and require the full use of the *tertium non datur* as well as impredicative definitions[84].

In 1956, Takeuti nevertheless published a partial result by showing that a certain system of ramified real numbers is consistent[85]. This was done by proving that the 'cut' can be eliminated from the underlying calculus of sequents, and the consistency then follows by a transfinite induction up to ω^ω. In a series of papers, Takeuti has also proved his 'fundamental conjecture', i.e., that every provable sequent in *GLC* is provable without cuts, for five different subsystems of GLC[86]. In his proofs, Takeuti incidentally uses certain metamathematical concepts closely allied to the concepts of the '*degree of a formula in a proof figure*', and the '*level*' and '*order*' of a sequent in a proof as defined by Gentzen in #8. A study of these proofs makes it clear to what extent Gentzen-type methods can be used profitably to push back the frontier of foundational knowledge. Schütte seems to have summed up the progress inherent in Takeuti's work appropriately when he concludes that 'it may be said that all these investigations constitute a remarkable advance in the research of the metamathematical properties of simple type theory, although it appears somewhat doubtful whether it will be possible to prove the fundamental conjecture of *GLC* generally'[87]. Takeuti has also generalized Gentzen's calculus *LK* to numerous mathematical systems[88], and, among others, obtained a metamathematical theorem on functions, and a result on Skolem's theorem in which he shows that 'if Gödel's axioms of set theory are consistent, then they are consistent with additional axioms of denumerability'. Further extensions of the *Hauptsatz* by suitable modification of the calculus *GLC* have also been shown to be possible.

Schütte himself has made significant contributions to the solution of the consistency problem of analysis both for ramified and for type-free analysis[89]. In the case of ramified analysis, the consistency of Schütte's system follows from the eliminability of the 'cut', i.e., an application of Gentzen's results of #3. Schütte has also succeeded in generalizing Gentzen's theorem on the derivability of 'restricted transfinite induction' in elementary number theory of #9[90] by proving the following theorem about formalized transfinite induction:

The derivation of the general transfinite induction up to α has as order at least (i) the last critical ε-number smaller than α (if such a number exists), (ii) the number α, if α is a critical ε-number.

This theorem represents a further refinement of earlier discoveries[91]. Schütte has also obtained various consistency results for type-free analysis, and here too the consistency follows from the eliminability of the 'cut'.

Many other important contributions to the solution of the consistency problem of analysis within metamathematics have been published, principally the results by Ackermann[92], one of whose contributions to the solution of consistency problems consisted in his adaptation of Gentzen's method of 'restricted transfinite induction' to proofs involving Hilbert's ε-symbol technique. In this way Ackermann was able to prove the consistency of number theory and of certain special type-free systems of analysis not involving the law of the excluded middle. The consistency followed from an application of 'restricted transfinite induction' extending over a wider segment of Cantor's second number class than that required for the consistency proof of elementary number theory. The segment involved consists of all ordinals less than the first critical ε-number, i.e., the first ε-number η such that $\varepsilon_\eta = \eta$. This result compares with Gentzen's anticipation that there exists 'a general affinity between formally delimited techniques of proof, their possibility of extension and newly arising incompletenesses, on the one hand, and the transfinite ordinal numbers of the second number class, transfinite induction and the constructive progression into the second number class on the other'[93].

Most researchers who have extended Gentzen's techniques of proof to the solution of consistency problems in analysis have based their logical calculi on Gentzen's calculus *LK*. In fact, Gentzen himself re-wrote his consistency proof for elementary number theory by couching it in terms of the calculus *LK*[94] and in this way was able to simplify the earlier proof considerably. This time the consistency follows from the nonderivability of the 'empty sequent', again proved by a restricted transfinite induction up to ε_0. Gentzen conjectured that 'it is reasonable to assume that by and large this correlation ((of transfinite ordinals)) is already fairly *optimal*, i.e., that we could not make do with essentially lower ordinal numbers. In particular, the totality of all our derivations cannot be handled by means of ordinal numbers all of which lie below a number smaller than ε_0. For transfinite induction up to such a number is itself provable in our formalism; a consistency proof carried out by means of this induction would therefore contradict Gödel's theorem (given, of course, that the other techniques of proof used, especially the correlation of ordinal numbers, have not assumed forms that are nonrepresentable in our number-theoretical formalism). By the same roundabout argument we can presumably also show that certain sub-classes of derivations cannot be handled by ordinal numbers below certain numbers of the form $\omega^{\cdot^{\cdot^{\cdot^\omega}}}$[95]. It is quite likely that one day a direct approach to the proof of such impossibility theorems will be found'[96].

In #9, Gentzen essentially answers these questions by giving a direct proof for the nonprovability of transfinite induction up to ε_0 in elementary number theory and by showing that it is impossible to prove still more restricted forms of transfinite induction to numbers below ε_0 in certain subsystems of formalized number theory. The results of #9 not only confirm Gödel's incompleteness theorem, but in the special case at hand, they illustrate the incompleteness of the number-theoretical formalism in a *direct* way[97].

Gentzen's reaction to the incompleteness of formal systems is that 'this is undoubtedly a very interesting, but certainly not an alarming result. We can paraphrase it by saying that for number theory no once-and-for-all sufficient system of forms of inference can be specified, but that on the contrary, new theorems can always be found whose proof requires a new form of inference. That this is so may not have been anticipated in the beginning, but it is certainly not implausible. The ((incompleteness)) theorem reveals of course a certain weakness of the axiomatic method. Since consistency proofs generally apply only to delimited systems of techniques of proof, these proofs must obviously be extended when an extension of the methods of proof takes place. It is remarkable that in the whole of existing mathematics only very few easily classifiable and constantly recurring forms of inference are used, so that an extension of these methods may be desirable in theory, but is insignificant in practice'[98]. In the case of the consistency proof for elementary number theory, for example, the consequences of the incompleteness have been minimized by the requirement that as a new form of inference is introduced into the formalism, a 'reduction procedure' for that form of inference is stipulated at the same time. Gentzen cites as an example of such a new form of inference an instance of transfinite induction up to a fixed number of the second number class. The introduction of a new form of inference into the number-theoretical formalism may of course necessitate a corresponding extension of the 'restricted transfinite induction' that enters into the proof of the finiteness of the reduction procedure. Gentzen's method of dealing with the inherent incompleteness of formalized mathematics by pairing the 'reduction rules' required for the forms of inference with 'restricted transfinite inductions' over different segments of the second number class is reminiscent of Weierstrass's method of overcoming the mystery of the 'infinitesimals' in the differential calculus by pairing 'epsilons' and 'deltas' and establishing functional dependences between them. This analogy makes Gentzen's argument extremely convincing that the formal incompleteness of mathematical systems constitutes a relatively minor obstacle in metamathematics. It must be mentioned, however, that some authors have evaluated the impact of Gödel's incompleteness theorem rather more negatively[99].

An entirely different problem is that of the significance of finitist metamathematical results for nonfinitist mathematics. #7 deals with this problem and, as Curry writes,

> 'the main theme of this ((Gentzen's)) expository article is to contrast the 'constructive' and '*an sich*' conception of infinity and then to defend the opinion that the Hilbert programme makes it possible for the two sides to agree on the retention of classical analysis in its present form'[100].

Although some writers have begun to distinguish meticulously between 'constructive' and 'finitist' methods, presumably because they do not consider such methods as 'restricted transfinite induction' to be in harmony with the notion of 'finitist' described earlier[101], Gentzen considers Brouwer's 'intuitionist' and Hilbert's 'finitist' methods as representing only two somewhat different instances of the 'constructive' point of view in mathematics[102]. The real issue does not therefore concern the question whether all 'constructive' methods are necessarily 'finitist' in the narrower sense, but rather whether 'nonconstructive' mathematics, based on the 'actualist' interpretation of infinity, can to some extent be justified by means of 'constructive' arguments. Gentzen here agrees with Hilbert that classical analysis, for example, has proven itself by the successes which it has scored in physics, and argues convincingly that nothing is lost by treating the real numbers as 'ideal elements'. On this basis classical analysis and presumably a considerable portion of axiomatic set theory could be retained. A constructive consistency proof for analysis would make it in fact quite safe to deal with extensive parts of nonconstructive mathematics 'as if' they were given 'constructively'. Gentzen illustrates the benefits which accrue from the idealization of experience by comparing 'constructive' analysis to the 'natural' geometry of Hjelmslev, while likening 'classical' analysis to 'pure' geometry[103]. As a result of its idealization, 'pure' geometry is a much simpler and considerably smoother theory than 'natural' geometry, since the latter is continually plagued by unpleasant exceptions. The same is true of constructive and nonconstructive analysis. In his review of #7, Rosser discusses the question of *as if* arguments in mathematics and states that

> 'the book (#7) is a well written summary of the present status of foundations, and contains one of the most lucid accounts of the Brouwer viewpoint that the present reviewer has seen. The distinction between

the Brouwer and Hilbert schools is presented from the point of view of their treatment of the infinite. For Brouwer, who always insists on finite constructibility, the infinite exists only in the sense that he can at any time take a larger (finite) set than any which he has taken hitherto. Hilbert would treat of infinite sets by the same methods used for finite sets, *as if* he could comprehend them in their entirety. Gentzen refers to this point of view as the *as if* point of view. He presents various paradoxes which arise when the *as if* method is used without proper care. This of course opens the question of what is 'proper care'. In the nature of things, the Brouwer method must fail to produce a paradox, since it never leaves the domain of the constructive finite. However, the Brouwer method does not produce sufficient mathematical theory for physical and engineering uses. So Brouwer's method must be described as 'excessive care'. A proposed way out of the difficulty is to base the *as if* method on an appropriate formal system, and use the Brouwer method to prove that the formal system is without contradiction'[104].

Rosser's assessment of Gentzen's arguments in #7 gives rise to several important considerations. Gentzen indeed advocates that nonconstructive methods, which he calls *an sich* methods, should be accepted in mathematics, provided that they can be justified by 'finitist' arguments. Yet, there is a clear distinction in German between the notion of *an sich* and that of *als ob*. Kant frequently speaks of *Dinge an sich*[105] and by this he means *things in themselves*, or *noumena*, which lie beyond the realm of experience but whose existence is a necessary presupposition. The expression *als ob*, however, is a neo-Kantian term on which Vaihinger, an eminent Kant scholar of recent times, has based his philosophy of the *als ob*[106]. It would seem appropriate, therefore, to render *als ob* by *as if* and search for a different English expression for *an sich*. The choice of terminology is made somewhat easier by Gentzen's own comment in his address to the Mathematical Congress in Paris[107] that the *konstruktive Auffassung* and the *an sich Auffassung* of infinity can be placed in a certain parallel with the philosophical views of *idealism* and *realism*. Gentzen's *an sich* thus has a realist status that corresponds in a certain sense to Lorenzen's notion of *das Aktual-Unendliche*[108]. The neologism *actualist* thus seems a suitable English translation of Gentzen's notion of *an sich* and it has been consistently used in this way throughout the present volume.

It is interesting that even Rosser's correct interpretation of Gentzen's *as if* attitude towards nonconstructive mathematics still exemplifies a considerable divergence from the basic tenets of the *as if* philosophy. Vaihinger argues that since reality cannot be truly known, man constructs systems of thought according to his requirements and then assumes that reality coincides with these constructs. He therefore acts 'as if' reality were what he assumes it to be. Waisman paraphrases the premisses on which Vaihinger's philosophy is based by stating that

'Vaihinger advances the opinion that our thinking is frequently guided by fictions, that is, by wittingly false assumptions, which, however, have held up by their results'[109].

This view still differs appreciably in its ontological assessment of the reality of mathematics from Gentzen's *as if* interpretation in the case of the continuum, for example, since here Gentzen concludes merely that 'whether the continuum should outwardly be regarded as a mere fiction, as an ideal construction, or whether it should be insisted upon that it possesses a reality independent of our methods of construction, in the sense of the actualist interpretation, is a purely theoretical question whose answer will probably remain a matter of taste; for practical mathematics it has hardly any further significance'[110]. Here Gentzen is not thinking of the continuum as a 'wittingly false' assumption. What he does have in mind is that an extreme formalist position is compatible with several different philosophical points of view. Quine characterizes such an extreme formalist position by stating that

'the formalist ((presumably Hilbert and his followers, including Gentzen)) keeps classical mathematics as a play of insignificant notations. This play of notations can still be of utility – whatever utility it has already shown itself to have as a crutch for physicists and technologists. But utility need not imply significance, in any literal linguistic sense. Nor need the marked success of mathematicians in spinning out theorems, and in finding objective bases for agreement with one another's results, imply significance. For an adequate basis for agreement among mathematicians can be found simply in the rules which govern the manipulation of the notations – these syntactic rules being, unlike the notations themselves, quite significant and intelligible'[111].

Quine is saying, in effect, that the formalist can, if he so desires, evade consistently the question of the ontological status of his 'notations' and Gentzen clearly concurs with this view. There is after all a categorical difference between the problem of existence *within* a framework (Bernays speaks of *bezogene Existenz*) and the problem of the existence *of the framework itself*[112]. Gentzen's argument is simply that his ideal *as if* point of view represents a minimal ontological commitment and thus forms a suitable basis for agreement among practicing mathematicians.

Gentzen envisages such an agreement along the lines suggested by Weyl, who writes that

> 'in studying mathematics for its own sake one should follow Brouwer and confine oneself to discernible truths into which the infinite enters only as an open field of possibilities; there can be no motive for exceeding these bounds. In the natural sciences, however, a sphere is touched upon which is no longer penetrable by an appeal to visible self-evidence, in any case; here cognition necessarily assumes a symbolic form. For this reason it is no longer necessary, as mathematics is drawn into the process of a theoretical reconstruction of the world by physics, to be able to isolate mathematics into a realm of the intuitively certain: On this higher plane, from which the whole of science appears as a unit, I agree with *Hilbert*'[113].

This position has also been endorsed on several occasions by Curry, who writes:

> 'I agree with Weyl and Gentzen that there are purposes for which intuitionist systems are acceptable, although they are not acceptable on empirical grounds, for application to physics. Again acceptability is a different question from truth; in fact, a formalist definition of mathematical truth is compatible with almost any position in regard to acceptability. In this sense, formalist mathematics is compatible with various philosophical views; it is an objective science which can form part of the data of philosophy'[114].

The philosophical question of what kind of 'actualist' sense can be ascribed to nonfinitist mathematics as a result of finitist discoveries, however, cannot be settled within metamathematics. Gentzen concludes #4 with the observation that his consistency proof for elementary number theory, for example, shows 'that it is possible to reason consistently *as if* everything in the infinite domains of objects were as actualisticly determined as in the finite domains. What the proof does *not* answer is to what extent anything 'real' corresponds to the actualist sense of a transfinite proposition apart from what its restricted finitist sense expresses'. In this connection the intuitionists have been particularly adamant in their insistence that actualist (nonconstructive) mathematics is utterly without sense and should be abandoned altogether. Their arguments seem to be based on the Kantian conviction that

> 'all concepts, and with them all principles, even such as are possible *a priori*, relate to empirical intuitions, that is, to the data for a possible experience. Apart from this relation they have no objective validity, and in respect of their representations ((Vorstellungen)) are a mere play of imagination or of understanding. Take, for instance, the concepts of mathematics . . .. Although all these principles, and the representation of the object with which this science occupies itself, are generated in the mind completely *a priori*, they would mean nothing, were we not always able to present their meaning in appearances, that is, in empirical objects. We therefore demand that a bare ((abgesondert)) concept be made sensible, that is, that an object corresponding to it be presented in intuition'[115].

Kant considered mathematics to meet this demand automatically since its concepts are usually illustrated by means of 'figures' and since mathematics thus produces 'appearances' present to the senses. In 1926, even Hilbert stated that

> 'as a condition for the use of logical inferences and the performance of logical operations, something must already be given to our faculty of representation ((in der Vorstellung)), certain extralogical concrete objects that are intuitively ((anschaulich)) present as immediate experience prior to all thought. If logical inference is to be reliable, it must be possible to survey these objects completely in all their parts, and the fact that they occur, that they differ from one another, and that they follow each other, or are concatenated, is immediately given intuitively, together with the objects, as something that neither can be reduced to anything else nor requires reduction. This is the basic philosophical position I consider requisite for mathematics and, in general, for all scientific thinking, understanding, and communication. And in mathematics, in particular, what we consider is the concrete signs themselves, whose shape, according to the conception we have adopted, is immediately clear and recognizable'[116].

Hilbert here states in embryonic form the 'finitist' view of formalist mathematics and in this assessment finds himself in agreement with Kant and the intuitionists. Gentzen's decisive contribution to mathematics lies in his success in developing and exploring the consequences of this 'finitist' point of view. He showed that the adoption of finitist principles secures extensive areas of mathematics against the destructive effect of the antinomies of set theory, and that these principles are also likely to provide an adequate basis for the *as if* interpretation of many of those levels of mathematics which can still consistently be regarded as idealizations of reality.

Montreal, June 1968   M.E.S.

NOTES TO THE INTRODUCTION

[0] P. Bernays, Contributions to logic and methodology, edited by A. Tymieniecka, North-Holland, 1965, chapter 1, page 5.
[1] J. Lambek, Deductive systems and categories, I, Math. Systems Theory, 2, No. 4, 1968, pp. 287–318; II, forthcoming in Proc. of the Seattle Conf. on Categories, 1968.
[2] A. Church, Introduction to mathematical logic, Princeton University Press, 1956, page 325, footnote 533.
[3] E. W. Beth, The foundations of mathematics, North-Holland, 1965, page 319.
[4] K. Schröter, Theorie des logischen Schliessens, I, II, Zeitschr. f. math. Logik, **1** (1955) pages 37–86 and **4** (1958) pages 10–65. For an English discussion of Schröter's papers the reader is referred to G. T. Kneebone, Mathematical logic, Van Nostrand, 1963, pages 382–384.
[5] A. Tarski, Logic, semantics, metamathematics, Oxford University Press, 1956, page 417, footnote, in which he states: 'After the original of this paper appeared in print, H. Scholz in his article 'Die Wissenschaftslehre Bolzanos, Eine Jahrhundert-Betrachtung', Abhandlungen der Fries'schen Schule, new series, vol. 6, pages 399–472 (see in particular page 472, footnote 58) pointed out a far-reaching analogy between this definition of consequence and the one suggested by B. Bolzano about a hundred years earlier.'
[6] Cf. J. van Heijenoort, From Frege to Gödel, Harvard University Press, 1967, pages 129–138.
[7] Cf. J. van Heijenoort, loc. cit., pages 367–392 and 464–479.
[8] H. Weyl, Die Stufen des Unendlichen, Jena, 1931.
[9] #4, article 9.3.
[10] An English translation of the relevant paper by Gödel has appeared in The undecidable, edited by M. Davis, Raven Press, 1965, pages 75–81.
[11] Cf. J. van Heijenoort, loc. cit., pages 414–437.
[12] #2, § 6.
[13] English translations of the relevant paper by Gödel have appeared in M. Davis, loc. cit., and J. van Heijenoort, loc. cit.
[14] Cf. #3, note 23, from which it appears that Gentzen is the originator of the symbol ∀ for the universal quantifier.
[15] Cf. Frege's *Begriffsschrift*, translated in J. van Heijenoort, loc. cit., pages 1–82; Russell and Whitehead's Principia Mathematica; and Hilbert and Ackermann's Grundzüge der theoretischen Logik, translated under the title of Mathematical logic, Chelsea, 1950.
[16] H. B. Curry, Foundations of mathematical logic, McGraw-Hill, 1963, pages 248–249.
[17] H. B. Curry, loc. cit., pages 184–185.
[18] #3, synopsis.
[19] S. Jaśkowski, On the rules of supposition in formal logic, Studia logica **1**, Warsaw

1934, reprinted in Polish logic, edited by S. MacCall, Oxford University Press, 1967, pages 232–258; cf. also D. Prawitz, loc. cit., page 98; and for a comprehensive historical survey cf. H. B. Curry, loc. cit. pages 245–253; also extremely pertinent at this point are Curry's Remarks on inferential deduction, Contributions to logic and methodology, loc. cit., chapter 2, where several important questions, including the recent 'dialogue approach' to logical calculi by Lorenzen are discussed.

[20] #8, introd. remarks, paragraph 4.

[21] W. Kneale, The privince of logic, Contemporary British philosophy, third series; cf. also the remarks made in the preface to the excellent introductory logic text by J. Anderson and H. Johnstone, Jr., entitled Natural deduction, Wadsworth, 1962, in which it is stated that 'this book is based upon Gentzen's techniques of natural deduction. Gentzen's techniques constitute a very natural approach to the study of the proofs occurring in axiom systems as well as a sound basis for the analysis of the properties of formula systems as such'.

[22] #4, articles 1.8 and 9.6.

[23] H. Leblanc, chapter 3 of Contributions to logic and methodology, loc. cit.

[24] D. Prawitz, Natural deduction, Stockholm Studies in Philosophy 3, 1965; cf. also the review of this monograph by R. Thomason in the J. Symb. Logic 32 (1967) pages 255–256.

[25] Cf. #3, p. 70 below; also pp. 88–89 below.

[26] The L-systems are developed in #3, section III.

[27] #3, article 2.3.

[28] The term 'extended Hauptsatz' appears to have been coined by S. C. Kleene in his Introduction to metamathematics, North-Holland, 1952, page 460; it has been erroneously translated back into German by some authors as the 'erweiterter Hauptsatz' (cf. E. W. Beth, loc. cit., page 267; H. B. Curry, Foundations of mathematical logic, page 351); In his Mathematical logic, Wiley, 1967, page 342, Kleene now refers to the theorem as the 'sharpened Hauptsatz', and in the present volume the translator follows this terminology.

[29] J. van Heijenoort, loc. cit., pages 525–581.

[30] #3, note 25.

[31] J. van Heijenoort, loc. cit., page 526.

[32] H. B. Curry, A theory of formal deducibility, Notre Dame mathematical lectures 6 Univ. of Notre Dame, 1950.

[33] H. B. Curry, J. Symb. Logic. 17 (1952) pages 249–265.

[34] C. I. Lewis and C. H. Langford, Symbolic logic, New York, 1932.

[35] M. Ohnishi and K. Matsumoto, Osaka J. Math. 9 (1957) pages 113–130 and 11 (1959) pages 115–120.

[36] S. Kanger, Provability in logic, Stockholm Studies in Philosophy, 1 1957.

[37] H. Rasiowa and R. Sikorski, On the Gentzen theorem, Fund. Math. 48 (1960) pages 57–69.

[38] W. Craig, Linear reasoning, A new form of the Herbrand-Gentzen theorem, J. Symb. Logic 22 (1957) pages 250–268; three uses of the Herbrand-Gentzen theorem in relating model theory and proof theory, ibid., pages 269–285.

[39] S. C. Kleene, Mathematical logic, Wiley, 1967.

[40] E. W. Beth, loc. cit., section 67, pages 186–201.

[41] E. W. Beth, loc. cit., page 200.

[42] S. C. Kleene, Mathematical logic, loc. cit., pages 283–295.

[43] E. W. Beth, loc. cit., section 94, pages 288–293; cf. also S. C. Kleene, Mathematical logic, pages 361–367 for a heuristic discussion of this theorem.

[44] W. Craig, loc. cit., page 269.

[45] O. Ketonen, Ajatus (Porvoo) 10 (1941) pages 77–92.

[46] An English translation of Gödel's result has appeared in J. van Heijenoort, loc. cit., pages 582–591.
[47] O. Ketonen, Ann. Acad. Sci. Fenn., Ser. A, I, Mathematica-physica **23**, Helsinki (1944), pages 71 seq.
[48] P. Bernays, J. Symb. Logic **10** No. 4, (1945) pages 127–130.
[49] #3, section IV, § 3.
[50] #3, section IV, § 3, article 3.4.
[51] #4, article 16.11.2.
[52] This observation was communicated to the editor by Prof. Bernays.
[53] For a discussion of the fan theorem (also known as the fundamental theorem on finitary spreads) cf. E. W. Beth, loc. cit., section 69, pages 194–196 and section 140, pages 427–433.
[54] #4, articles 14.5 and 16.2.
[55] #4, article 2.3.
[56] Hilbert and Bernays, Grundlagen der Mathematik, Vol. 1, Springer, 1934, page 32: 'Die ausgeführte Betrachtung der Anfangsgründe von Zahlentheorie und Algebra diente dazu, uns das direkte inhaltliche, in Gedanken-Experimenten an anschaulich vorgestellten Objekten sich vollziehende und von axiomatischen Annahmen freie Schliessen in seiner Anwendung und Handhabung vorzuführen. Diese Art des Schliessens wollen wir, um einen kurzen Ausdruck zu haben, als das '*finite*' Schliessen und ebenso auch die diesem Schliessen zugrunde liegende methodische Einstellung als die 'finite' Einstellung oder den 'finiten' Standpunkt bezeichnen. Im gleichen Sinne wollen wir von finiten Begriffsbildungen und Behauptungen sprechen, indem wir allemal mit dem Worte 'finit' zum Ausdruck bringen, dass die betreffende Überlegung, Behauptung oder Definition sich an die Grenzen der grundsätzlichen Vorstellbarkeit von Objekten sowie der grundsätzlichen Ausführbarkeit von Prozessen hält und sich somit im Rahmen konkreter Betrachtung vollzieht.'
[57] #4, article 16.1.
[58] S. C. Kleene, Introduction to metamathematics, North-Holland, 1952, page 479.
[59] Cf. Revue internationale de philosophie **28** (1954), page 19.
[60] M. Black, Mind, new series, **49** (1940), page 247. Black here introduces a series of 'patterns' and uses them to show that Gentzen's 'transfinite ordinals' are entirely independent of Cantor's ordinal number theory.
[61] #4, §§ 10 and 11.
[62] #4, article 1.8.
[63] #4, § 11.
[64] This short address has been omitted from the present volume because it overlaps entirely with #4 and #6. Several details, however, have been woven into this introduction, whenever appropriate. Gentzen's text has appeared under the title Unendlichkeitsbegriff und Widerspruchsfreiheit der Mathematik, in IXe Congrès Int. d. Phil. VI, logique et mathématiques, Paris, 1937.
[65] Cf. J. van Heijenoort, loc. cit., pages 446–463.
[66] #4, § 17.
[67] #4, article 13.4. The concept is illustrated in article 13.8. In article 13.90, Gentzen shows why the consistency of elementary number theory follows from the reducibility of the derivable sequents to reduced form.
[68] The expression 'restricted transfinite induction' is here used in the sense of Gentzen's 'Anfangsfall der transfiniten Induktion' and conforms with current English usage; cf. for example, R. L. Goodstein, On the restricted ordinal theorem, J. Symb. Logic **9**, No. 2, (1944) pages 33–41.
[69] Cf. note 64, page 12, of this Introduction.
[70] For further studies of the constructivity of segments of Cantors' second number class

cf. also A. Church, The constructive second number class, Bull. Amer. Math. Soc. **44** (1938); S. C. Kleene, On notation for ordinal numbers, J. Symb. Logic **3** (1938); W. Ackermann, Konstruktiver Aufbau eines Abschnitts der zweiten Cantorschen Zahlenklasse, Math. Zeitschrift, **53**, No. 5, (1951) pages 403–413, where Ackermann explains that in view of their metamathematical applications the relevant ordinal numbers must be developed constructively, and considers his construction in agreement with the conditions imposed by Church and Kleene in loc. cit., viz., that of any two numbers it can be decided recursively whether they are equal, or which is the greater and which is the smaller, that the first ordinal number is known explicitly and of any ordinal number it can be decided whether or not it has an immediate predecessor. If an immediate predecessor exists, it can be uniquely specified; if not, a sequence of ordinal numbers can be produced recursively such that the given ordinal number is the limit number of that sequence. It can furthermore be shown that 'restricted transfinite induction' holds up to that ordinal number. These criteria coincide essentially with those stipulated by Gentzen in #8, § 4, article 4.1.

[71] Cf. #4, article 17.1; #8, article 4.4; #6, pp. 232–233; #7, §§ 1 and 3.

It seems appropriate to mention at this point that in a letter to Paul Bernays dated September 24th, 1938, László Kalmár presented a consistency proof for elementary number theory by adapting Gentzen's results in #4 and #8 to the ordinary formalization of the propositional and predicate calculi as presented in volume I of *Hilbert and Bernays*, §§ 3–4. This proof has been included in the new edition of *Hilbert and Bernays*. An English translation of Kalmár's proof was recently prepared and is at present available in note form.

[72] F. B. Fitch, The consistency of the ramified Principia, J. Symb. Logic **3** (1938) pages 140–149.

[73] P. Lorenzen, Algebraische und logistische Untersuchungen über freie Verbände, J. Symb. Logic **16**, No. 2, (1951) pages 81–106.

[74] H. Wang, J. Symb. Logic **16**, No. 4, (1951) pages 269–272.

[75] #3 of the present volume.

[76] Cf. J. van Heijenoort, loc. cit., pages 525–581.

[77] The translation of this quotation is by the editor.

[78] G. Takeuti, On a generalized logic calculus, Japan. J. Math. **23** (1953) pages 39–96.

[79] D. Hilbert and W. Ackermann, loc. cit., English edition, pages 87–88.

[80] P. Bachmann, Jahrbuch über die Fortschritte der Mathematik, 1936, pages 43–44; the translation of the quotation is by the editor.

[81] D. Hilbert and W. Ackermann, loc. cit., English edition, page 88.

[82] H. A. Schmidt, Zentralblatt Math. **15**, No. 5, page 193.

[83] A. Tarski, Einige Betrachtungen über die Begriffe der ω-Widerspruchsfreiheit und der ω-Vollständigkeit, Monatsh. Math. Phys. **40** (1933) pages 97–112; English translation by J. H. Woodger in Logic, semantics, metamathematics, Oxford Univeristy Press, 1956, pages 279–295.

[84] K. Schütte, J. Symb. Logic **22**, No. 4, (1957) pages 351–352.

[85] G. Takeuti, Construction of ramified real numbers, Ann. Japan Ass. Philos. Sci. **1**, No. 1, (1956) pages 41–61.

[86] G. Takeuti, On the fundamental conjecture of *GLC*, ibid. **7** (1955) pages 249–275; **8** (1956) pages 54–64, 145–155, and **10** (1958) pages 121–134.

[87] K. Schütte, J. Symb. Logic **24** (1959) page 64.

[88] G. Takeuti, J. Math. Soc. Japan **8**, **9**, **10**.

[89] K. Schütte, Beweistheoretische Untersuchungen der verzweigten Analysis, Math. Ann. **124** (1952) pages 123–147; Ein widerspruchsloses System der Analysis auf typenfreier Grundlage, Math. Zeitschrift **61** (1954) pages 160–179; for a comprehensive summary of the main metamathematical results up to 1960 cf. K. Schütte, Beweis-

theorie, Springer, 1960.
[90] K. Schütte, Math. Ann. **124** (1952) pages 123–147.
[91] K. Schütte, Beweistheoretische Erfassung der unendlichen Induktion in der Zahlentheorie, Math. Ann. **122**, No. 5, (1951) pages 369–389.
[92] W. Ackermann, Zur Widerspruchsfreiheit der Zahlentheorie, Math. Ann. **117** (1940) pages 162–194; Widerspruchsfreier Aufbau der Logik, J. Symb. Logic **15** (1950) pages 403–413; Math. Zeitschrift **55** (1951–52) pages 364–384, ibid. **57** (1952–53) pages 155–166.
[93] Cf. #9, page 308.
[94] Cf. #8.
[95] Cf. G. Takeuti, Construction of ramified real numbers, in which the consistency of a certain subsystem of analysis is proved by cut elimination using a transfinite induction up to ω^ω.
[96] Cf. #8, article 4.4.
[97] Cf. Paul Bernays's review of #9 in the J. Symb. Logic **9** (1944) pages 70–72, for a summary discussion.
[98] Cf. #7, § 2; cf. also note 66.
[99] Cf. A. Mostowski, The present state of investigations on the foundations of mathematics, Rozprawy Mat. **9** (1955), 48 pages.
[100] H. B. Curry, Zentralblatt Math. **19**, No. 3, (1938) page 97.
[101] Cf. page 9.
[102] Cf. #6.
[103] J. Hjelmslev, Die natürliche Geometrie, Hamb. Math. Einzelschriften **1**, 1923.
[104] J. B. Rosser, Bull. Amer. Math. Soc., Nov. 1939, pages 812–813.
[105] Cf. I. Kant, Kritik der reinen Vernunft, Felix Meiner, 1956, page 289, where Kant speaks of 'Dinge überhaupt und an sich selbst' and which N. Kemp Smith, Macmillan, 1956, page 259, translated as 'things in general and in themselves'; in the original handwritten manuscript, Kant had actually used the phrase 'Gegenstände, die uns in keiner Anschauung gegeben werden, mithin nicht sinnliche Gegenstände' (objects which are not given to us by intuition, hence not sensible objects); it seems reasonable that Gentzen's 'an sich' should be taken in this sense.
[106] Cf. H. Vaihinger, Die Philosophie des Als-Ob, 1911; an English edition was published in 1924 under the title of 'The Philosophy of the As-If'; for a very stimulating and pertinent discussion of the philosophical questions involved in the 'as if' attitude to mathematics cf. also A. Fraenkel and Y. Bar-Hillel, Foundations of set theory, North-Holland Publ. Co., 1958, pages 332–347.
[107] Cf. foot 64 of this Introduction.
[108] Cf. P. Lorenzen, Das Aktual-Unendliche in der Mathematik, Philosophia Naturalis, Vol. IV (1959), pages 1–11.
[109] Cf. the preface to F. Waisman's Introduction to mathematical thinking, Harper Torchbooks.
[110] Cf. the end of #7.
[111] W. V. O. Quine, On what there is, From a logical point of view, Harper Torchbooks, page 15.
[112] Cf. E. W. Beth, loc. cit., page 364.
[113] Cf. #7, § 4.
[114] H. B. Curry, Remarks on the definition and nature of mathematics, in Philosophy of mathematics, edited by P. Benacerraf and H. Putnam, Prentice Hall, 1964, page 156; cf. also H. B. Curry, Outlines of a formalist philosophy of mathematics, North-Holland, 1951, pages 59–64.
[115] I. Kant, Critique of pure reason, translated by N. Kemp Smith, loc. cit., pages 259–260; cf. Kritik der reinen Vernunft, loc. cit., page 290 of the German edition.
[116] D. Hilbert, On the infinite, J. van Heijenoort, loc. cit., page 376.

1. ON THE EXISTENCE OF INDEPENDENT AXIOM SYSTEMS FOR INFINITE SENTENCE SYSTEMS

In the following, I intend to discuss the question raised in a paper by P. Hertz[1]:

Are there infinite closed sentence systems without an independent axiom system? (Terminology vid. seq.)

I shall prove two main results, viz.:
1. There are infinite closed sentence systems for which no independent axiom system exists.
2. For every denumerably infinite closed *linear* sentence system, it is possible to state an independent axiom system.

The proofs of these theorems form the contents of sections II and III of the present paper; section I is devoted to the definition of the terminology used and to the proofs of several lemmas.

A knowledge of Hertz's papers is not assumed.

SECTION I. NOTATIONS AND SOME LEMMAS

The notations used agree largely with those of Hertz. I shall define them once more in order to be independent of Hertz's papers and also because the choice of somewhat simpler rules of inference has necessitated a reformulation of the definitions. For the benefit of readers familiar with Hertz's papers, I have marked those expressions whose meanings I have retained by an asterisk when first introducing them.

§ 1. The 'sentences' [2]

A *sentence** has the form
$$u_1 u_2 \ldots u_\nu \to v.$$
The u's and v's are called *elements**. We might think of them as events,

and the 'sentence' then reads: The happening of the events $u_1, \ldots, u_v$ causes the happening of v.

The 'sentence' may also be understood thus: A domain of elements containing the elements $u_1, \ldots, u_v$ also contains the element v.

The elements may furthermore be thought of as properties and the 'sentence' can then be interpreted thus: An object with the properties $u_1, \ldots, u_v$ also has the property v.

Or we imagine the elements to stand for 'propositions', in the sense of the propositional calculus, and the 'sentence' then reads: If the propositions $u_1, \ldots, u_v$ are true, then proposition v is also true.

Our considerations do not depend on any particular kind of informal interpretation of the 'sentences', since we are concerned only with their formal structure.

We assume the u's to be distinct. The order of their arrangement is unimportant, i.e., sentences which differ only with respect to the order of the u's, are considered identical.

It may happen, however, that v is identical with one of the u's; such a sentence is called *trivial**.

If we denote a system of finitely many elements, which is called a *complex**, by a capital letter, we can also write a 'sentence' thus:

$$K \to v,$$

where $K = (u_1, u_2, \ldots, u_v)$. This complex is called the *antecedent** and the element v the *succedent** of the sentence.

A complex shall not be empty unless this has been explicitly specified.

LM denotes the set-theoretic union of L and M.

A sentence with only one antecedent element is called *linear**, and a linear trivial sentence, which therefore has the form $v \to v$, *tautologous**.

§ 2. The 'proof' of a 'sentence' from other 'sentences'

From some 'sentences' we can 'prove' other 'sentences' by applying certain 'forms of inference' to them. E.g., from $u \to v$ and $v \to w$ we can infer: $u \to w$.

An individual inference consists of a number of sentences called *premisses**, and a further sentence, called the *conclusion**, which is deduced from the premisses. We shall introduce two such forms of inference and shall show in § 4 that, informally interpreted, they are correct and sufficient:

the 'thinning', which has one premiss, and the 'cut', which has two premisses.

1. A *thinning* (Hertz calls it an 'immediate inference') has the form:

$$\frac{L \to v}{ML \to v} \begin{array}{l}\text{Premiss}\\ \text{Conclusion}\end{array}.$$

M may be empty. – We shall also use the expressions: 'to thin a sentence', 'a thinned sentence of ...'. Informally, we may interpret a thinning as an adjunction of assumptions (M).

2. A *cut* has the form:

$$\begin{array}{cc}\textit{Lower sentence:} & \textit{Upper sentence:}\\ L \to u & Mu \to v\\ \hline & LM \to v\\ & \textit{Conclusion of the cut}\end{array} \begin{array}{l}\text{Premisses}\\ \text{Conclusion}\end{array}.$$

M may be empty. u must not occur in M. u is called the *cut element*. We shall also use the phrase: 'To cut a sentence (called the lower sentence) with another sentence (called the upper sentence)'.

It may happen that, given two sentences $\mathfrak{p}$ and $\mathfrak{q}$, it is impossible to cut either $\mathfrak{p}$ with $\mathfrak{q}$ or $\mathfrak{q}$ with $\mathfrak{p}$, or that $\mathfrak{p}$ can be cut with $\mathfrak{q}$ but not $\mathfrak{q}$ with $\mathfrak{p}$; finally, it may be possible to cut both $\mathfrak{p}$ with $\mathfrak{q}$ and $\mathfrak{q}$ with $\mathfrak{p}$. (In the latter case the conclusion of the cut is trivial both ways, as is easily seen.) Informally, a cut may be interpreted as the replacement of one assumption (u) by another assumption (L), where the latter must include the former.

A cut of two linear sentences has necessarily the form

$$\frac{u \to v \quad v \to w}{u \to w}.$$

The conclusion is again linear.

By a *proof* of a sentence q from the sentences $\mathfrak{p}_1, \ldots, \mathfrak{p}_\nu$ ($\nu \geq 0$) we shall henceforth mean an ordered succession of inferences (i.e., thinnings and cuts)[3] arranged in such a way that the conclusion of the last inference is q and that its premisses are either premisses of the $\mathfrak{p}$'s or tautologies, or coincide with the conclusions of earlier inferences.

The fact that we consider tautologous sentences as proved will later on (cf. § 4) turn out to have been justified. With this exception, our definition of a proof expresses precisely what it informally suggests.

A sentence q is called *provable* from the sentences $\mathfrak{p}_1, \ldots, \mathfrak{p}_\nu$, if there exists a proof for q from $\mathfrak{p}_1, \ldots, \mathfrak{p}_\nu$.

§ 3. The equivalence of our concept of provability with that of P. Hertz
(This paragraph will not be needed below.)

In place of the 'cut', P. Hertz uses a somewhat more complicated form of inference, the 'syllogism' (cf. H. 4, pp. 462–463). The cut is a special syllogism. Since our other definitions coincide with those of Hertz, the equivalence of our concept of provability with that of P. Hertz is assured if we can show that the syllogism may be transformed into a proof in our sense. This will now be done.

The syllogism runs:

$$\left. \begin{array}{c} L_1 \to u_1 \\ \vdots \\ L_\nu \to u_\nu \end{array} \right\} \text{Lower sentences}$$

$$\frac{Mu_1 \ldots u_\nu \to v}{L_1 \ldots L_\nu M \to v} \quad \begin{array}{l} \text{Upper sentence} \\ \text{Conclusion} \end{array}$$

M may be empty.

We begin by transforming it into a proof consisting of a succession of simpler syllogisms:

$$\frac{L_2 \to u_2 \quad \dfrac{L_1 \to u_1 \quad Mu_1 \ldots u_\nu \to v}{L_1 Mu_2 \ldots u_\nu \to v}}{}$$

$$\frac{L_\nu \to u_\nu \quad L_{\nu-1} \ldots L_1 Mu_\nu \to v}{L_\nu \ldots L_1 M \to v}.$$

Every single one of these syllogisms has already the form of a cut, viz.,

$$\frac{L_\rho \to u_\rho \quad L_{\rho-1} \ldots L_1 Mu_\rho \ldots u_\nu \to v}{L_\rho L_{\rho-1} \ldots L_1 Mu_{\rho+1} \ldots u_\nu \to v}.$$

According to our definition, this expression fails to be a cut only if u_ρ occurs in $L_{\rho-1} \ldots L_1 Mu_{\rho+1} \ldots u_\nu$. In this case, however, the conclusion is a thinning of the upper sentence.

This shows that Hertz's syllogism may be broken up into cuts and thinnings, which proves our assertion.

§ 4. The correctness and completeness of the forms of inference. The normal proof
(Cf. H.3)

Our formal definition of provability and, more generally, our choice of the forms of inference will seem appropriate only if it is certain that a sentence q is 'provable' from the sentences $\mathfrak{p}_1, \ldots, \mathfrak{p}_\nu$ if and only if it represents informally a consequence of the $\mathfrak{p}$'s. We shall be able to show that this is indeed so as soon as we have fixed the meaning of the still somewhat vague notion of 'consequence', in accordance with a particular informal interpretation of our 'sentences' as follows:

We first introduce an auxiliary concept:

We say that a complex of elements *satisfies** (cf. H.3) a given sentence if it either does not contain all antecedent elements of the sentence, or alternatively, contains all of them and also the succedent of that sentence. – In order that a complex should *not* satisfy a sentence, it is therefore necessary and sufficient that it contains the antecedent of that sentence, but not its succedent.

We now look at the complex K of all (finitely many) elements of $\mathfrak{p}_1, \ldots, \mathfrak{p}_\nu$ and q, and call q a *consequence* of $\mathfrak{p}_1, \ldots, \mathfrak{p}_\nu$[4] if (and only if) every subcomplex of K which satisfies the sentences $\mathfrak{p}_1 \ldots \mathfrak{p}_\nu$, also satisfies q.

(If we are working within the framework of the propositional calculus, then the place of a subcomplex of K is taken by an assignment of truth-values to the elements of K in which the elements of this subcomplex take the value 'true' and the others the value 'false'. Under this interpretation, all proofs that follow can be carried out analogously. Our concept of consequence also proves adequate for the other informal interpretations of the 'sentences' mentioned above.)

That part of our assertion, made above, which deals with what might be called the 'informal correctness' of our forms of inference, can now be formulated as follows:

THEOREM I.[5] If a sentence q is 'provable' from the sentences $\mathfrak{p}_1, \ldots, \mathfrak{p}_\nu$, then it is a 'consequence' of them.

The theorem is proved[5] in five simple steps

1. A tautologous sentence is a 'consequence' of any sentence.

This is so since every tautologous sentence is satisfied by any given complex.

2. A thinning of a sentence is a consequence of that sentence.

The thinning schema runs:
$$\frac{L \to v}{ML \to v}.$$

A complex satisfying the sentence $L \to v$ either does not contain the whole antecedent L, nor therefore the whole antecedent ML of the thinned sentence, and hence satisfies that sentence; or it contains both L and v, in which case it also satisfies the thinned sentence, since it contains the succedent v of that sentence.

3. The conclusion of a cut of two sentences is a consequence of these sentences.

The cut schema runs:
$$\frac{L \to u \qquad Mu \to v}{LM \to v}.$$

Thus a complex which does not satisfy the conclusion of the cut, would have to contain LM and not v. If it also contained u, then it would not satisfy the upper sentence; if it did not contain u, then it would not satisfy the lower sentence $L \to u$. A complex satisfying both premises thus also satisfies the conclusion of the cut.

4. A sentence which results by a thinning or a cut from (one or two) premises which are themselves consequences of $\mathfrak{p}_1, \ldots, \mathfrak{p}_\nu$, is also a consequence of $\mathfrak{p}_1, \ldots, \mathfrak{p}_\nu$.

This follows since, by 2) and 3), the sentence is a consequence of its premises, and thus a complex which did not satisfy it would have to fail to satisfy at least one of the premises and hence could not satisfy all $\mathfrak{p}$'s.

5. Suppose now that a proof for q from $\mathfrak{p}_1, \ldots, \mathfrak{p}_\nu$ is given. Each one of the sentences $\mathfrak{p}$ is a trivial consequence of $\mathfrak{p}_1, \ldots, \mathfrak{p}_\nu$, and so, by 1), is every tautologous sentence; from 4) it then follows that every sentence of the proof and q in particular, is a consequence of $\mathfrak{p}_1, \ldots, \mathfrak{p}_\nu$.

The proof of the other part of our assertion, which might be called the 'informal completeness' of our forms of inference, is more difficult and is expressed as follows: If a sentence q is a 'consequence' of the sentences $\mathfrak{p}_1, \ldots, \mathfrak{p}_\nu$, then it is also 'provable' from them[6].

§ 4, THE CORRECTNESS AND COMPLETENESS OF THE FORMS OF INFERENCE

In view of its later applications, we shall prove this theorem in a considerably stronger form by showing that for every 'proof' there in fact exists a proof in 'normal form', in a sense which we shall describe below.

We begin by pointing out that if q is trivial and if t is the tautologous sentence with the same succedent, then q is immediately provable from $\mathfrak{p}_1, \ldots, \mathfrak{p}_\nu$ by the proof

$$\frac{t}{q}.$$

From now on, we can therefore confine our attention to nontrivial sentences q.

By a *normal proof* of a nontrivial sentence q from the sentences $\mathfrak{p}_1, \ldots, \mathfrak{p}_\nu$ ($\nu \geqq 1$) we mean a proof which can be written in the following form:

$$\frac{\dfrac{\dfrac{\dfrac{\mathfrak{r}_0 \quad \mathfrak{s}_0}{\mathfrak{r}_1 \quad \mathfrak{s}_1}}{\vdots}}{\dfrac{\mathfrak{r}_{\rho-1} \quad \mathfrak{s}_{\rho-1}}{\mathfrak{s}_\rho}}}{q},$$

where $\rho \geqq 0$. (If $\rho = 0$, we mean of course the form $\dfrac{\mathfrak{s}_0}{q}$.) Here

$$\frac{\mathfrak{r}_\lambda \quad \mathfrak{s}_\lambda}{\mathfrak{s}_{\lambda+1}}$$

always denotes a cut with $\mathfrak{r}_\lambda$ for its lower sentence, $\mathfrak{s}_\lambda$ for its upper sentence, and $\mathfrak{s}_{\lambda+1}$ for its conclusion.

Suppose now that no trivial sentence occurs. The *initial sentences* $\mathfrak{s}_0, \mathfrak{r}_0, \ldots, \mathfrak{r}_{\rho-1}$ form part of the $\mathfrak{p}$'s. (Not all of the sentences $\mathfrak{p}$ need occur, and the same sentence may appear several times.)

We notice: In a normal proof a single thinning is carried out at the end (which may of course leave the sentence $\mathfrak{s}_\rho$ unchanged) and before it occur only cuts (possibly none); the *lower sentence* of a cut must here always be one of the $\mathfrak{p}$'s. The sentences $\mathfrak{s}_0, \ldots, \mathfrak{s}_\rho, q$ all have the same succedent (cf. the cut schema). In the entire proof no elements other than those of $\mathfrak{p}_1, \ldots, \mathfrak{p}_\nu$ and q occur, since a cut cannot give rise to elements which did not occur in its upper or lower sentences[7].

Our assertion may now be formulated as follows:

THEOREM II.[8] If a non-trivial sentence q is a 'consequence' of the sentences $\mathfrak{p}_1, \ldots, \mathfrak{p}_\nu$, then there exists a 'normal proof' for q from $\mathfrak{p}_1, \ldots, \mathfrak{p}_\nu$.

PROOF. Suppose q is of the form

$$L \to v.$$

We examine the system $\mathfrak{S}$ of all those sentences which have v for their succedent, are nontrivial, and for which there exists a normal proof from $\mathfrak{p}_1, \ldots, \mathfrak{p}_\nu$ without a thinning (i.e., a proof which is of the above normal form, but without the thinning at the end). (In particular, all of the $\mathfrak{p}$'s which are of this form themselves belong to $\mathfrak{S}$.) The system $\mathfrak{S}$ is finite since with the finitely many elements of the $\mathfrak{p}$'s no more than finitely many sentences can be formed and since no cut ever gives rise to new elements.

If there is a sentence in $\mathfrak{S}$ whose antecedent is entirely contained in L (possibly identical with L), then q is a thinning of that sentence and our assertion is proved.

Suppose that this is not the case. Then we shall derive a contradiction from the assumption that q is a consequence of $\mathfrak{p}_1, \ldots, \mathfrak{p}_\nu$, by producing a subcomplex N of the complex K of the elements of $\mathfrak{p}_1, \ldots, \mathfrak{p}_\nu, q$, which satisfies the sentences $\mathfrak{p}_1, \ldots, \mathfrak{p}_\nu$, but not the sentence q. N is formed by constructing a sequence of complexes

$$M_1, M_2, \ldots, M_\sigma = N$$

of which each results from its predecessor by the adjunction of a further element. For M_1 we take L, the antecedent of q. Suppose now that M_μ has been determined, we then obtain $M_{\mu+1}$ in the following way: From among the $\mathfrak{p}$'s we choose a sentence which does not satisfy the complex M_μ. (If there is no such sentence, vid. seq.) The succedent of that sentence, therefore, does not belong to M_μ (although its entire antecedent does). This sentence we then adjoin to M_μ and thus obtain $M_{\mu+1}$. After finitely many adjunctions we obviously get to the point where the last complex M_σ, which we then take as N, satisfies all $\mathfrak{p}$'s. This is bound to happen since the $\mathfrak{p}$'s have only finitely many elements and since the complex consisting of all of these elements certainly satisfies every $\mathfrak{p}$.

If we are now able to show that the complex N obtained in this way does not contain the succedent v of q, then we have finished since N does not satisfy q, although it satisfies all the $\mathfrak{p}$'s. For this purpose, we prove by induction the following assertion:

Each M satisfies all sentences of the system $\mathfrak{S}$ and does not contain the element v. From this it follows of course at once that v does not occur in N.

The assertion holds first of all for M_1, i.e., L. L does not contain v, otherwise q would be trivial. If the complex L did not satisfy a sentence of $\mathfrak{S}$, then it would contain the entire antecedent of that sentence. This case was ruled out earlier.

Now we assume that the assertion holds for M_τ and is to be proved for $M_{\tau+1}$. Suppose that $M'_{\tau+1} = uM_\tau$ and that $O \to u$ is that sentence among the $\mathfrak{p}$'s which gave rise to the adjunction of u. I.e., M_τ does not satisfy it. The antecedent O, therefore, belongs entirely to M_τ. The element v does not occur in M_τ (by the induction hypothesis) and therefore not in O. We have $u \neq v$, for otherwise $O \to u$ would be a sentence of $\mathfrak{S}$ which is not satisfied by M_τ. Hence $M_{\tau+1}$ does not contain the element v either.

Now suppose that there exists a sentence in $\mathfrak{S}$ which $M_{\tau+1}$ does not satisfy. Since the sentence is still satisfied by M_τ, it must obviously have the form:

$$Pu \to v.$$

The complex P may be empty; suppose that u does not occur in it. The antecedent Pu belongs to $M_{\tau+1}$ and therefore P belongs to M_τ. Now we consider the sentence

$$OP \to v$$

which results from $O \to u$ and $Pu \to v$ by a cut. This sentence belongs to $\mathfrak{S}$, since for $Pu \to v$ (as a sentence of $\mathfrak{S}$) there exists a normal proof from the $\mathfrak{p}$'s without a thinning, and by adjoining this cut in which $O \to u$, as a sentence of $\mathfrak{p}$, is the lower sentence, a normal proof for $OP \to v$ without a thinning results. Furthermore, $OP \to v$ is nontrivial, since v occurs in neither O nor P.

We thus have a sentence of $\mathfrak{S}$ which is not satisfied by M_τ (O and P belong to M_τ, but v does not). Hence we have a contradiction and theorem II is proved.

The only results of this paragraph needed later are two simple corollaries of theorems I and II, which we shall formulate as follows:

THEOREM III. *If a nontrivial sentence* q *is provable from the sentences* $\mathfrak{p}_1, \ldots, \mathfrak{p}_\nu$, *then there exists a normal proof for* q *from* $\mathfrak{p}_1, \ldots, \mathfrak{p}_\nu$.

This follows at once from theorems I and II together. The theorem can also be obtained directly without reference to the notion of consequence by taking an arbitrary proof and transforming it step by step into a normal proof. The reason for the approach chosen in this paper is that it involves little extra effort and yet provides us with important additional results, viz., the correctness and completeness of our forms of inference.

THEOREM IV. If a non-tautologous *linear* sentence q is provable from the *linear* sentences $\mathfrak{p}_1, \ldots, \mathfrak{p}_\nu$, then among the $\mathfrak{p}$'s there exists a sequence of sentences of the form

$$u \to v_1, v_1 \to v_2, \ldots, v_\lambda \to w \qquad (\lambda \geq 0),$$

where $u \to w = q$, and where $v_1, \ldots, v_\lambda, u, w$ are pairwise distinct elements.

By theorem III there after all exists a normal proof for q from $\mathfrak{p}_1, \ldots, \mathfrak{p}_\nu$. This proof must obviously look like this:

$$\frac{\dfrac{u \to x_1 \quad \dfrac{x_1 \to x_2 \quad x_2 \to w}{x_1 \to w}}{u \to w} \quad \cdots \quad \dfrac{x_{\rho-1} \to x_\rho \quad x_\rho \to w}{}}{u \to w}$$

Only linear sentences can occur in it, since q and the $\mathfrak{p}$'s are linear and since a cut of two linear sentences results in another linear sentence. Already the initial sentences

$$u \to x_1, x_1 \to x_2, \ldots, x_{\rho-1} \to x_\rho, x_\rho \to w$$

of the $\mathfrak{p}$'s almost yield our assertion; yet the elements $u, x_1, \ldots, x_\rho, w$ need not all be distinct. If several of these elements are identical, then we omit from the sequence of sentences those that lie between the first and last occurrence of the same element. In this way we eventually obtain a sequence of sentences in which only neighbouring elements are equal, and this yields our assertion. (It should be noted that since $u \neq w$, our procedure never leads to the cancellation of *all* sentences.)

§ 5. Systems of sentences

A system of sentences is called *closed under cuts*, if every possible cut of two sentences of the system results in another sentence of the system.

A *closed sentence system** for a given domain of elements is a system of sentences which consists of the elements of the domain together with all those sentences that are provable from sentences of the system (where thinnings may be carried out only with elements of the domain).

Any closed sentence system obviously contains all trivial sentences that can be formed from the elements of the domain. (After all, they are already provable from 0 sentences.)

THEOREM V. If, in a system $\mathfrak{S}$, closed under cuts, we incorporate all possible thinnings of the sentences of the system by means of elements of the system, as well as all trivial sentences (made up from the elements of the domain), we obtain a closed sentence system $\overline{\mathfrak{S}}$.

This follows at once from theorem III. For suppose that q is a non-trivial sentence provable from $\overline{\mathfrak{S}}$ (i.e., from sentences of $\overline{\mathfrak{S}}$). Then it is of course also provable from $\mathfrak{S}$, and the proof may be carried out, according to theorem III, by means of a single thinning at the end and by using only sentences from $\mathfrak{S}$ as initial sentences. Since the cuts of these sentences result in sentences of $\mathfrak{S}$, therefore q is a sentence of $\overline{\mathfrak{S}}$.

A sentence system is called *linear*, if every sentence of the system is a thinning of another *linear* sentence also belonging to that system.

An *axiom system** for a closed sentence system is a system of sentences belonging to the sentence system proper and from which all other sentences of the sentence system can be proved.

A sentence system is called *independent** if none of its sentences are provable from other sentences. (Such a system contains therefore no trivial sentences.)

In this paper we are particularly interested in the independence of *axiom systems*.

§ 6. The independence of axiom systems

In this paragraph we shall state certain results without exact proof; they will not be used later, but they facilitate the understanding of sections II and III. These results are proved in H.4 in the places referred to below.

For every *finite* closed sentence system there exists an independent axiom system. (H. 4, p. 466.)[9]

This is so since an independent axiom system results if all sentences which are provable from others are systematically eliminated.

This cancellation procedure may break down if the sentence system is *infinite*. It may happen, for example, that this process leads to the cancellation of all sentences. (H. 4, pp. 470–471.) There may, nevertheless, exist an independent axiom system. (H. 4, p. 471.)

Very instructive is the following example (given by Hertz) of a system $\mathfrak{A}$ of linear sentences closed under cuts for which no independent axiom system of linear sentences exists. (The minor differences between Hertz's definition of an axiom system and ours, introduced here only for closed sentence systems, should present no difficulty for the reader.) On the other hand, the linear closed sentence system $\overline{\mathfrak{A}}$ (cf. § 5) which results from the inclusion of the thinned and trivial sentences (using only the given domain of objects) does possess an independent axiom system (which does not, of course, consist only of linear sentences).

The system $\mathfrak{A}$ consists of the sentences:

$\mathfrak{A}_1$: $a_1 \to a_\nu$ $(\nu > 1)$
$\mathfrak{A}_2$: $a_\lambda \to a_\mu$ $(\lambda > \mu > 1)$.

(H. 4, pp. 469–470, 482–483.)

The impossibility of an independent axiom system stems from the following two facts (proved by Hertz):

1. The sentences of $\mathfrak{A}_2$ must be provable from every axiom system of $\mathfrak{A}$ without the use of $\mathfrak{A}_1$-sentences.

2. From one $\mathfrak{A}_1$-sentence and the $\mathfrak{A}_2$-sentences all $\mathfrak{A}_1$-sentences with a smaller ν are provable; but none with a larger ν.

The impossibility follows from these two facts since at least two $\mathfrak{A}_1$-sentences must occur among the axioms, and the sentence with a smaller ν would then be provable from other axioms; this makes the axiom system dependent.

An entirely analogous situation arises in the example of a *closed* sentence system without an independent axiom system described in section II below.

If the system $\mathfrak{A}$ is extended to a linear closed sentence system $\overline{\mathfrak{A}}$, however, then that system, as already mentioned, possesses an independent axiom system. Such a system is, for example, (H. 4, p. 482):

$$a_1 \to a_2 \qquad a_3 \to a_2$$
$$a_2 a_1 \to a_3 \qquad a_4 \to a_3$$
$$a_3 a_2 a_1 \to a_4 \qquad a_5 \to a_4$$
$$\cdots \qquad \cdots$$

The independence is achieved by including, in place of the $\mathfrak{A}_1$-sentences, thinnings of these sentences from which the associated $\mathfrak{A}_1$-sentence is in each case provable only with the help of earlier axioms (in the schema, these axioms occur on the left above the sentence concerned), so that we can now no longer prove earlier axioms from later ones (with the help of the $\mathfrak{A}_2$-sentences.)

The example that will be given in section II is constructed in such a way that this device cannot be applied. The sentence system concerned fails to be linear and indeed cannot be linear, since linear closed sentence systems always possess an independent axiom system, as will be shown in section III. There we shall also employ, among other things, the device just mentioned of taking as axioms certain thinnings of the linear sentences in order to achieve independence.

§ 7. 'Maximal nets'
(This paragraph will be needed only in section III.)
(Cf. H. 1, articles 10–16.)

For a given closed linear sentence system $\overline{\mathfrak{S}}$, we introduce the following concepts:

A non-tautologous linear sentence $u \to v$ of $\overline{\mathfrak{S}}$ with the property that $v \to u$ also belongs to $\overline{\mathfrak{S}}$, is called a *net sentence*.

An element occurring in a net sentence is called a *net element*.

The set of all elements v, for a given net element u, for which $u \to v$ and $v \to u$ are sentences of $\overline{\mathfrak{S}}$, is called a *maximal net*[*10]. (A maximal net contains at least two elements.)

THEOREM VI. *Every net element belongs to exactly one maximal net.*

The fact that a net element belongs to at least one maximal net is obvious. Suppose now that u belongs to two maximal nets and that these nets are determined by v and w respectively. (One of these may be identical with u.) Hence $u \to v$, $v \to u$, $u \to w$, $w \to u$ are sentences of $\overline{\mathfrak{S}}$. Suppose that x is an element which belongs to only one of the two maximal nets, say it belongs to the one determined by v. In that case $v \to x$, $x \to v$ are sentences of $\overline{\mathfrak{S}}$. The following sentences are thus obtained by cuts from these and the preceding sentences: $w \to v$, $w \to x$ and $v \to w$, $x \to w$. These conclusions of cuts belong to $\overline{\mathfrak{S}}$, since that system was assumed to be closed. Hence x also belongs to the maximal net determined by w, in contradiction to our assumption.

COROLLARIES:

The elements of a net sentence both belong to the same maximal net (by theorem VI).

If u and v are two distinct elements of the same maximal net, then $u \to v$ and $v \to u$ are sentences of $\overline{\mathfrak{S}}$. (This follows trivially.)

A linear sentence of $\overline{\mathfrak{S}}$ in which one element is a net element and in which the other is not an element of the same maximal net is called a *semi-net sentence*.

SECTION II. AN INFINITE CLOSED SENTENCE SYSTEM POSSESSING NO INDEPENDENT AXIOM SYSTEM

§ 1. Construction of the paradigm $\overline{\mathfrak{A}}$

The elements of the system are $b, c, a_1, a_2, \ldots$ (all mutually distinct).

We begin by considering a system $\mathfrak{A}$ of sentences that are divided into two classes:

$\mathfrak{A}_1$: $a_\nu b \to c$ for arbitrary ν,

$\mathfrak{A}_2$: $a_\lambda \to a_\mu$ for $\lambda < \mu$.

(This subdivision corresponds to that into $\mathfrak{A}_1$- and $\mathfrak{A}_2$-sentences in the system stated in section I, § 6.)

$\mathfrak{A}$ is closed under cuts.

PROOF. A cut of two $\mathfrak{A}_1$-sentences is not possible since c does not occur in any antecedent.

A cut of two $\mathfrak{A}_2$-sentences has the form

$$\frac{a_\rho \to a_\sigma \quad a_\sigma \to a_\tau}{a_\rho \to a_\tau},$$

where $\rho < \sigma$, $\sigma < \tau$, hence $\rho < \tau$, i.e., the conclusion of the cut belongs again to $\mathfrak{A}_2$.

A cut of an $\mathfrak{A}_1$- and an $\mathfrak{A}_2$-sentence has the form (c cannot be a cut element since it does not occur in an $\mathfrak{A}_2$-sentence):

$$\frac{a_\rho \to a_\sigma \quad a_\sigma b \to c}{a_\rho b \to c}.$$

The conclusion of the cut belongs again to $\mathfrak{A}_1$.

This shows that $\mathfrak{A}$ is closed under cuts.

We now adjoin to $\mathfrak{A}_1$ all nontrivial thinnings of $\mathfrak{A}_1$-sentences (formed with elements of $\mathfrak{A}$), this system we call $\mathfrak{B}_1$; similarly, we denote the system which results from $\mathfrak{A}_2$ by the inclusion of all non-trivial thinnings of $\mathfrak{A}_2$-sentences (formed with elements from $\mathfrak{A}$) by $\mathfrak{B}_2$.

The systems $\mathfrak{B}_1$ and $\mathfrak{B}_2$ have no sentences in common since all $\mathfrak{B}_1$-sentences, but none of the $\mathfrak{B}_2$-sentences, have c for their succedent. $\mathfrak{B}_1$ and $\mathfrak{B}_2$, together with the trivial sentences formed from the elements of $\mathfrak{A}$, now constitute a closed infinite sentence system $\overline{\mathfrak{A}}$. (Cf. theorem V, section I, § 5.)

We intend to show that for this system $\overline{\mathfrak{A}}$ no independent axiom system exists.

In order to facilitate the understanding of the following, the reader should first read the proof in § 3, since then the purpose of the various lemmas proved in § 2 will become clear.

§ 2. Some lemmas about the system $\overline{\mathfrak{A}}$

The smallest subscript of the a's in the antecedent of a $\mathfrak{B}_1$-sentence will be called the 'degree' of that sentence. (E.g. $a_3 a_7 a_2 b \to c$ has degree 2.)

LEMMA 1. A thinning of a $\mathfrak{B}_1$-sentence by elements of the system results in a $\mathfrak{B}_1$-sentence whose degree is no larger than that of the original sentence or it results in a trivial sentence, but never in a $\mathfrak{B}_2$-sentence.

LEMMA 2. A thinning of a $\mathfrak{B}_2$-sentence by elements of the system results in another $\mathfrak{B}_2$-sentence or in a trivial sentence, but never in a $\mathfrak{B}_1$-sentence.

Both assertions (lemmas 1 and 2) are easily recognized as true.

LEMMA 3. A cut of two $\mathfrak{B}_1$-sentences is not possible since both have c for their succedent and neither one has c in its antecedent.

LEMMA 4. A cut of two $\mathfrak{B}_2$-sentences results in a $\mathfrak{B}_2$-sentence or in a trivial sentence.

This is so since the conclusion of the cut belongs to $\overline{\mathfrak{A}}$ and can therefore not have c for its succedent.

LEMMA 5. A cut of a $\mathfrak{B}_1$-sentence with a $\mathfrak{B}_2$-sentence (which is therefore the upper sentence of the cut, cf. section I, § 2) results in a $\mathfrak{B}_2$-sentence or in a trivial sentence (since c is not the succedent of the conclusion of the cut).

A cut of a $\mathfrak{B}_2$-sentence with a $\mathfrak{B}_1$-sentence results either in a trivial sentence or in a $\mathfrak{B}_1$-sentence (since c remains the succedent) whose degree is no greater than the degree of the upper sentence.

PROOF (of the latter assertion). Suppose the degree of the upper sentence is μ, i.e., a_μ is the a with the smallest subscript occurring in the sentence. If the degree of the conclusion of the cut were greater than that of the upper sentence, then a_μ would have to be the cut element. This means that the lower sentence would have the form:

$$(b)(c)a_{v_1} \ldots a_{v_\sigma} \to a_\mu$$

(b and c may or may not occur). This sentence is a thinning of an $\mathfrak{A}_2$-sentence and as such has the form:

$$a_{v_\rho} \to a_\mu$$

From the definition of the $\mathfrak{A}_2$-sentences it then follows that $v_\rho < \mu$. On the other hand, a_{v_ρ} also occurs in the conclusion of the cut, which therefore has a degree smaller than μ, contrary to assumption.

LEMMA 6. *If in a normal proof composed of sentences from $\overline{\mathfrak{A}}$, a $\mathfrak{B}_1$-sentence occurs as an initial sentence, then b occurs in the antecedent of the conclusion of the proof.*

PROOF. For suppose that the $\mathfrak{B}_1$-sentence is the sentence $\mathfrak{r}_v$ or $\mathfrak{z}_0$ in the schema of the normal proof (section I, § 4); in that case the b occurring in the antecedent of the $\mathfrak{B}_1$-sentence can never disappear from the sequence of sentences $\mathfrak{r}_v \mathfrak{z}_{v+1} \ldots \mathfrak{z}_\rho \mathfrak{q}$ or $\mathfrak{z}_0 \ldots \mathfrak{z}_\rho \mathfrak{q}$; this follows from the fact that b can never be a cut element, since it does not occur as a succedent in any non-trivial sentence of $\overline{\mathfrak{A}}$.

LEMMA 7. *From all sentences of $\mathfrak{A}_2$ and a single $\mathfrak{B}_1$-sentence all other $\mathfrak{B}_1$-sentences of equal or lower degree are provable.*

PROOF. Suppose the $\mathfrak{B}_1$-sentence is

$$a_{v_1} \ldots a_{v_\rho} b \to c,$$

where v_1 is exactly the degree of the sentence, i.e., the smallest subscript.

Now, the sentences

$$a_{v_1} \to a_{v_2}$$
$$a_{v_1} \to a_{v_3}$$
$$\vdots$$
$$a_{v_1} \to a_{v_\rho}$$

are sentences from $\mathfrak{A}_2$. If we cut the first one with the $\mathfrak{B}_1$-sentence, the second one with the conclusion of the first cut, the third one with the conclusion of the second cut, etc., we finally obtain

$$a_{v_1} b \to c.$$

(If $\rho = 1$, then this expression occurs at the beginning and the above argument becomes redundant.)

This sentence is the (uniquely determined) $\mathfrak{A}_1$-sentence of the same degree as the original $\mathfrak{B}_1$-sentence. From it all other $\mathfrak{A}_1$-sentences of lower degree now follow easily:

For suppose $\mu < v_1$, then $a_\mu \to a_{v_1}$ is an $\mathfrak{A}_2$-sentence, and its cut with the $\mathfrak{A}_1$-sentence yields: $a_\mu b \to c$, the $\mathfrak{A}_1$-sentence of degree μ.

The $\mathfrak{B}_1$-sentences of the same degree follow from the $\mathfrak{A}_1$-sentences by thinnings, and this completes the proof of lemma 7.

§ 3. Proof of the nonexistence of an independent axiom system for $\overline{\mathfrak{A}}$

This result now follows quickly, since the above lemmas already contain the crucial steps. (Cf. Section I, § 6.)

We suppose that there exists an independent axiom system $\mathfrak{C}$ for $\overline{\mathfrak{A}}$. Let $\mathfrak{C}_2$ be the system of those axioms which are $\mathfrak{B}_2$-sentences.

We claim that all $\mathfrak{A}_2$-sentences are already provable from $\mathfrak{C}_2$. This follows at once from § 2, lemma 6: Every $\mathfrak{A}_2$-sentence was after all assumed to be provable from the axioms, this proof can be brought into normal form (theorem III) and, by § 2, lemma 6, no $\mathfrak{B}_1$-sentence can thus be an initial sentence, since b does not occur in the conclusion.

No $\mathfrak{B}_1$-sentence is provable from $\mathfrak{C}_2$. (By theorem III and § 2, lemmas 2 and 4).

Hence the axiom system $\mathfrak{C}$ must contain at least one $\mathfrak{B}_1$-sentence.

From theorem III and § 2, lemmas 3, 4, 5, 1 and 2 it now follows further: From $\mathfrak{C}_2$ and a $\mathfrak{B}_1$-axiom no $\mathfrak{B}_1$-sentence of greater degree (than that of the axiom) is provable. On the other hand, there are $\mathfrak{B}_1$-sentences of arbitrarily large degree. Hence there must exist a second $\mathfrak{B}_1$-sentence of a larger degree than that of the first axiom. ($\mathfrak{C}$ must actually contain infinitely many $\mathfrak{B}_1$-sentences; a fact which we do not need.)

From § 2, lemma 7, it therefore follows:

The first $\mathfrak{B}_1$-axiom is provable from the second one, together with the axioms $\mathfrak{C}_2$ (since all $\mathfrak{A}_2$-sentences follow from the latter axioms).

Hence the axiom system is not independent.

SECTION III. CONSTRUCTION OF AN INDEPENDENT AXIOM SYSTEM FOR A GIVEN DENUMERABLY INFINITE CLOSED LINEAR SENTENCE SYSTEM

§ 1. Outline of procedure

Suppose that an arbitrary denumerably infinite closed linear sentence system $\overline{\mathfrak{S}}$ is given.

We shall construct a subsystem $\mathfrak{T}$ of $\overline{\mathfrak{S}}$ (in § 2) which we shall prove to be an axiom system of $\mathfrak{S}$ (§ 3) and, moreover, an independent one (§ 4).

In developing $\mathfrak{T}$ we start with the system $\mathfrak{S}$ of the non-tautologous linear sentences of $\overline{\mathfrak{S}}$. This is an axiom system of $\overline{\mathfrak{S}}$, although in general of course not independent.

We begin by selecting axioms from the net sentences of $\mathfrak{S}$ (section I, § 7); these already form part of $\mathfrak{T}$ (§ 2.1).

Since all elements of a maximal net may in some sense be regarded as equivalent (after all, any element may be replaced by any other element in a sentence without disturbing the membership of that sentence in $\overline{\mathfrak{S}}$) it therefore suffices for our purpose to select from every maximal net a representative element and to consider only those semi-net sentences of $\mathfrak{S}$ whose net elements are representatives[11]. (§ 2.2, step 1.)

Steps 2 and 3 (§ 2.2) which then follow, represent the nucleus of the procedure. In step 3, we include certain thinnings as axioms (in $\mathfrak{T}$) in place of the linear sentences with which we had started; the method is similar to that described in the example in section I, § 6. Step 3, however, is not sufficient to guarantee the independence of the axioms; hence step 2, in which sentences are cancelled and where their order is rearranged.

§ 2. Construction of the axiom system $\mathfrak{T}$

1. Selection of axioms from the net sentences.

Consider all net sentences occurring in a fixed predetermined enumeration of the sentences of $\mathfrak{S}$; these form a sequence $\mathfrak{P}_1$. This sequence is examined step by step and modified as follows:

If a sentence is provable from a preceding sentence (in the already modified sequence), then it is omitted. If this is not the case, then the sentence, call it $u \to v$, is retained and next to it we write the sentence $v \to u$ which, by the definition of net sentences, also belongs to $\mathfrak{P}_1$.

In this way a new sequence $\mathfrak{P}_2$ results, which already forms part of $\mathfrak{T}$.

2. Construction of the other axioms.

Step 1. (Selection of representatives from the semi-net sentences.)

To every maximal net (relative to $\overline{\mathfrak{S}}$) we assign that one of its elements as a 'representative element' which occurs first in the enumeration of the sentences of $\mathfrak{S}$. Then we go through the sequence of sentences of $\mathfrak{S}$ and omit every semi-net sentence containing a net element which is not the representative of its maximal net. We also eliminate all net sentences. (These have already been dealt with separately in 1.) The sentences of $\mathfrak{S}$ that remain form a sequence which we shall call $\mathfrak{Q}_1$.

Step 2. (Cancellation and rearrangement of sentences.)

We go through the sequence $\mathfrak{Q}_1$ and modify it as follows: Suppose we have reached a sentence $u \to v$. If it is provable from preceding sentences in the modified sequence, then it is omitted. If this is not the case and if, among the earlier sentences (in the new sequence), there exists a sentence of the form $u \to w$ and if, furthermore, there exists in $\mathfrak{Q}_1$ a sentence of the form $w \to v$, then we first replace $u \to v$ by $w \to v$. If there are several such possibilities, we choose the one determined by the first preceding sentence with that property.

This newly obtained sentence is then examined for the same property and may itself be replaced by a further sentence, etc. In this way no sentence ever occurs which is provable entirely from sentences which precede it in the sequence (obtained by modifying $\mathfrak{Q}_1$); after all, the same argument would then obviously also apply to the sentence that was replaced, etc., up to $u \to v$ itself which, by assumption, is not provable from preceding sentences.

After a finite number of replacements, this procedure leads to a sentence which no longer possesses the property concerned and that sentence is then retained (this may already be $u \to v$ itself).

If this were not the case, then an already occurring sentence would recur at some point during the replacements, e.g., $w \to v$. (This is so since the succedent (v) always remains the same and since, by the rule of replacement, the antecedent is always the same as the succedent of one of the finitely many earlier sentences (in the sequence).) Say that the sequence of successive replacements from $w \to v$ up to the first reoccurrence of $w \to v$ runs:

$$w \to v, x_1 \to v, \ldots, x_\nu \to v, w \to v$$

($\nu \geq 1$, for $w \to v$ cannot be replaced by itself immediately, since $w \to w$ does not occur in $\mathfrak{Q}_1$). I.e., the following sentences, among others, occur earlier in the sequence:

$$w \to x_1, x_1 \to x_2, \ldots, x_{\nu-1} \to x_\nu, x_\nu \to w.$$

From these (without the first one) the sentence $x_1 \to w$ is now provable; hence it, in addition to $w \to x_1$, also belongs to $\mathfrak{S}$; from this it follows that both sentences are net sentences. It was assumed, however, that no net sentence occurs in $\mathfrak{Q}_1$; we thus have a contradiction.

The sequence which results from $\mathfrak{Q}_1$ by step 2 will be called $\mathfrak{Q}_2$.

Step 3. (Thinning of certain sentences.)

We go through the sequence $\mathfrak{Q}_2$ step by step and modify it as follows: Suppose we reach a sentence $u \to v$. If, among the earlier sentences in $\mathfrak{Q}_2$ (this time they are therefore not 'in the already modified sequence'), there is none with u as an antecedent element, then $u \to v$ remains unchanged. If there are such sentences, let these be the following:

$$u \to w_1, \ldots, u \to w_\nu \qquad (\nu \geq 1).$$

In that case $u \to v$ is replaced by the thinned sentence:

$$uw_1 \ldots w_\nu \to v.$$

(No w is identical with v, since $u \to v$ would otherwise occur twice in $\mathfrak{Q}_2$ and this is impossible by step 2. Hence $uw_1 \ldots w_\nu \to v$ is a nontrivial sentence of $\overline{\mathfrak{S}}$.)

The sequence which results in this way we call $\mathfrak{Q}_3$.

The sequences $\mathfrak{P}_2$ and $\mathfrak{Q}_3$, together, form the system $\mathfrak{T}$. The obvious result: The sentences of $\mathfrak{T}$ belong to $\overline{\mathfrak{S}}$.

§ 3. The system $\mathfrak{T}$ is an axiom system of $\overline{\mathfrak{S}}$

LEMMA 1. Every net sentence of $\mathfrak{S}$ is provable from $\mathfrak{P}_2$.

This is trivial, since the only sentences that were eliminated from the net sentences ($\mathfrak{P}_1$) were those provable from sentences that had already been taken as axioms.

LEMMA 2. *Every non-net sentence of $\mathfrak{S}$ is provable from $\mathfrak{Q}_1$ together with $\mathfrak{P}_2$.*

This needs to be shown only for the semi-net sentences eliminated in step 1. (This is, of course, due to the 'equivalence' between the net elements and their representatives expressed by the net sentences (cf. § 1).)

Suppose that $u \to v$ is such a semi-net sentence, and that u is a net element different from the 'representative' of its net, and that v is neither a net element nor a representative. Suppose also that w is a representative of the maximal net to which u belongs ($w \neq u, v$). Now $u \to w$ and $w \to u$ are net sentences and therefore, by lemma 1, provable from $\mathfrak{P}_2$. The sentence $w \to v$ is a consequence of $w \to u$ and $u \to v$, is therefore a sentence of $\mathfrak{S}$ and moreover, a semi-net sentence occurring in $\mathfrak{Q}_1$. On the other hand, $u \to v$ is a consequence of $u \to w$ and $w \to v$, and $u \to v$ is thus provable from $\mathfrak{Q}_1$ and $\mathfrak{P}_2$.

If either the succedent or both elements are not representatives of their nets, our reasoning is quite analogous.

LEMMA 3. *Every sentence of $\mathfrak{Q}_1$ is provable from $\mathfrak{Q}_2$.*

This is an immediate consequence of the procedure developed in step 2. For every sentence that was here eliminated was provable from an earlier sentence in $\mathfrak{Q}_2$, possibly with the help of a sentence by which it had been replaced.

LEMMA 4. *Every sentence of $\mathfrak{Q}_2$ is provable from $\mathfrak{Q}_3$.*

For the first sentence of $\mathfrak{Q}_2$ this is trivial (since that sentence is also the first sentence of $\mathfrak{Q}_3$). Suppose the lemma has been proved for the first u sentences of $\mathfrak{Q}_2$. If the $\mu+1$-th sentence was not thinned in step 3, then the lemma is also proved for that sentence. If the sentence was thinned, suppose that it has the form $u \to v$ and that, in step 3, it was transformed into
$$uw_1 \ldots w_\nu \to v \quad (\nu \geq 1).$$
Then
$$u \to w_1, \ldots, u \to w_\nu$$
are earlier sentences in $\mathfrak{Q}_2$ (among the first μ). By the induction hypothesis,

these sentences are already provable from $\mathfrak{Q}_3$. From them, together with

$$uw_1 \ldots w_\nu \to v,$$

the sentence $u \to v$ obviously follows by ν cuts. (Analogous to section II, § 2, lemma 7.)

This proves our assertion.

LEMMA 5. From lemmas 2, 3, 4 and 1 it follows that every sentence of $\mathfrak{S}$ is provable from $\mathfrak{T}$. The same, therefore, also holds for every sentence of $\overline{\mathfrak{S}}$.

§ 4. The system $\mathfrak{T}$ is independent

LEMMA 1. No sentence of $\mathfrak{P}_2$ is provable from the remaining axioms.

PROOF. Suppose that $u \to v$ is a sentence of $\mathfrak{P}_2$ which is provable from the other sentences of $\mathfrak{P}_2$ and $\mathfrak{Q}_3$. In place of the sentences of $\mathfrak{Q}_3$ we can take the corresponding sentences in $\mathfrak{Q}_2$ (i.e., those which have resulted from them in step 3), since the former sentences follow from the latter by thinnings. We are therefore dealing entirely with linear sentences.

By theorem IV, these sentences contain a sequence of sentences of the form

$$u \to w_1, w_1 \to w_2, \ldots, w_{\nu-1} \to w_\nu, w_\nu \to v$$

($\nu \geq 1$). Since, in addition to $u \to v$, the sentence $v \to u$ also belongs to $\overline{\mathfrak{S}}$, by being a net sentence, so do therefore the sentences provable from it and the above list of provable sentences, viz.,

$$w_1 \to u, w_2 \to w_1, \ldots, w_\nu \to w_{\nu-1}, v \to w_\nu.$$

Hence all these sentences are net sentences, i.e.,

$$u \to w_1, w_1 \to w_2, \ldots, w_\nu \to v$$

belong to $\mathfrak{P}_2$ (and none to $\mathfrak{Q}_2$). It follows from the construction of $\mathfrak{P}_2$ that in it all sentences occur in pairs. Of the $\nu+2$ pairs

$$u \to v, v \to u; u \to w_1, w_1 \to u; \ldots; w_\nu \to v, v \to w_\nu,$$

occurring in $\mathfrak{P}_2$, we consider the last one. This pair contradicts the construction of $\mathfrak{P}_2$, since both of its sentences are provable from the preceding pairs.

LEMMA 2. No sentence of $\mathfrak{Q}_3$ is provable from the remaining axioms.

§ 4, THE SYSTEM $\mathfrak{T}$ IS INDEPENDENT

PROOF. Let $\mathfrak{q}_{2\nu}$, $\mathfrak{q}_{3\nu}$, and $\mathfrak{p}_{2\nu}$ denote the ν-th sentence in the sequences $\mathfrak{Q}_2$, $\mathfrak{Q}_3$, and $\mathfrak{P}_2$, resp. This means that $\mathfrak{q}_{3\nu}$ is always a thinning of $\mathfrak{q}_{2\nu}$, by virtue of step 3.

Suppose, therefore, that $\mathfrak{q}_{3\nu}$ is provable from other sentences of $\mathfrak{T}$. Then every sentence of $\mathfrak{S}$ is provable from $\mathfrak{T}$ without the use of axiom $\mathfrak{q}_{3\nu}$; for example, the sentence $\mathfrak{q}_{2\nu}$. Suppose it has the form $u \to v$ and that it is provable from the axioms

$$\mathfrak{p}_{2\lambda}, \ldots, \mathfrak{p}_{2\mu}, \mathfrak{q}_{3\rho}, \ldots, \mathfrak{q}_{3\sigma},$$

and that all of these axioms are required for the proof (this can of course always be accomplished). (Suppose also that $\mathfrak{q}_{3\nu}$ does not occur among these axioms.)

As in lemma 1, we conclude that $u \to v$ is provable from the linear sentences

$$\mathfrak{p}_{2\lambda}, \ldots, \mathfrak{p}_{2\mu}, \mathfrak{q}_{2\rho}, \ldots, \mathfrak{q}_{2\sigma}$$

(here it is no longer mandatory that all sentences are required for the proof), and that from these sentences we may single out a sequence of sentences of the form

$$u \to w_1, w_1 \to w_2, \ldots, w_\tau \to v$$

(theorem IV). In these sentences we replace all occurring net elements by their representatives. This means that the sentences of $\mathfrak{Q}_2$ are left unchanged and those of $\mathfrak{P}_2$ become tautologous sentences. (The reason for this is that both elements of a $\mathfrak{P}_2$-sentence always belong to the same maximal net and hence have the same representative.) If the tautologous sentences are eliminated, then another sequence of sentences of the form

$$u \to x_1, x_1 \to x_2, \ldots, x_\kappa \to v$$

obviously results, consisting entirely of sentences of $\mathfrak{Q}_2$. (We have $\kappa \geq 1$, for if the sequence were to consist only of $u \to v$, then this sentence would occur twice in $\mathfrak{Q}_2$, which is impossible by step 2.)

We can suppose, without loss of generality, that $u \to x_1$ is the sentence $\mathfrak{q}_{2\rho}$. Since the sentence $x_1 \to v$ ($x_1 \neq v$) is provable, it is a sentence of $\mathfrak{S}$ and, in particular, of $\mathfrak{Q}_1$. From this it follows by step 2, that $u \to x_1$ does not occur *before* $u \to v$ in $\mathfrak{Q}_2$, since $u \to v$ would then have been replaced by another sentence (initially by $x_1 \to v$). $u \to x_1$ therefore occurs *later* in $\mathfrak{Q}_2$ than $u \to v$. Hence the sentence $\mathfrak{q}_{3\rho}$ which resulted from $u \to x_1$ in step 3, runs:

$$u \ldots v \ldots \to x_1,$$

by the construction of step 3. For it we put

$$vK \to x_1.$$

According to assumption, the sentence $u \to v$ ($\mathfrak{q}_{2\nu}$) is now provable from

$$\mathfrak{p}_{2\lambda}, \ldots, \mathfrak{p}_{2\mu}, \mathfrak{q}_{3\rho}, \ldots, \mathfrak{q}_{3\sigma},$$

but not without $vK \to x_1$($\mathfrak{q}_{3\rho}$). This cannot be so. A sentence with v in the antecedent can, in no case, be an integral component of a proof for a sentence with v for its succedent. The following argument explains why this is so:

By theorem III there would have to exist a normal proof for $u \to v$ in which $vK \to x_1$ occurs among the initial sentences. In that case the sentences $\mathfrak{F}$ in the schema of the normal proof (cf. Section I, § 4) would all have the succedent v. Hence $vK \to x_1$ would have to belong to the $\mathfrak{r}$'s. It would then be cut with one of the $\mathfrak{F}$'s. Since v could not be the cut element, v would therefore occur in the antecedent and succedent of the conclusion of the cut, the latter would thus be trivial; however, trivial sentences are not allowed to occur in normal proofs.

We therefore have a contradiction, and the independence of the axiom system $\mathfrak{T}$ is proved.

2. ON THE RELATION BETWEEN INTUITIONIST AND CLASSICAL ARITHMETIC

INTRODUCTION

By classical arithmetic we mean the theory of the natural numbers as it is built up from the axioms of Peano, together with classical predicate logic (called 'restricted predicate calculus' in Hilbert-Ackermann)[12], and the introduction of recursive definitions.

Intuitionist arithmetic differs from classical arithmetic, purely externally, by accepting only part of classical predicate logic as admissible. Intuitionist predicate logic may be extended to classical predicate logic by including, for example, the law of the excluded middle ($\mathfrak{A}$ is true or $\mathfrak{A}$ is false) or, alternatively, the law of double negation (if $\mathfrak{A}$ is not false, then $\mathfrak{A}$ is true).

In the following we intend to show that the applications of the law of double negation in proofs of classical arithmetic can in many instances be eliminated. The most important consequences that follow are these:

THEOREM VI. If intuitionist arithmetic is consistent, then classical arithmetic is also consistent.

THEOREM IV[13]. Every definite proposition of arithmetic which does not involve the concepts 'or' and 'there is', and is classically provable is also intuitionistically provable.

A definition of the expressions used will follow below.

The proofs of these and the remaining theorems will be carried out intuitionistically. All of them are reasonably straightforward and require mainly facility in the use of the calculus of intuitionist predicate logic.

§ 1. Terminology and notations

We shall distinguish the following symbols and combinations of symbols (expressions):

1.1. *Logical symbols*: & and, ∨ or, ⊃ if ... then ..., ¬ not, ($\mathfrak{x}$) for all $\mathfrak{x}$, E($\mathfrak{x}$) there is an $\mathfrak{x}$. The last two symbols are called *quantifiers*.

1.21. *Symbol for definite objects*: 1.

1.22. *Object variables*: $a, b, c, \ldots$.

1.3. *Symbols for definite functions*: ′ (the successor function, with one argument place); + , · (with two argument places).

1.41. *Symbols for definite predicates*: = , < (with two argument places).

1.42. *Propositional variables*: $A, B, C, \ldots$.

1.5. German letters will serve as syntactic variables, i.e., as variables for our deliberations *about* arithmetic.

1.6. The concept of an expression involving objects, briefly called a *term* (defined inductively):

1.61. Object variables and symbols for definite objects are terms.

1.621. If $\mathfrak{f}$ is a term, so is $\mathfrak{f}'$.

1.622. If $\mathfrak{f}$ and $\mathfrak{g}$ are terms, then so are $\mathfrak{f}+\mathfrak{g}$ and $\mathfrak{f} \cdot \mathfrak{g}$.

1.7. Example of a term: $(a+1'')' \cdot x'$. (The brackets serve to make the structure of the term unambiguous.)

The numerals 2, 3, 4, ... are abbreviations for the terms $1', 1'', 1''', \ldots$.

1.8. The concept of a propositional expression, briefly called a *formula* (defined inductively) – cf. H.-A.[14], p. 52 –:

1.81. A symbol for a definite predicate with terms in the argument places is a formula (e.g.: $x' < y+1$).

A propositional variable, with several terms behind it, is a formula (e.g.: $Fx \cdot y'$). (Informally, such an expression stands for an arbitrary proposition in which the mentioned objects occur.)

It is also permissible for *no* term to stand behind the propositional variable.

Formulae of the kinds introduced so far will be called *elementary formulae*, and the terms will be called the arguments of the predicate symbol or of the propositional variable.

1.821. If $\mathfrak{A}$ is a formula, so is ¬$\mathfrak{A}$.

1.822. If $\mathfrak{A}$ and $\mathfrak{B}$ are formulae, so are $\mathfrak{A} \& \mathfrak{B}$, $\mathfrak{A} \vee \mathfrak{B}$, $\mathfrak{A} \supset \mathfrak{B}$.

1.823. If $\mathfrak{A}$ is a formula and $\mathfrak{x}$ is an object variable occurring within the scope of a quantifier in $\mathfrak{A}$, then ($\mathfrak{x}$)$\mathfrak{A}$ and (E$\mathfrak{x}$)$\mathfrak{A}$ are also formulae.

1.9. The formulae $\mathfrak{A}$ and $\mathfrak{B}$ in 1.821–1.823 are called the *scopes* of the logical symbols involved.

Brackets and dots serve to display unambiguously the scopes of the logical symbols in a formula.

Example of a formula:

$(Ax'+1, a \cdot \& : \cdot (x) : x'' < 1 \cdot \supset Fb) \supset \neg (Ez)A.$

The dots are to be understood thus: The scope of a logical symbol extends on a given side up to the point where a larger number of dots occurs than the number of dots next to the symbol on that side. (If no such larger number of dots occurs, the scope naturally extends to the beginning or end of the formula.)

For greater clarity we also use brackets ( ) in place of dots: each pair of brackets encloses a scope.

An object variable in a formula is said to have a *bound* occurrence if it stands within the scope of a quantifier with the same object variable (or if it stands in the quantifier itself); otherwise it is said to have a *free* occurrence.

§ 2. The formal structure of arithmetic

Arithmetic is a system of 'true' propositions, some of which are axioms, and from which the other propositions are obtained by repeated application of certain inferences.

In formalized arithmetic (which we must consider in place of informal arithmetic, in order to be able to reason *about* arithmetic) we have a corresponding system of *true formulae*, some of which are *axiom formulae* and from which the others are obtained by repeated application of certain *operational rules*.

The formal counterpart of a proof is a sequence of formulae each of which is either an axiom formula or results from earlier formulae by the application of an operational rule. This formal counterpart of a proof is called a proof figure or, briefly, a *derivation*. The last formula in the sequence is called the *endformula* of the derivation. A formula is called derivable if there exists a derivation of which it is the endformula. (This derivation is then also called a 'derivation of the formula'.)

The choice of the axiom formulae and operational rules is largely arbitrary. In this paper we shall stipulate the following system:

2.1. Statement of the axiom formulae.

2.11. $\neg \neg A \cdot \supset A$. (Formal counterpart of the 'law of double negation'.)

By introducing this axiom formula at the beginning, we obtain the following arrangement:

All axiom formulae and operation rules that now follow constitute together the formalism of *intuitionist arithmetic*. From it the formalism of *classical arithmetic* results by the inclusion of the axiom formula $\neg \neg A \cdot \supset A$.

Formulae which are derivable without the use of the latter axiom formula are called *intuitionistically true*; those derivable in the entire formalism, *classically true*.

2.12. Axiom formulae of intuitionist propositional logic.

We adopt Heyting's axiom formulae[15]:

1. $A \supset \cdot A \mathbin{\&} A$,
2. $A \mathbin{\&} B \cdot \supset \cdot B \mathbin{\&} A$,
3. $A \supset B \cdot \supset : A \mathbin{\&} C \cdot \supset \cdot B \mathbin{\&} C$,
4. $A \supset B \cdot \mathbin{\&} \cdot B \supset C : \supset \cdot A \supset C$,
5. $B \supset \cdot A \supset B$,
6. $A \mathbin{\&} \cdot A \supset B : \supset B$,
7. $A \supset \cdot A \vee B$,
8. $A \vee B \cdot \supset \cdot B \vee A$,
9. $A \supset C \cdot \mathbin{\&} \cdot B \supset C : \supset : A \vee B \cdot \supset C$,
10. $\neg A \cdot \supset \cdot A \supset B$,
11. $A \supset B \cdot \mathbin{\&} \cdot A \supset \neg B : \supset \neg A$.

2.13. Axiom formulae for 'all' and 'there is' (H.-A. p. 53):

1. $(x)Fx \cdot \supset Fy$,
2. $Fy \supset (Ex)Fx$.

2.14. Axiom formulae for the natural numbers (according to Herbrand)[16]:

1. $x = x$,
2. $x = y \cdot \supset \cdot y = x$,
3. $x = y \cdot \mathbin{\&} \cdot y = z : \supset \cdot x = z$,
4. $\neg \cdot x' = 1$,
5. $x = y \cdot \supset \cdot x' = y'$,
6. $x' = y' \cdot \supset \cdot x = y$,
7. $F1 \mathbin{\&} : (x) \cdot Fx \supset Fx' : \cdot \supset (x)Fx$
   (axiom formula for complete induction),
8. $\neg \neg \cdot x = y : \supset \cdot x = y$
   (This special case of the law of double negation is intuitionistically true.).

Any other intuitionistically true axiom formulae are admitted provided that (as before) they do not contain the symbols $\vee$ and **E**.

For example, the following axiom formulae for addition are admissible:
$$x+1 = x', \quad x+y' = (x+y)'.$$
Multiplication, exponentiation, the predicate 'smaller than', and others may be introduced by means of similar axiom formulae. All we require further is that for predicates the law of double negation is intuitionistically valid; for 'smaller than', for example, we require that
$$\neg\neg \cdot x < y : \supset \cdot x < y$$
is an axiom formula.

It is easily seen that these and similar derived concepts customary in arithmetic fulfil the stated conditions. ($\neg\neg \cdot x < y : \supset \cdot x < y$ e.g., is indeed intuitionistically valid.)

2.2. Statement of the operational rules (following H.-A., p. 53–54):

2.21. If $\mathfrak{A}$ and $\mathfrak{A} \supset \mathfrak{B}$ are true formulae, so is $\mathfrak{B}$.

2.22. Rules for 'all' and 'there is':

If $\mathfrak{A} \supset \mathfrak{B}$ is a true formula and if $\mathfrak{x}$ is an object variable with no free occurrence in $\mathfrak{A}$ and no bound occurrence in $\mathfrak{B}$, then $\mathfrak{A} \supset (\mathfrak{x})\mathfrak{B}$ is a true formula.

If $\mathfrak{B} \supset \mathfrak{A}$ is a true formula and $\mathfrak{x}$ is an object variable with no free occurrence in $\mathfrak{A}$ and no bound occurrence in $\mathfrak{B}$, then $(E\mathfrak{x})\mathfrak{B} \cdot \supset \mathfrak{A}$ is also a true formula.

2.23. Rules of replacement:

2.231. For bound object variables:

From a true formula another true formula results if an object variable is replaced in a quantifier and throughout the scope of that quantifier, by another object variable, provided that the replacement variable occurs neither in the scope of that quantifier nor in another quantifier of the formula in which the replacement takes place.

2.232. For free object variables:

From a true formula another true formula results if every free occurrence of an object variable is replaced by one and the same term provided that this term contains no object variable occurring in a quantifier in whose scope the replacement takes place.

2.233. For propositional variables:

From a true formula $\mathfrak{A}$ another true formula results by the following replacement:

We replace all occurrences of a propositional variable $\mathfrak{B}$, together with all of its arguments, in all places in $\mathfrak{A}$ where the variable $\mathfrak{B}$ occurs with the same number of arguments. Let us call that number v (v may be 0). For the replacement we use a formula $\mathfrak{S}$ which does not contain an object variable with a bound occurrence in the scope of a quantifier of $\mathfrak{A}$ within which the replacement takes place. $\mathfrak{S}$ may, in particular, contain v object variables designated by $\mathfrak{x}_1, \ldots, \mathfrak{x}_v$, but these variables must not be bound. (Not all of these variables need to occur, it may happen that none of them occurs, cf. 1.81.) By $\mathfrak{S}^{\mathfrak{b}}(\mathfrak{S}^{\mathfrak{x}_1 \ldots \mathfrak{x}_v}_{\mathfrak{y}_1 \ldots \mathfrak{y}_v})$ we mean that formula which results from $\mathfrak{S}$ by the replacement of every occurrence of $\mathfrak{x}_\mu$ by the term $\mathfrak{y}_\mu$. The replacement as a whole may be described thus: If, in the place where it is to be replaced, the propositional variable has for its arguments the terms $\mathfrak{z}_1, \ldots, \mathfrak{z}_v$, then we substitute for it (and for its arguments) the formula

$$\mathfrak{S}^{\mathfrak{b}}(\mathfrak{S}^{\mathfrak{x}_1 \ldots \mathfrak{x}_v}_{\mathfrak{z}_1 \ldots \mathfrak{z}_v}).$$

2.3. Some true formulae of intuitionist propositional logic.

The following true formulae, which we shall need for the proofs in §§ 3 and 4, are derivable from the axiom formulae 2.12. by means of the operational rules 2.21 and 2.233 (applied only to formulae without terms). The formulae are taken from Heyting's paper and the numbers behind them refer to that paper[17].

1.	$A \& B \cdot \supset A,$	2.2
2.	$A \supset B \cdot \& \cdot C \supset D : \supset : A \& C \cdot \supset \cdot B \& D,$	2.23
3.	$A \supset B \cdot \& A \supset C : \supset : A \supset \cdot B \& C,$	2.24
4.	$A \& B \cdot \supset C : \supset : A \supset \cdot B \supset C,$	2.27
5.	$A \supset \cdot B \supset C : \supset : B \supset \cdot A \supset C,$	2.271
6.	$A \supset B \cdot \supset : B \supset C \cdot \supset \cdot A \supset C,$	2.29
7.	$A \supset B \cdot \supset : \neg B \cdot \supset \neg A,$	4.2
8.	$A \supset B \cdot \supset : \neg \neg A \cdot \supset \neg \neg B,$	4.22
9.	$A \supset \neg \neg A,$	4.3
10.	$\neg \neg \neg A \cdot \supset \neg A,$	4.32
11.	$\neg \neg \cdot A \& B : \supset : \neg \neg A \cdot \& \neg \neg B.$	4.61

§ 3. The intuitionist validity of the law of double negation

Preliminary remarks about the connections between the theorems which follow: Theorem II contains the essential basis of our results theorem I, serves as a lemma for its proof. Theorem II says that the law of double negation is to a large extent intuitionistically valid; this fact is then used (in § 4) in order to transform an arbitrary classical proof into an intuitionist

§ 3, THE INTUITIONIST VALIDITY OF THE LAW OF DOUBLE NEGATION

proof, although this transformation causes certain changes in the conclusion of the proof. Theorem III indicates what can be achieved in this direction. From theorem III then follow almost at once a number of individual results (in § 5), among them the two theorems mentioned in the introduction.

3.1. THEOREM I. If $\mathfrak{A}$, $\mathfrak{B}$, $\mathfrak{C}$ are any given formulae, $\mathfrak{x}$ any given object variable not occurring as a bound variable in $\mathfrak{A}$, and if $\neg\neg\mathfrak{A}\cdot\supset\mathfrak{A}$ as well as $\neg\neg\mathfrak{B}\cdot\supset\mathfrak{B}$ are intuitionistically true formulae, then $(\neg\neg\cdot\mathfrak{A}\&\mathfrak{B})\supset(\mathfrak{A}\&\mathfrak{B})$, $(\neg\neg\cdot\mathfrak{C}\supset\mathfrak{A})\supset(\mathfrak{C}\supset\mathfrak{A})$, $(\neg\neg\cdot\neg\mathfrak{C})\supset(\neg\mathfrak{C})$ and $(\neg\neg(\mathfrak{x})\mathfrak{A})\supset((\mathfrak{x})\mathfrak{A})$ are also intuitionistically true formulae.

3.11. The informal sense of this theorem can be formulated as follows:

If the law of double negation is intuitionistically valid for certain propositions, then it also holds for the conjunction of two such propositions, for the implication of an arbitrary proposition with such a proposition, for the negation of an arbitrary proposition of this kind, as well as for the universal generalization of such a proposition.

3.2. PROOF of theorem I. We must deal with four individual propositions, each of which is proved separately.

3.21. From $\neg\neg\mathfrak{A}\cdot\supset\mathfrak{A}$ and $\neg\neg\mathfrak{B}\cdot\supset\mathfrak{B}$ we derive $(\neg\neg\cdot\mathfrak{A}\&\mathfrak{B})\supset(\mathfrak{A}\&\mathfrak{B})$ intuitionistically as follows (the numbers refer to § 2):

From $\neg\neg\mathfrak{A}\cdot\supset\mathfrak{A}$ and $\neg\neg\mathfrak{B}\cdot\supset\mathfrak{B}$ follows $(\neg\neg\mathfrak{A}\cdot\&\neg\neg\mathfrak{B})\supset(\mathfrak{A}\&\mathfrak{B})$ by 2.32 and 2.34. This, together with 2.3.11 and 2.36, yields

$$(\neg\neg\cdot\mathfrak{A}\&\mathfrak{B})\supset(\mathfrak{A}\&\mathfrak{B}).$$

3.22. From $\neg\neg\mathfrak{A}\cdot\supset\mathfrak{A}$ we wish to derive $(\neg\neg\cdot\mathfrak{C}\supset\mathfrak{A})\supset(\mathfrak{C}\supset\mathfrak{A})$. This is done as follows: 2.126, together with 2.34, yields

$$\mathfrak{C}\supset(\mathfrak{C}\supset\mathfrak{A}\cdot\supset\mathfrak{A}).$$

According to 2.38, it holds that

$$(\mathfrak{C}\supset\mathfrak{A}\cdot\supset\mathfrak{A})\supset(\neg\neg\cdot\mathfrak{C}\supset\mathfrak{A}:\supset\neg\neg\mathfrak{A}).$$

Both together yield (2.36)

$$\mathfrak{C}\supset(\neg\neg\cdot\mathfrak{C}\supset\mathfrak{A}:\supset\neg\neg\mathfrak{A}).$$

From this follows (2.35)

$$(\neg\neg\cdot\mathfrak{C}\supset\mathfrak{A})\supset(\mathfrak{C}\supset\neg\neg\mathfrak{A}).$$

From $\neg\neg\mathfrak{A}\cdot\supset\mathfrak{A}$ we obtain (2.35, 2.36)

$$(\mathfrak{C}\supset\neg\neg\mathfrak{A})\supset(\mathfrak{C}\supset\mathfrak{A}),$$

hence (2.36)
$$(\neg\neg\cdot\mathfrak{C}\supset\mathfrak{A})\supset(\mathfrak{C}\supset\mathfrak{A}).$$

3.23. $(\neg\neg\neg\mathfrak{C})\supset(\neg\mathfrak{C})$ holds by 2.3.10. (This case offers nothing new and is included in theorem I only for the sake of completeness.)

3.24. From $\neg\neg\mathfrak{A}\cdot\supset\mathfrak{A}$ we wish to derive: $\neg\neg(\mathfrak{x})\mathfrak{A}\cdot\supset\cdot(\mathfrak{x})\mathfrak{A}$.

2.131 yields (2.231, 2.232, 2.233) $(\mathfrak{x})\mathfrak{A}\cdot\supset\mathfrak{A}$ (permissible, since $\mathfrak{x}$ was not allowed to have a bound occurrence in $\mathfrak{A}$).

Using 2.38, we obtain $\neg\neg(\mathfrak{x})\mathfrak{A}\cdot\supset\neg\neg\mathfrak{A}$; using $\neg\neg\mathfrak{A}\cdot\supset\mathfrak{A}$, we further obtain (2.36)
$$\neg\neg(\mathfrak{x})\mathfrak{A}\cdot\supset\mathfrak{A},$$
and from this, by 2.22:
$$\neg\neg(\mathfrak{x})\mathfrak{A}\cdot\supset(\mathfrak{x})\mathfrak{A}.$$

(The application of the operational rule is permissible since $\mathfrak{x}$ was assumed to have no bound occurrence in $\mathfrak{A}$ and quite obviously has no free occurrence in $\neg\neg(\mathfrak{x})\mathfrak{A}$.)

3.3. THEOREM II. *If $\mathfrak{A}$ is a formula without the symbols $\vee$ and E, and if all of its elementary formulae are prefixed by $\neg$, then $\neg\neg\mathfrak{A}\cdot\supset\mathfrak{A}$ is an intuitionistically true formula.*

3.4. The proof of this theorem follows easily from an application of theorem I. The formula $\mathfrak{A}$ is, after all, made up of formulae of the form
$$\mathfrak{f}=\mathfrak{g},\ \mathfrak{h}<\mathfrak{i},\ldots,\ \text{also}\ \neg\mathfrak{Ef}_1\ldots\mathfrak{f}_\nu$$

($\mathfrak{f}, \mathfrak{g}, \mathfrak{h}, \mathfrak{i}, \mathfrak{f}_1, \ldots, \mathfrak{f}_\nu$ designate terms; $\mathfrak{E}$ designates a propositional variable) and the logical symbols &, $\supset$, $\neg$, $(\mathfrak{x})$ (1.8).

For any one of these formulae, let us call it $\mathfrak{D}$, $\neg\neg\mathfrak{D}\cdot\supset\mathfrak{D}$ is intuitionistically true (2.14 together with 2.232, 2.3.10 together with 2.233). According to theorem I, this property is inherited by the subformulae of $\mathfrak{A}$, as they are combined in the construction of $\mathfrak{A}$, and is finally passed on to $\mathfrak{A}$ itself, i.e., $\neg\neg\mathfrak{A}\cdot\supset\mathfrak{A}$ is an intuitionistically true formula.

§ 4. Transformation of proofs of classical arithmetic into proofs of intuitionist arithmetic

4.1. THEOREM III. *A proof figure of classical arithmetic with the endformula $\mathfrak{E}$ is transformable into a proof figure of intuitionist arithmetic with the endformula $\mathfrak{E}^*$, where $\mathfrak{E}^*$ results from $\mathfrak{E}$ in the following way: Each sub-*

formula of $\mathfrak{E}$ which has the form $\mathfrak{C} \vee \mathfrak{D}$ is replaced by $\neg : \neg \mathfrak{C} \cdot \& \neg \mathfrak{D}$, each subformula which has the form $(E\mathfrak{x})\mathfrak{C}$ is replaced by $\neg (\mathfrak{x}) \neg \mathfrak{C}$; and each elementary formula containing a propositional variable is replaced by the same formula, prefixed by two negation symbols.

4.11. In classical logic, the formula $\mathfrak{E}^*$ which thus results is, as is well-known, equivalent to $\mathfrak{E}$. (Cf., for example, H.-A., pp. 5, 7, 46, 61.)

The somewhat tedious arguments required for the proof of theorem III contain nothing new; all we need to do is to carry them out rigorously for the case at hand.

4.2. PROOF of theorem III. The transformation of the classical proof figure – we think of this figure as given – is carried out in four steps (4.21–4.24).

4.21. First of all we can see to it that every formula of the proof figure, with the exception of the endformula $\mathfrak{E}$, is used *exactly once to obtain a new formula* (by means of an operational rule).

This is done by omitting one by one those formulae that are not used to obtain another formula, except the endformula, and by writing down correspondingly often those formulae that are used more than once, together with the formulae required for their derivation.

4.22. Elimination of the symbols $\vee$ and E from the proof figure.

The derivation (= proof figure) is now transformed thus: Wherever a formula of the form $\mathfrak{A} \vee \mathfrak{B}$ occurs as a component of a derivation formula, it is replaced by $\neg : \neg \mathfrak{A} \cdot \& \neg \mathfrak{B}$, and every occurrence of a formula of the form $(E\mathfrak{x})\mathfrak{A}$ is replaced by $\neg (\mathfrak{x}) \neg \mathfrak{A}$. (The order in which these replacements are carried out is obviously immaterial.)

We must now examine to what extent the new figure which has resulted in this way has remained a correct derivation and, where this is not the case, modify the figure accordingly. For this purpose we examine, first, the application of the operational rules and, second, the axiom formulae (4.221, 4.222).

In the course of the transformation we shall adjoin to the proof figure several true formulae from 2.3, together with their derivations, and we note that these formulae, as may be seen from Heyting's paper, are derivable without the use of the symbols $\vee$ and E, so that these symbols in fact no longer occur in the proof figure. (This, incidentally, is not crucial.) We assume that these derivations have already been modified in such a way by the procedure described at 4.21 that every formula, with the exception of the endformula, is used exactly once to obtain another formula.

4.221. All applications of operational rules in the proof figure have, as is easily seen, remained correct in the replacement described, except for those places in which a rule for 'there is' (2.22, second part) was applied.

An instance of this kind, which had the form $\mathfrak{B} \supset \mathfrak{A}$, $(E\mathfrak{x})\mathfrak{B} \cdot \supset \mathfrak{A}$ before the replacement, now has the form $\mathfrak{B}^* \supset \mathfrak{A}^*$, $\neg (\mathfrak{x}) \neg \mathfrak{B}^* \cdot \supset \mathfrak{A}^*$, where $\mathfrak{x}$ has no free occurrence in $\mathfrak{A}^*$ and no bound occurrence in $\mathfrak{B}^*$. This place is modified as follows:

From $\mathfrak{B}^* \supset \mathfrak{A}^*$ we obtain $\neg \mathfrak{A}^* \cdot \supset \neg \mathfrak{B}^*$ by using the true formula 2.37 (which must be adjoined to the proof figure, together with its derivation); from this we obtain $\neg \mathfrak{A}^* \cdot \supset (\mathfrak{x}) \neg \mathfrak{B}^*$, by the rule for 'all', and this, by means of 2.37, yields $\neg (\mathfrak{x}) \neg \mathfrak{B}^* \cdot \supset \neg \neg \mathfrak{A}^*$, which, in turn, yields $\neg (\mathfrak{x}) \neg \mathfrak{B}^* \cdot \supset \mathfrak{A}^*$, by the application of $\neg \neg \mathfrak{A}^* \supset \mathfrak{A}^*$ (2.11) and (2.36), and the gap which had developed in the proof figure has therefore once again been filled.

4.222. The replacement of the symbols ∨ and E may also have caused changes in certain axiom formulae, i.e., in those which had contained a ∨ or E. The axiom formulae affected are 2.127, 2.128, 2.129, and 2.132. (The arithmetic axiom formulae (2.14) do not – this was explicitly stipulated at 2.14 – contain ∨ and E.) The formulae which have resulted from the above four axiom formulae are now derived as follows:

4.222.1. The axiom formula 2.127 became

$$A \supset \neg : \neg A \cdot \& \neg B.$$

By 2.31 it holds that

$$\neg A \cdot \& \neg B : \supset \neg A;$$

from this, with the help of 2.37, we obtain

$$\neg \neg A \supset : \cdot \neg : \neg A \cdot \& \neg B;$$

it also holds that (2.39)

$$A \supset \neg \neg A,$$

and both together yield (2.36) the formula to be derived.

4.222.2. The axiom formula 2.128 became

$$\neg : \neg A \cdot \& \neg B : \cdot \supset : \cdot \neg : \neg B \cdot \& \neg A.$$

From 2.122

$$\neg B \cdot \& \neg A : \supset : \neg A \cdot \& \neg B$$

results by replacement and from this, with the help of 2.37, the desired result follows.

4.222.3. The axiom formula 2.129 became

$$(A \supset C \cdot \& \cdot B \supset C) \supset (\neg : \neg A \cdot \& \neg B : \cdot \supset C).$$

It can be derived thus: 2.37 yields

$$A \supset C \cdot \supset : \neg C \cdot \supset \neg A \quad \text{and} \quad B \supset C \cdot \supset : \neg C \cdot \supset \neg B.$$

From this, by 2.32 and 2.34, follows

$$(A \supset C \cdot \& \cdot B \supset C) \supset (\neg C \cdot \supset \neg A : \& : \neg C \cdot \supset \neg B).$$

By 2.33 it further holds that

$$(\neg C \cdot \supset \neg A : \& : \neg C \cdot \supset \neg B) \supset (\neg C \cdot \supset : \neg A \cdot \& \neg B),$$

and by 2.37

$$(\neg C \cdot \supset : \neg A \cdot \& \neg B) \supset (\neg : \neg A \cdot \& \neg B : \cdot \supset \neg \neg C).$$

If we include (2.11) $\neg \neg C \supset C$, then repeated applications of 2.36 (and a single application of 2.35) finally yield the formula to be derived.

4.222.4. The axiom formula 2.132 became

$$Fy \supset \neg (x) \neg Fx.$$

From $(x) \neg Fx \cdot \supset \neg Fy$ (2.13) an application of 2.37 yields

$$\neg \neg Fy \cdot \supset \neg (x) \neg Fx,$$

and from this, with the help of $Fy \supset \neg \neg Fy$ (2.39) and 2.36, follows the formula to be derived.

4.23. Now we shall modify the proof figure in such a way that replacements of propositional variables (2.233) take place only immediately after the application of the axiom formulae, i.e., before one of the remaining operational rules (2.21, 2.22, 2.231, 2.232) is applied. (As before, it remains permissible to carry out *several* such replacements in succession immediately after the application of an axiom formula.)

4.231. Preparatory step: All places of the proof figure in which a replacement of a propositional variable takes place are modified as follows:

Suppose the replacement transforms a formula $\mathfrak{F}$ into a formula $\mathfrak{G}$. The formula $\mathfrak{S}$ (2.233) which was used as a replacement, is itself replaced by a formula $\mathfrak{T}$ which, in place of the object variables of $\mathfrak{S}$, contains other object variables not yet occurring in the proof figure. Once $\mathfrak{T}$ has been incorporated into the formula $\mathfrak{F}$, we carry out several replacements of object variables according to 2.231 and 2.232, so that finally the formula 𝔊 once again results.

4.232. Step-by-step reversal of the order of the replacements of propositional variables:

A replacement of a propositional variable which takes place *after* an application of one of the *remaining* operational rules is moved *ahead*, so that the same replacement now occurs before that application instance. By 'the same replacement' we mean: For the same propositional variable with the same number of arguments we substitute the same formula 𝔖 (2.233) into that formula, or into the two formulae (in the case of 2.21) to which the other operational rule is applied. (If the propositional variable is missing from one of these formulae, then the replacement becomes of course redundant.) The replacement is correct by virtue of the preparatory step 4.231. The application of the operational rule, which now occurs *after* the replacement (or after the two replacements, in the case 2.21), is also correct, as a consideration of all four cases (2.21 to 2.232) easily shows. (On the basis of the preparatory step, the only new object variables that could have been introduced at different places by the substitution of 𝔖 are at most object variables that did not yet occur in the formulae involved in the application of the operational rule.)

By repeating this reversal of order sufficiently often, it can be achieved that the replacements of proposition variables eventually occur only after the applications of the axiom formulae.

4.24. The prefixing of two negation symbols in the case of propositional variables that are not being replaced:

After having completed the replacements of propositional variables, we carry out several further replacements of this kind in such a way that every elementary formula with a propositional variable that still occurs is replaced by the same formula prefixed by two negation symbols. In the remainder of the proof figure, every occurrence of that elementary formula is then prefixed by two negation symbols. The proof figure remains correct.

4.25. Now the proof of theorem III is easily completed:

We have achieved that in the proof figure under discussion the axiom formula $\neg \neg A \cdot \supset A$ (as well as all other axiom formulae, a fact which will not be used here) is used only in the following form: after the axiom formula there occur replacements of propositional variables and this leads to a formula in which every elementary formula with a propositional variable is prefixed by $\neg$ and in which the symbols $\vee$ and E no longer occur. This formula has (in the case of the axiom formula $\neg \neg A \cdot \supset A$) the form $\neg \neg \mathfrak{A} \cdot \supset \mathfrak{A}$. (After all, every replacement necessarily preserves this form.)

This formula is intuitionistically derivable by virtue of theorem II. Thus the axiom formula $\neg \neg A \cdot \supset A$ can now be eliminated (as an axiom formula) from the proof figure by writing down its intuitionist derivation wherever the formula occurs, so that finally an intuitionist proof figure results. In steps 4.22 and 4.24, its endformula was subjected to precisely the modifications stated in theorem III. This proves the theorem.

§ 5. Consequences

5.1. THEOREM IV. A proof figure in classical arithmetic whose endformula contains no propositional variables, and which does not contain the symbols ∨ and E, can be transformed into a proof figure in intuitionist arithmetic with the same endformula.

5.11. The informal sense of this theorem may be formulated thus: Every definite proposition of arithmetic which does not contain the concepts 'or' and 'there is' and is classically provable, is also intuitionistically provable.

By a 'definite proposition' we mean a proposition whose formula contains no propositional variables. 'Provability' should always be understood as provability in our formal system of arithmetic. Proofs using techniques from analysis, for example, are not included.

5.2. Theorem IV follows at once from theorem III.

5.3. THEOREM V. For every formula (of arithmetic) there exists a classically equivalent formula which is intuitionistically derivable if and only if the former is classically derivable.

5.31. The informal sense of this theorem may be formulated thus: For every proposition of arithmetic there exists a classically equivalent proposition which is intuitionistically provable if and only if the former is classically provable.

5.4. Theorem V also follows at once from theorem III. The formulae $\mathfrak{E}$ and $\mathfrak{E}^*$ in theorem III are classically equivalent and the true formulae $\mathfrak{E} \supset \mathfrak{E}^*$ and $\mathfrak{E}^* \supset \mathfrak{E}$ which, together, express this equivalence, are derivable in our logical formalism (§ 2 without 2.14). (This derivation will not be carried out; it is an easy consequence of H.-A. and Heyting's paper.) Thus, if $\mathfrak{E}$ is classically derivable, $\mathfrak{E}^*$ is intuitionistically derivable by theorem III, and if $\mathfrak{E}^*$ is intuitionistically derivable, then $\mathfrak{E}$ is classically derivable with the help of $\mathfrak{E}^* \supset \mathfrak{E}$.

5.5. THEOREM VI. *If intuitionist arithmetic is consistent, then classical arithmetic is also consistent.*

5.51. The consistency of arithmetic means formally that there is no derivation with an endformula of the form $\mathfrak{A} \,\&\, \neg\, \mathfrak{A}$, where $\mathfrak{A}$ stands for an arbitrary formula.

5.6. In order to prove theorem VI we assume such a proof figure to exist in classical arithmetic. Then, according to theorem III, we can find an intuitionist proof figure whose endformula is of the form $\mathfrak{A}^* \,\&\, \neg\, \mathfrak{A}^*$. Hence intuitionist arithmetic is not consistent either.

5.7. THEOREM VII. *For every purely logical formula there exists a classically equivalent formula which is derivable in intuitionist predicate logic if and only if the former is derivable in classical predicate logic.*

5.71. By a purely logical formula we mean a formula which (besides logical symbols) contains only variables (object and propositional variables), hence no symbols for definite objects, for definite functions, nor for definite predicates.

By a proof figure in classical predicate logic we mean a proof figure in which the axiom formulae 2.11 to 2.13 and all operational rules (2.2) may be applied, but where the operational rules may be used only to produce new purely logical formulae (such as the axiom formulae mentioned). This means that for free object variables only other object variables may be substituted (2.232); for propositional variables only purely logical formulae may be substituted (2.233).

The same holds for a proof figure of intuitionist predicate logic, except that here the axiom formula 2.11 may not be used.

5.8. The proof of theorem VII is obtained, as was that of theorem V, merely by specializing theorems I to III and their proofs to derivations in predicate logic. A brief reflection shows that this procedure is legitimate.

From theorem VII it follows, for example, that the 'decision problem' for classical predicate logic would be solved, if it were solved for intuitionist predicate logic. (The converse does not follow.)

§ 6. The consistency of arithmetic. The redundancy of negation in intuitionist arithmetic

6.1. A few remarks concerning the question of the consistency of arithmetic:
If intuitionist arithmetic is accepted as consistent, then the consistency of

classical arithmetic is also guaranteed by theorem VI. If a narrower view is taken as a starting point, such as Hilbert's 'finitist' view outlined in his paper 'On infinity'[18], then we are still left with the task of proving the consistency of intuitionist arithmetic from this point of view. Whether this can be done at all is rather doubtful since Gödel[19] has shown that the consistency of classical arithmetic (more precisely: an equivalent arithmetical proposition) cannot be proved within arithmetic itself (given that arithmetic is consistent). This does not contradict theorem VI since that theorem merely traces the consistency of classical arithmetic back to that of intuitionist arithmetic and the latter still remains unproved. Applying Gödel's result, we conclude that the consistency of *intuitionist* arithmetic cannot be proved within *classical* arithmetic (assuming that arithmetic is consistent).

6.2. Redundancy of negation in intuitionist arithmetic.

In intuitionist arithmetic we can dispense with negation altogether by defining $\neg \mathfrak{A}$ as an abbreviation for $\mathfrak{A} \supset 1 \neq 1$, and by considering 'non-equality' as a primitive predicate with the following axiom formulae: $x \neq y \cdot \supset : \neg \cdot x = y$ and $\neg \cdot x = y : \supset x \neq y$ (where $\neg \cdot x = y$ stands only for $x = y \cdot \supset \cdot 1 \neq 1$). This interpretation of negation is correct since $\neg A \cdot \supset (A \supset \neg \cdot 1 = 1)$ and $(A \supset \neg \cdot 1 = 1) \supset \neg A$ are intuitionistically derivable formulae (2.12.10 and 2.141, 2.125, 2.12.11 together with 2.34).

The axiom formula 2.12.11 thus becomes derivable (by means of 2.33, 2.126), and the other axiom formula for negation, 2.12.10, becomes equivalent to $1 \neq 1 \supset B$. This formula may be adopted as an arithmetical formula, although it can actually be omitted altogether since $1 \neq 1 \supset \mathfrak{A}$ is always derivable if $\mathfrak{A}$ contains no propositional variable.

6.21. The latter result follows thus: It holds, first of all, that

$$1 \neq 1 \cdot \supset : B \supset \cdot 1 \neq 1$$

(2.125), i.e., $1 \neq 1 \cdot \supset \neg B$; in particular, that $1 \neq 1 \cdot \supset : \neg \neg \cdot x = y$, hence (together with 2.148) $1 \neq 1 \cdot \supset \cdot x = y$. In the same way we derive $1 \neq 1 \cdot \supset \cdot x < y$. Furthermore, an arbitrary formula $\mathfrak{A}$ without propositional variables (1.8) is composed of elementary formulae $\mathfrak{f} = \mathfrak{g}, \mathfrak{h} < \mathfrak{i}, \ldots,$ (where $\mathfrak{f}, \mathfrak{g}, \mathfrak{h}, \mathfrak{i}$ are terms) together with logical symbols. The formula $1 \neq 1 \supset \mathfrak{C}$ is already valid if $\mathfrak{f} = \mathfrak{g}, \mathfrak{h} < \mathfrak{i}, \ldots,$ stands for $\mathfrak{C}$ (by virtue of the preceding together with 2.232) and, in the construction of $\mathfrak{A}$, the validity of the formula $1 \neq 1 \supset \mathfrak{C}$ carries over to the various subformulae of $\mathfrak{A}$ and finally to $\mathfrak{A}$ itself (in accordance with 2.33 for &; 2.127 for ∨; 2.125 for ⊃; 2.22 for $(\mathfrak{x})$; 2.132 for $(E\mathfrak{x})$).

3. INVESTIGATIONS INTO LOGICAL DEDUCTION

SYNOPSIS

The investigations that follow concern the domain of *predicate logic* (H.-A.[20] call it the 'restricted predicate calculus'). It comprises the types of inference that are continually used in all parts of mathematics. What remains to be added to these are axioms and forms of inference that may be considered as being proper to the particular branches of mathematics, e.g., in elementary number theory the axioms of the natural numbers, of addition, multiplication, and exponentiation, as well as the inference of complete induction; in geometry the geometric axioms.

In addition to *classical logic* I shall also deal with *intuitionist logic* as formalized, for example, by Heyting[21].

The present investigations into classical and intuitionist predicate logic fall essentially into two only loosely connected parts.

1. My starting point was this: The formalization of logical deduction, especially as it has been developed by Frege, Russell, and Hilbert, is rather far removed from the forms of deduction used in practice in mathematical proofs. Considerable formal advantages are achieved in return.

In contrast, I intended first to set up a formal system which comes as close as possible to actual reasoning. The result was a '*calculus of natural deduction*' ('*NJ*' for intuitionist, '*NK*' for classical predicate logic). This calculus then turned out to have certain special properties; in particular, the 'law of the excluded middle', which the intuitionists reject, occupies a special position.

I shall develop the calculus of natural deduction in section II of this paper together with some remarks concerning it.

2. A closer investigation of the specific properties of the natural calculus finally led me to a very general theorem which will be referred to below as the '*Hauptsatz*'.

The *Hauptsatz*[22] says that every purely logical proof can be reduced to a definite, though not unique, normal form. Perhaps we may express the essential properties of such a normal proof by saying: it is not roundabout. No concepts enter into the proof other than those contained in its final result, and their use was therefore essential to the achievement of that result.

The *Hauptsatz* holds both for classical and for intuitionist predicate logic.

In order to be able to enunciate and prove the *Hauptsatz* in a convenient form, I had to provide a logical calculus especially suited to the purpose. For this the natural calculus proved unsuitable. For, although it already contains the properties essential to the validity of the *Hauptsatz*, it does so only with respect to its intuitionist form, in view of the fact that the law of excluded middle, as pointed out earlier, occupies a special position in relation to these properties.

In section III of this paper, therefore, I shall develop a new calculus of logical deduction possessing all the desired properties in both their intuitionist and their classical forms ('*LJ*' for intuitionist, '*LK*' for classical predicate logic). The *Hauptsatz* will then be enunciated and proved by means of that calculus.

The *Hauptsatz* permits of a variety of *applications*. To illustrate this I shall develop a decision procedure (IV, § 1) for intuitionist propositional logic in section IV, and shall in addition give a new proof of the consistency of classical arithmetic without complete induction (IV, § 3).

Sections III and IV may be read independently of section II.

3. Section I contains the terminology and notations used in this paper.

In section V, I prove the *equivalence* of the logical calculi *NJ*, *NK*, and *LJ*, *LK*, developed in this paper, by means of a calculus modelled on the formalisms of Russell, Hilbert, and Heyting (and which may easily be compared with them). ('*LHJ*' for intuitionist, '*LHK*' for classical predicate logic.)

SECTION I. TERMINOLOGY AND NOTATIONS

To the concepts 'object', 'function', 'predicate', 'proposition', 'theorem', 'axiom', 'proof', 'inference', etc., in logic and mathematics there correspond, in the formalization of these disciplines, certain symbols or combinations of symbols. We divide these into:

1. *Symbols.*

2. *Expressions*, i.e., finite sequences of symbols.
3. *Figures*, i.e., finite sets of symbols, with some ordering.

Symbols count as special cases of expressions and figures, expressions as special cases of figures.

In this paper we shall consider symbols, expressions, and figures of the following kind:

1. *Symbols.*

These divide into constant symbols and variables.

1.1. Constant symbols:

Symbols for definite objects: 1, 2, 3, ...

Symbols for definite functions: +, −, ·.

Symbols for definite propositions: $\vee$ ('the true proposition'), $\wedge$ ('the false proposition').

Symbols for definite predicates: =, <.

Logical symbols:[23] & 'and', $\vee$ 'or', $\supset$ 'if ... then', $\supset \subset$ 'is equivalent to', $\neg$ 'not', $\forall$ 'for all', $\exists$ 'there is'.

We shall also use the terms: conjunction symbol, disjunction symbol, implication symbol, equivalence symbol, negation symbol, universal quantifier, existential quantifier.

Auxiliary symbols:) , (, → .

1.2. Variables:

Object variables. These we divide into *free* object variables: $a, b, c, \ldots, m$ and *bound* object variables: $n, \ldots, x, y, z$.

Propositional variables: $A, B, C, \ldots$.

An arbitrary number of variables will be assumed to be available; if the alphabet is insufficient, we adjoin numerical subscripts, e.g., a_7, C_3.

1.3. German and Greek letters serve as 'syntactic variables', i.e., not as symbols of the logic formalized, but as variables of our deliberations *about* that logic. Their meanings are explained as they are used.

2. *Expressions.*

2.1. The concept of a propositional expression, called a 'formula' for short (defined inductively):

(The concept of a formula is ordinarily used in a more general sense; the special case defined below might thus perhaps be described as a 'purely logical formula'.)

2.11. A symbol for a definite proposition (i.e., the symbols $\vee$ and $\wedge$) is a formula.

A propositional variable followed by a number (possibly zero) of free object variables is a formula, e.g., *Abab*.

The object variables are called the *arguments* of the propositional variables.

Formulae of the two kinds mentioned are also called *elementary formulae*.

2.12. If $\mathfrak{A}$ is a formula, then $\neg\, \mathfrak{A}$ is also a formula.

If $\mathfrak{A}$ and $\mathfrak{B}$ are formulae, then $\mathfrak{A}\,\&\,\mathfrak{B}$, $\mathfrak{A} \vee \mathfrak{B}$, $\mathfrak{A} \supset \mathfrak{B}$ are formulae.

(We shall not introduce the symbol $\supset \subset$ into our presentation; it is in fact superfluous, since $\mathfrak{A} \supset \subset \mathfrak{B}$ may be regarded as an abbreviation for $(\mathfrak{A} \supset \mathfrak{B})\,\&\,(\mathfrak{B} \supset \mathfrak{A})$.

2.13. A formula not containing the bound object variable $\mathfrak{x}$ yields another formula, if we prefix either $\forall \mathfrak{x}$ or $\exists \mathfrak{x}$. At the same time we may substitute $\mathfrak{x}$ in a number of places for a free object variable occurring in the formula.

2.14. Brackets (or parentheses) are to be used to show the structure of a formula unambiguously. Example of a formula:

$$\exists x\, (((\neg\, Abxa) \vee Bx) \supset (\forall z\, (A\,\&\,B)))$$

By special convention the number of brackets may be reduced, but (with one exception, *vide* 2.4) no use will be made of this, since we do not have to write down many formulae.

2.2. The number of logical symbols occurring in a formula is called the *degree of the formula*. (Thus an elementary formula is of degree 0.)

The logical symbol of a nonelementary formula that has been adjoined last in the construction of the formula according to 2.12 and 2.13, is called the *terminal symbol of the formula*.

Formulae that may have arisen in the course of the construction of a formula according to 2.12 and 2.13, including the formula itself, are called *subformulae*.

Example: the subformulae of $A\,\&\,\forall x\, Bxa$ are A, $\forall x\, Bxa$, $A\,\&\,\forall x\, Bxa$ as well as all formulae of the form $B\mathfrak{a}a$, where $\mathfrak{a}$ represents any free object variable (this variable may also be a, for example). The degree of $A\,\&\,\forall x\, Bxa$ is 2, the terminal symbol is $\&$.

2.3. The concept of a *sequent*:

(This concept will not be used until section III, and it is only then that the purpose of its introduction becomes clear.)

A sequent is an expression of the form

$$\mathfrak{A}_1, \ldots, \mathfrak{A}_\mu \rightarrow \mathfrak{B}_1, \ldots, \mathfrak{B}_\nu,$$

where $\mathfrak{A}_1, \ldots, \mathfrak{A}_\mu, \mathfrak{B}_1, \ldots, \mathfrak{B}_\nu$ may represent any formula whatever. (The $\rightarrow$, like commas, is an auxiliary symbol and not a logical symbol.)

The formulae $\mathfrak{A}_1, \ldots, \mathfrak{A}_\mu$ form the *antecedent*, and the formulae

$\mathfrak{B}_1, \ldots, \mathfrak{B}_\nu$, the *succedent* of the sequent. Both expressions may be empty.

2.4. The sequent $\mathfrak{A}_1, \ldots, \mathfrak{A}_\mu \rightarrow \mathfrak{B}_1, \ldots, \mathfrak{B}_\nu$ has exactly the same informal meaning as the formula

$$(\mathfrak{A}_1 \,\&\, \ldots \,\&\, \mathfrak{A}_\mu) \supset (\mathfrak{B}_1 \vee \ldots \vee \mathfrak{B}_\nu).$$

(By $\mathfrak{A}_1 \,\&\, \mathfrak{A}_2 \,\&\, \mathfrak{A}_3$ we mean $(\mathfrak{A}_1 \,\&\, \mathfrak{A}_2) \,\&\, \mathfrak{A}_3$, likewise for $\vee$.)

If the antecedent is empty, the sequent reduces to the formula $\mathfrak{B}_1 \vee \ldots \vee \mathfrak{B}_\nu$.

If the succedent is empty, the sequent means the same as the formula $\neg\, (\mathfrak{A}_1 \,\&\, \ldots \,\&\, \mathfrak{A}_\mu)$ or $(\mathfrak{A}_1 \,\&\, \ldots \,\&\, \mathfrak{A}_\mu) \supset \wedge$.

If both the antecedent and the succedent of the formula are empty, the sequent means the same as $\wedge$, i.e., a false proposition.

Conversely, to every formula there corresponds an equivalent sequent, e.g., the sequent whose antecedent is empty and whose succedent consists precisely of that formula.

The formulae making up a sequent are called *S-formulae* (i.e., sequent formulae). By this we intend to indicate that we are not considering the formula by itself, but as it appears in the sequent. Thus we say, for example:

'A formula occurs in several places in a sequent as an *S*-formula', which may also be expressed as follows:

'Several distinct *S*-formulae (which shall simply mean: having distinct occurrences in the sequent) are formally identical'.

3. *Figures*

We require inference figures and proof figures.

Such figures consist of formulae or sequents, as the case may be. In what follows (3.1 to 3.3, 3.5) we shall be speaking only of formulae, but whatever is said applies analogously to sequents; all we need to do is to replace the word 'formula', wherever it occurs, by the word 'sequent'.

3.1. An *inference figure* may be written in the following way:

$$\frac{\mathfrak{A}_1, \ldots, \mathfrak{A}_\nu}{\mathfrak{B}} \quad (\nu \geqq 1),$$

where $\mathfrak{A}_1, \ldots, \mathfrak{A}_\nu, \mathfrak{B}$ are formulae. $\mathfrak{A}_1, \ldots, \mathfrak{A}_\nu$ are then called the *upper formulae* and $\mathfrak{B}$ the *lower formula* of the inference figure. (The concepts of the upper sequents and of the lower sequent of an inference figure consisting of sequents are to be understood correspondingly.)

We shall have to consider only particular inference figures and they will be stated for each calculus as they arise.

3.2. A *proof figure*, called a *derivation* for short, consists of a number of formulae (at least one), which combine to form inference figures in the following way: Each formula is a lower formula of at most one inference figure; each formula (with the exception of exactly one: the *endformula*) is an upper formula of at least one inference figure; and the system of inference figures is noncircular, i.e., there is in the derivation no cycle (no sequence whose last member is again succeeded by its first member) of formulae such that each member is an upper formula of an inference figure whose lower formula is the next formula in the sequence.

3.3. The formulae of a derivation that are not lower formulae of an inference figure are called *initial formulae* of the derivation.

A derivation is in 'tree form' if each one of its formulae is an upper formula of *at most one* inference figure.

Thus all formulae except the endformula are upper formulae of *exactly one* inference figure.

We shall have to treat only of derivations in tree form.

The formulae which compose a derivation so defined are called *D-formulae* (i.e., derivation formulae). By this we wish to indicate that we are not considering merely the formula as such, but also its position in the derivation. In this sense we shall be using, for example, expressions such as: 'A formula occurs in a derivation as a *D-formula*'. 'Two distinct *D*-formulae (i.e., formulae occurring merely in distinct places in the derivation) are formally identical, viz., identical to the same formula'.

Thus by '$\mathfrak{A}$ is the *same D*-formula as $\mathfrak{B}$' we mean that $\mathfrak{A}$ and $\mathfrak{B}$ are not only formally identical, but occur also in the same place in the derivation. We shall use the words 'formally identical' to indicate identity of form regardless of place.

For object variables, however, we shall not introduce a special term that would associate the variable with a specific place of occurrence in the formula. Thus we say, e.g.: 'The *same* object variable occurs in two distinct *D*-formulae.'

3.4. The inference figures of the derivation are called *D-inference figures* (i.e., derivation inference figures).

In a derivation consisting of sequents the *S*-formulae of the *D*-sequents are called *D-S-formulae* (i.e., derivation sequent formulae).

3.5. A *path* in a derivation is (following Hilbert) a sequence of *D*-formulae whose first formula is an initial formula and whose last formula is the endformula, and of which each formula except the last is an upper formula of a *D*-inference figure whose lower formula is the next formula in the path.

We say that 'a *D*-formula stands *above* (*below*) another *D*-formula' if there exists a path in which the former occurs before (after) the latter.

We are here thinking of the fact that a derivation is written in tree form with the initial formulae above and the endformula below. (Examples may be found in II, § 4.)

Furthermore, we say that 'a D-inference figure occurs above (below) a D-formula', if all formulae of the inference figure occur above (below) that D-formula.

A derivation with the endformula $\mathfrak{A}$ is also called a 'derivation of $\mathfrak{A}$'.

The initial formulae of a derivation may be *basic formulae* or *assumption formulae*; more about their nature will have to be said as we reach the different calculi.

SECTION II. THE CALCULUS OF NATURAL DEDUCTION

§ 1. Examples of natural deduction

We wish to set up a formalism that reflects as accurately as possible the actual logical reasoning involved in mathematical proofs.

By means of a number of examples we shall first of all show what form deductions tend to take in practice and shall examine, for this purpose, three 'true formulae' and try to see their truth in the most natural way possible.

1.1. *First example*:

$(X \vee (Y \& Z)) \supset ((X \vee Y) \& (X \vee Z))$ is to be established as a true formula (H.-A., p. 28, formula 19).

The argument runs as follows: Suppose that either X or $Y \& Z$ holds. We distinguish the two cases: 1. X holds, 2. $Y \& Z$ holds. In the first case it follows that $X \vee Y$ holds, and also $X \vee Z$; hence $(X \vee Y) \& (X \vee Z)$ also holds. In the second case $Y \& Z$ holds, which means that both Y and Z hold. From Y follows $X \vee Y$; from Z follows $X \vee Z$. Thus $(X \vee Y) \& (X \vee Z)$ again holds. The latter formula has thus been derived, generally, from $X \vee (Y \& Z)$, i.e., $(X \vee (Y \& Z)) \supset ((X \vee Y) \& (X \vee Z))$ holds.

1.2. *Second example*:

$(\exists x \forall y \, Fxy) \supset (\forall y \exists x \, Fxy)$.

(H.-A., formula 36, p. 60). The argument runs as follows: Suppose there is an x such that for all y Fxy holds. Let a be such an x. Then for all y:

Fay holds. Now let *b* be an arbitrary object. Then *Fab* holds. Thus there is an *x*, viz., *a*, such that *Fxb* holds. Since *b* was arbitrary, our result therefore holds for all objects, i.e., for all *y* there is an *x* such that *Fxy* holds. This yields our assertion.

1.3. *Third example*:
($\neg \exists x\, Fx) \supset (\forall y\, \neg Fy)$ is to be established as intuitionistically true. We reason as follows: Assume there is no *x* for which *Fx* holds. From this we wish to infer: For all *y*, $\neg Fy$ holds. Now suppose *a* is some object for which *Fa* holds. It then follows that there is an *x* for which *Fx* holds, viz., *a* is such an object. This contradicts our hypothesis that $\neg \exists x\, Fx$. We have therefore a contradiction, i.e., *Fa* cannot hold. But since *a* was completely arbitrary, it follows that for all *y*, $\neg Fy$ holds. Q.E.D.

We intend now to integrate proofs of the kind carried out in these three examples into an exactly defined calculus (in § 4, we shall show how these examples are presented in that calculus).

§ 2. Construction of the Calculus *NJ*

2.1. We intend now to present a calculus for 'natural' intuitionist derivations of true formulae. The restriction to intuitionist reasoning is only provisional; we shall explain below (cf. § 5) our reasons for doing so and shall show in what way the calculus has to be extended for classical reasoning (by including the law of the excluded middle).

Externally, the essential difference between '*NJ*-derivations' and derivations in the systems of Russell, Hilbert, and Heyting is the following: In the latter systems true formulae are derived from a sequence of 'basic logical formulae' by means of a few forms of inference. Natural deduction, however, does not, in general, start from basic logical propositions, but rather from *assumptions* (cf. examples in § 1) to which logical deductions are applied. By means of a later inference the result is then again made independent of the assumption.

Calculi of the former kind will be referred to as *logistic* calculi.

2.2. After this preliminary remark we define the concept of an *NJ-derivation* as follows:
(Examples in § 4.)
An *NJ*-derivation consists of formulae arranged in tree form (I3.3).
(By demanding that the formulae are arranged in tree form we are

deviating somewhat from the analogy with actual reasoning. This is so, since in actual reasoning we necessarily have (1) a linear sequence of propositions due to the linear ordering of our utterances, and (2) we are accustomed to applying repeatedly a result once it has been obtained, whereas the tree form permits only of a single use of a derived formula. These two deviations permit us to define the concept of a derivation in a more convenient form and are not essential.)

The initial formulae of the derivation are *assumption formulae*. Each of these is adjoined to precisely one *D*-inference figure (and in fact occurs 'above' (I.3.5) the lower formula of that figure, as will be explained more fully below).

All formulae occurring below an assumption formula, but still above the lower formula of the *D*-inference figure to which that assumption formula was adjoined, the assumption formula itself included, are said to *depend* on that assumption formula. (Thus the inference makes all succeeding propositions independent of the assumption which is correlated with it.)

According to what we have said the endformula of the derivation depends on no assumption formula.

2.21. We shall now state the permissible *inference figures*.

The inference figure schemata below are to be understood in the following way:

We obtain an *NJ*-inference figure from one of the schemata by replacing $\mathfrak{A}, \mathfrak{B}, \mathfrak{C}, \mathfrak{D}$ by arbitrary formulae; and $\forall \mathfrak{x} \mathfrak{F} \mathfrak{x}$ ($\exists \mathfrak{x} \mathfrak{F} \mathfrak{x}$) by an arbitrary formula containing $\forall (\exists)$ for its terminal symbol, where $\mathfrak{x}$ designates the bound object variable belonging to that terminal symbol; and $\mathfrak{F}\mathfrak{a}$ by the formula obtained from $\mathfrak{F}\mathfrak{x}$ by replacing the bound variable $\mathfrak{x}$, wherever it occurs, by the free object variable $\mathfrak{a}$.

(For $\mathfrak{a}$ we may, for instance, take a variable already occurring in $\mathfrak{F}\mathfrak{x}$. For the inference figures $\forall$–I and $\exists$–E, this possibility will, however, be excluded by the restrictions on variables which follow below, but it remains for $\forall$–E and $\exists$–I. Nor need $\mathfrak{x}$ occur at all in $\mathfrak{F}\mathfrak{x}$, in which case $\mathfrak{F}\mathfrak{a}$ is, of course, identical with $\mathfrak{F}\mathfrak{x}$. – $\mathfrak{F}\mathfrak{a}$ is obviously always a subformula of $\forall \mathfrak{x} \mathfrak{F}\mathfrak{x}$ ($\exists \mathfrak{x} \mathfrak{F}\mathfrak{x}$), according to the definition of a subformula in I.2.2.)

Symbols written in square brackets have the following meaning: An arbitrary number (possibly zero) of formulae of this form, all formally identical, may be adjoined to the inference figure as assumption formulae. They must then be initial formulae of the derivation and occur, moreover, in those paths of the proof to which the particular upper formula of the inference figure belongs. (I.e., that upper formula above which the square bracket occurs in the scheme. This formula may itself be an assumption formula.)

The adjunction of the respective assumption formulae to a *D*-inference figure in a derivation must in some way be made explicit such as by appropriately numbering these assumption formulae (cf. the examples in § 4).

The designations of the various inference figure schemata: &–*I*, &–*E*, etc., stand for the following: An inference figure formed according to a particular schema is an 'introduction' (*I*) or an 'elimination' (*E*) of the conjunction (&), the disjunction (∨), the universal quantifier (∀), the existential quantifier (∃), the implication (⊃), or of the negation (¬). More about this in § 5.

The inference figure schemata:

$$
\begin{array}{cccc}
\&\text{--}I & \&\text{--}E & \vee\text{--}I & \vee\text{--}E \\
& & & [\mathfrak{A}]\ [\mathfrak{B}] \\
\dfrac{\mathfrak{A}\quad \mathfrak{B}}{\mathfrak{A}\,\&\,\mathfrak{B}} & \dfrac{\mathfrak{A}\,\&\,\mathfrak{B}}{\mathfrak{A}}\ \dfrac{\mathfrak{A}\,\&\,\mathfrak{B}}{\mathfrak{B}} & \dfrac{\mathfrak{A}}{\mathfrak{A}\vee\mathfrak{B}}\ \dfrac{\mathfrak{B}}{\mathfrak{A}\vee\mathfrak{B}} & \dfrac{\mathfrak{A}\vee\mathfrak{B}\quad \mathfrak{C}\quad \mathfrak{C}}{\mathfrak{C}}
\end{array}
$$

$$
\begin{array}{cccc}
\forall\text{--}I & \forall\text{--}E & \exists\text{--}I & \exists\text{--}E \\
& & & [\mathfrak{F}\mathfrak{a}] \\
\dfrac{\mathfrak{F}\mathfrak{a}}{\forall\mathfrak{x}\,\mathfrak{F}\mathfrak{x}} & \dfrac{\forall\mathfrak{x}\,\mathfrak{F}\mathfrak{x}}{\mathfrak{F}\mathfrak{a}} & \dfrac{\mathfrak{F}\mathfrak{a}}{\exists\mathfrak{x}\,\mathfrak{F}\mathfrak{x}} & \dfrac{\exists\mathfrak{x}\,\mathfrak{F}\mathfrak{x}\quad \mathfrak{C}}{\mathfrak{C}}
\end{array}
$$

$$
\begin{array}{cccc}
\supset\text{--}I & \supset\text{--}E & \neg\text{--}I & \neg\text{--}E \\
[\mathfrak{A}] & & [\mathfrak{A}] & \\
\dfrac{\mathfrak{B}}{\mathfrak{A}\supset\mathfrak{B}} & \dfrac{\mathfrak{A}\quad \mathfrak{A}\supset\mathfrak{B}}{\mathfrak{B}} & \dfrac{\wedge}{\neg\mathfrak{A}} & \dfrac{\mathfrak{A}\quad \neg\mathfrak{A}\quad \wedge}{\wedge\quad \mathfrak{D}}
\end{array}
$$

The free object variable of a ∀–*I* or ∃–*E*, designated by 𝔞 in the respective schema, is called the *eigenvariable*. (This, of course, presupposes that there is such a variable, i.e., that the bound object variable designated by 𝔵 occurs in the formula designated by 𝔉𝔵.)

Restrictions on variables:

An *NJ*-derivation is subject to the following restriction (for the significance of this restriction cf. § 3):

The eigenvariable of an ∀–*I* must not occur in the formula designated in the schema by ∀𝔵 𝔉𝔵; nor in any assumption formula upon which that formula depends.

The eigenvariable of an ∃–E must not occur in the formula designated in the schema by $\exists\mathfrak{x}\,\mathfrak{Fx}$; nor in an upper formula designated by $\mathfrak{C}$; nor in any assumption formula upon which that formula depends, with the exception of the assumption formulae designated by $\mathfrak{Fa}$ in the schema of the ∃–E.

This concludes the definition of the '*NJ*-derivation'.

§ 3. Informal sense of *NJ*-inference figures

We shall explain the informal sense of a number of inference figure schemata and thus try to show how the calculus in fact reflects 'actual reasoning'.

⊃–*I*: Expressed in words, this schema corresponds to the following inference: If $\mathfrak{B}$ has been proved by means of assumption $\mathfrak{A}$, we have (this time without the assumption): from $\mathfrak{A}$ follows $\mathfrak{B}$. (Further assumptions may, of course, have been made and the result still continues to depend on them.)

∨–*E* ('Distinction of cases'): If $\mathfrak{A} \vee \mathfrak{B}$ has been proved, we can distinguish two cases: What we first assume is that $\mathfrak{A}$ holds and derive, let us say, $\mathfrak{C}$ from it. If it is then possible to derive $\mathfrak{C}$ also by assuming that $\mathfrak{B}$ holds, then $\mathfrak{C}$ holds generally, i.e., it is now independent of both assumptions (cf.1.1).

∀–*I*: If $\mathfrak{Fa}$ has been proved for an 'arbitrary $\mathfrak{a}$', then $\forall\mathfrak{x}\,\mathfrak{Fx}$ holds. The presupposition that $\mathfrak{a}$ is 'completely arbitrary' can be expressed more precisely as: $\mathfrak{Fa}$ must not depend on any assumption in which the object variable $\mathfrak{a}$ occurs. And this, together with the obvious requirement that *every occurrence* of $\mathfrak{a}$ in $\mathfrak{Fa}$ must be replaced by an $\mathfrak{x}$ in $\mathfrak{Fx}$, constitutes precisely that part of the 'restrictions on variables' which applies to the schema of the ∀–*I*.

∃–*E*: We have $\exists\mathfrak{x}\,\mathfrak{Fx}$. We say: Suppose $\mathfrak{a}$ is an object for which $\mathfrak{F}$ holds, i.e., we assume that $\mathfrak{Fa}$ holds. (It is, of course, obvious that for $\mathfrak{a}$ we must take an object variable which does not yet occur in $\exists\mathfrak{x}\,\mathfrak{Fx}$.) If, on this assumption, we then prove a proposition $\mathfrak{C}$ which no longer contains $\mathfrak{a}$ and does not depend on any other assumption containing $\mathfrak{a}$, we have proved $\mathfrak{C}$ independently of the assumption $\mathfrak{Fa}$. We have here stated the part of the 'restrictions on variables' that concerns the ∃–*E*. (A certain analogy exists between the ∃–*E* and the ∨–*E* since the existential quantifier is indeed the generalization of ∨, and the universal quantifier the generalization of &.)

¬–*E*: $\mathfrak{A}$ and ¬$\mathfrak{A}$ signifies a contradiction and as such cannot hold true (law of contradiction). This is formally expressed by the inference figure $\neg\neg$-E, where $\curlywedge$ designates 'the contradiction', 'the false'.

$\neg$-I: (*Reductio ad absurdum.*) If we can derive any false proposition ($\curlywedge$) on an assumption $\mathfrak{A}$, then $\mathfrak{A}$ is not true, i.e., $\neg\,\mathfrak{A}$ holds.

The schema $\dfrac{\curlywedge}{\mathfrak{D}}$ expresses the fact that if a false proposition holds, any arbitrary proposition also holds.

The interpretation of the remaining inference figure schemata should be straightforward.

§ 4. The three examples of § 1 written as *NJ*-derivations

First example (1.1):

$$\dfrac{\overset{2}{X\vee(Y\,\&\,Z)}\quad \dfrac{\dfrac{\overset{1}{X}}{X\vee Y}\vee\text{-}I \quad \dfrac{\overset{1}{X}}{X\vee Z}\vee\text{-}I}{(X\vee Y)\,\&\,(X\vee Z)}\&\text{-}I \quad \dfrac{\dfrac{\dfrac{\overset{1}{Y\,\&\,Z}}{Y}\&\text{-}E}{X\vee Y}\vee\text{-}I \quad \dfrac{\dfrac{\overset{1}{Y\,\&\,Z}}{Z}\&\text{-}E}{X\vee Z}\vee\text{-}I}{(X\vee Y)\,\&\,(X\vee Z)}\&\text{-}I}{\dfrac{(X\vee Y)\,\&\,(X\vee Z)}{(X\vee(Y\,\&\,Z))\supset((X\vee Y)\,\&\,(X\vee Z))}\supset\text{-}I_2}\vee\text{-}E_1.$$

In this example the tree form must appear somewhat artificial since it does not bring out the fact that it is *after* the enunciation of $X\vee(Y\,\&\,Z)$ that we distinguish the cases $X,\,Y\,\&\,Z$.

Second example (1.2):

$$\dfrac{\overset{2}{\exists x\,\forall y\,Fxy}\quad \dfrac{\dfrac{\dfrac{\overset{1}{\forall y\,Fay}}{Fab}\forall\text{-}E}{\exists x\,Fxb}\exists\text{-}I}{\forall y\,\exists x\,Fxy}\forall\text{-}I}{\dfrac{\forall y\,\exists x\,Fxy}{(\exists x\,\forall y\,Fxy)\supset(\forall y\,\exists x\,Fxy)}\supset\text{-}I_2}\exists\text{-}E_1.$$

If we were using a *linear* arrangement, then the assumption of the $\exists$-E would here also follow naturally behind the upper formula on the left, as was the case in our treatment of that example in § 1.

Third example (1.3):

$$\cfrac{\cfrac{\cfrac{\cfrac{\cfrac{Fa}{\exists x\, Fx}\ \exists\text{-}I^{2} \quad \neg\, \exists x\, Fx^{1}}{\curlywedge}\ \neg\text{-}E}{\neg\, Fa}\ \neg\text{-}I_2}{\forall y\, \neg\, Fy}\ \forall\text{-}I}{(\neg\, \exists x\, Fx) \supset (\forall y\, \neg\, Fy)}\ \supset\text{-}I_1.$$

§ 5. Some remarks concerning the calculus *NJ*. The calculus *NK*

5.1. The calculus *NJ* lacks a certain formal elegance. This has to be put against the following advantages:

5.11. A close affinity to actual reasoning, which had been our fundamental aim in setting up the calculus. The calculus lends itself in particular to the formalization of mathematical proofs.

5.12. In most cases the derivations for true formulae are *shorter* in our calculus than their counterparts in the logistic calculi. This is so primarily because in logistic derivations one and the same formula usually occurs a number of times (as part of other formulae), whereas this happens only very rarely in the case of *NJ*-derivations.

5.13. The designations given to the various inference figures (2.21) make it plain that our calculus is remarkably *systematic*. To every logical symbol &, ∨, ∀, ∃, ⊃, ¬, belongs precisely one inference figure which 'introduces' the symbol – as the terminal symbol of a formula – and one which 'eliminates' it. The fact that the inference figures &–*E* and ∨–*I* each have two forms constitutes a trivial, purely external deviation and is of no interest. The introductions represent, as it were, the 'definitions' of the symbols concerned, and the eliminations are no more, in the final analysis, than the consequences of these definitions. This fact may be expressed as follows: In eliminating a symbol, we may use the formula with whose terminal symbol we are dealing only 'in the sense afforded it by the introduction of that symbol'. An example may clarify what is meant: We were able to introduce the formula 𝔄 ⊃ 𝔅 when there existed a derivation of 𝔅 from the assumption formula 𝔄. If we then wished to use that formula by eliminating the ⊃-symbol (we could, of course, also use it to form longer formulae, e.g., (𝔄 ⊃ 𝔅) ∨ ℭ, ∨–*I*), we could do this precisely by inferring 𝔅 directly, once 𝔄 has been proved, for what 𝔄 ⊃ 𝔅 attests is just the existence of a derivation of $\mathfrak{B}$ from $\mathfrak{A}$. Note that in saying this we need not go into the 'informal sense' of the $\supset$-symbol.

By making these ideas more precise it should be possible to display the E-inferences as unique functions of their corresponding I-inferences, on the basis of certain requirements.

5.2. It is possible to eliminate the *negation* from our calculus by regarding $\neg\,\mathfrak{A}$ as an abbreviation for $\mathfrak{A} \supset \wedge$. This is permissible, since by replacing every $\neg\,\mathfrak{A}$ by $\mathfrak{A} \supset \wedge$, and thus removing all $\neg$-symbols from an NJ-derivation, we obtain another NJ-derivation (the inference figures $\neg\text{-}I$ and $\neg\text{-}E$ then become special cases of the $\supset\text{-}I$ and the $\supset\text{-}E$) and vice versa: If, in an NJ-derivation, we replace every occurrence of $\mathfrak{A} \supset \wedge$ by $\neg\,\mathfrak{A}$, another NJ-derivation results.

The inference figure schema $\dfrac{\wedge}{\mathfrak{D}}$ occupies a special place among the schemata: It does not belong to a logical symbol, but to the propositional symbol $\wedge$.

5.3. The '*law of the excluded middle*' and the *calculus NK*.

From the calculus NJ we obtain a complete classical calculus NK by including the 'law of the excluded middle' (*tertium non datur*), i.e.: In addition to the assumption formulae we now also allow 'basic formulae' of the form $\mathfrak{A} \vee \neg\,\mathfrak{A}$, where $\mathfrak{A}$ stands for any arbitrary formula.

We have thus granted to the law of the excluded middle, in a purely external way, a special position, and we have done this because we considered that formulation the 'most natural'. It would be perfectly feasible to introduce a new inference figure schema, say $\dfrac{\neg\,\neg\,\mathfrak{A}}{\mathfrak{A}}$ (a schema analogous to the one formed by Hilbert and Heyting), in place of the basic formula schema $\mathfrak{A} \vee \neg\,\mathfrak{A}$. However, such a schema still falls outside the framework of the NJ-inference figures, because it represents a new elimination of the negation whose admissibility does not follow at all from our method of introducing the $\neg$-symbol by the $\neg\text{-}I$.

SECTION III. THE DEDUCTIVE CALCULI *LJ*, *LK* AND THE HAUPTSATZ

§ 1. The calculi *LJ* and *LK* (logistic intuitionist and classical calculi)

1.1. Preliminary remarks concerning the construction of the calculi *LJ* and *LK*.

What we want to do is to formulate a deductive calculus (for predicate logic) which is 'logistic' on the one hand, i.e., in which the derivations do not, as in the calculus *NJ*, contain assumption formulae, but which, on the other hand, takes over from the calculus *NJ* the division of the forms of inference into introductions and eliminations of the various logical symbols.

The most obvious method of converting an *NJ*-derivation into a logistic one is this: We replace a *D*-formula $\mathfrak{B}$, which depends on the assumption formulae $\mathfrak{A}_1, \ldots, \mathfrak{A}_\mu$, by the new formula $(\mathfrak{A}_1 \& \ldots \& \mathfrak{A}_\mu) \supset \mathfrak{B}$. This we do with all *D*-formulae.

We thus obtain formulae which are already true *in themselves*, i.e., whose truth is no longer *conditional* on the truth of certain assumption formulae. This procedure, however, introduces new logical symbols & and $\supset$, necessitating additional inference figures for & and $\supset$, and thus upsets the systematic character of our method of introducing and eliminating symbols. For this reason we have introduced the concept of a *sequent* (I.2.3). Instead of a formula $(\mathfrak{A}_1 \& \ldots \& \mathfrak{A}_\mu) \supset \mathfrak{B}$, e.g., we therefore write the sequent

$$\mathfrak{A}_1, \ldots, \mathfrak{A}_\mu \to \mathfrak{B}.$$

The informal meaning of this sequent is no different from that of the above formula; the expressions differ merely in their formal structure (cf. I. 2.4).

Even now new inference figures are required that cannot be integrated into our system of introductions and eliminations; but we have the advantage of being able to reserve them special places within our system, since they no longer refer to logical symbols, but merely to the structure of the sequents. We therefore call these 'structural inference figures', and the others 'operational inference figures'.

In the *classical* calculus *NK* the law of the excluded middle occupied a special place among the forms of inference (II.5.3), because it could not be integrated into our system of introductions and eliminations. In the classical logistic calculus *LK* about to be presented, this characteristic is removed. This is made possible by the admission of sequents with *several* formulae in the succedent, whereas the transition from the calculus *NJ* just described led only to sequents with *one* formula in the succedent. (For the informal meaning of sequents in general cf. I.2.4.) The symmetry thus obtained is more suited to classical logic. On the other hand, the restriction to at most one formula in the succedent will be retained for the intuitionist calculus *LJ*. (Cf. below. – An empty succedent means the same as if $\wedge$ stood in the succedent.)

We have thus outlined a number of points that underlie the construction of the calculi that follow. Their form is largely determined, however, by considerations connected with the '*Hauptsatz*' (§ 2) whose proof follows later. That form cannot therefore be justified more fully at this stage.

1.2. We now define the concepts of an '*LK*-derivation' and an '*LJ*-derivation' as follows:

An *LJ*- or *LK*-derivation consists of sequents arranged in tree form (I.3.3).

The *initial sequents* of the derivation are basic sequents of the form $\mathfrak{D} \to \mathfrak{D}$, where $\mathfrak{D}$ may be an arbitrary formula.

Each *inference figure* of the derivation results from one of the schemata below by a substitution of the following kind (cf. II.2.21):

Replace $\mathfrak{A}, \mathfrak{B}, \mathfrak{D}, \mathfrak{E}$ by an arbitrary formula; for $\forall \mathfrak{x} \mathfrak{F} \mathfrak{x}$ ($\exists \mathfrak{x} \mathfrak{F} \mathfrak{x}$) put an arbitrary formula having $\forall (\exists)$ for its terminal symbol, where $\mathfrak{x}$ designates the associated bound object variable; for $\mathfrak{F}\mathfrak{a}$ put that formula which is obtained from $\mathfrak{F}\mathfrak{x}$ by replacing every occurrence of the bound object variable $\mathfrak{x}$ by the free object variable $\mathfrak{a}$.

For $\Gamma, \Delta, \Theta, \Lambda$ put arbitrary (possibly empty) sequences of formulae separated by commas.

The following restriction is furthermore placed on *LJ-inference figures* (this is the *only* respect in which the concepts of an *LJ*- and an *LK*-derivation differ):

'In the succedent of each *D*-sequent no more than one *S*-formula may occur'.

The designations of the various schemata for operational inference figures &–*IS*, &–*IA*, etc., are intended to mean: An inference figure formed according to the schema is an introduction (I) in the succedent (S) or antecedent (A) of the conjunction (&), the disjunction ($\lor$), the universal quantifier ($\forall$), the existential quantifier ($\exists$), the negation ($\neg$), or the implication ($\supset$).

The inference figure schemata

1.21. Schemata for structural inference figures:

Thinning:

$$\text{in the antecedent} \qquad \text{in the succedent}$$
$$\frac{\Gamma \to \Theta}{\mathfrak{D}, \Gamma \to \Theta}, \qquad \frac{\Gamma \to \Theta}{\Gamma \to \Theta, \mathfrak{D}};$$

Contraction:

$$\text{in the antecedent} \qquad \text{in the succedent}$$
$$\frac{\mathfrak{D}, \mathfrak{D}, \Gamma \to \Theta}{\mathfrak{D}, \Gamma \to \Theta}, \qquad \frac{\Gamma \to \Theta, \mathfrak{D}, \mathfrak{D}}{\Gamma \to \Theta, \mathfrak{D}};$$

Interchange:

$$\text{in the antecedent} \qquad \text{in the succedent}$$
$$\frac{\Delta, \mathfrak{D}, \mathfrak{E}, \Gamma \to \Theta}{\Delta, \mathfrak{E}, \mathfrak{D}, \Gamma \to \Theta}, \qquad \frac{\Gamma \to \Theta, \mathfrak{E}, \mathfrak{D}, \Lambda}{\Gamma \to \Theta, \mathfrak{D}, \mathfrak{E}, \Lambda};$$

Cut:

$$\frac{\Gamma \to \Theta, \mathfrak{D} \qquad \mathfrak{D}, \Delta \to \Lambda}{\Gamma, \Delta \to \Theta, \Lambda}.$$

1.22. Schemata for operational inference figures:

$$\&\text{–}IS: \quad \frac{\Gamma \to \Theta, \mathfrak{A} \qquad \Gamma \to \Theta, \mathfrak{B}}{\Gamma \to \Theta, \mathfrak{A} \& \mathfrak{B}},$$

$$\&\text{–}IA: \quad \frac{\mathfrak{A}, \Gamma \to \Theta}{\mathfrak{A} \& \mathfrak{B}, \Gamma \to \Theta} \qquad \frac{\mathfrak{B}, \Gamma \to \Theta}{\mathfrak{A} \& \mathfrak{B}, \Gamma \to \Theta},$$

$$\vee\text{–}IA: \quad \frac{\mathfrak{A}, \Gamma \to \Theta \qquad \mathfrak{B}, \Gamma \to \Theta}{\mathfrak{A} \vee \mathfrak{B}, \Gamma \to \Theta},$$

$$\vee\text{–}IS: \quad \frac{\Gamma \to \Theta, \mathfrak{A}}{\Gamma \to \Theta, \mathfrak{A} \vee \mathfrak{B}} \qquad \frac{\Gamma \to \Theta, \mathfrak{B}}{\Gamma \to \Theta, \mathfrak{A} \vee \mathfrak{B}},$$

$$\forall\text{–}IS: \quad \frac{\Gamma \to \Theta, \mathfrak{F}\mathfrak{a}}{\Gamma \to \Theta, \forall \mathfrak{x}\, \mathfrak{F}\mathfrak{x}},$$

$$\exists\text{–}IA: \quad \frac{\mathfrak{F}\mathfrak{a}, \Gamma \to \Theta}{\exists \mathfrak{x}\, \mathfrak{F}\mathfrak{x}, \Gamma \to \Theta}.$$

Restrictions on variables: The object variable in the last two schemata, which is designated by $\mathfrak{a}$ and is called the *eigenvariable* of the $\forall\text{–}IS$ ($\exists\text{–}IA$), must not occur in the lower sequent of the inference figure (i.e., not in Γ, Θ, and $\mathfrak{F}\mathfrak{x}$).

$$\forall\text{–}IA: \quad \frac{\mathfrak{F}\mathfrak{a}, \Gamma \to \Theta}{\forall \mathfrak{x}\, \mathfrak{F}\mathfrak{x}, \Gamma \to \Theta},$$

$$\neg\text{–}IS: \quad \frac{\mathfrak{A}, \Gamma \to \Theta}{\Gamma \to \Theta, \neg \mathfrak{A}},$$

$\exists$-IS: $\dfrac{\Gamma \to \Theta, \mathfrak{F}\mathfrak{a}}{\Gamma \to \Theta, \exists\mathfrak{x}\, \mathfrak{F}\mathfrak{x}}$,

$\neg$-IA: $\dfrac{\Gamma \to \Theta, \mathfrak{A}}{\neg\, \mathfrak{A}, \Gamma \to \Theta}$,

$\supset$-IS: $\dfrac{\mathfrak{A}, \Gamma \to \Theta, \mathfrak{B}}{\Gamma \to \Theta, \mathfrak{A} \supset \mathfrak{B}}$,

$\supset$-IA: $\dfrac{\Gamma \to \Theta, \mathfrak{A} \quad \mathfrak{B}\, \Delta \to \Lambda}{\mathfrak{A} \supset \mathfrak{B}, \Gamma, \Delta \to \Theta, \Lambda}$.

1.3. Example of an *LJ*-derivation (using II.1.3):

$$\dfrac{\dfrac{Fa \to Fa}{Fa \to \exists x\, Fx}\,\exists\text{-}IS \quad \dfrac{\dfrac{\exists x\, Fx \to \exists x\, Fx}{\neg\, \exists x\, Fx, \exists x\, Fx \to}\,\neg\text{-}IA}{\exists x\, Fx, \neg\, \exists x\, Fx \to}\,\text{Interchange}}{\dfrac{\dfrac{Fa, \neg\, \exists x\, Fx \to}{\neg\, \exists x\, Fx \to \neg\, Fa}\,\neg\text{-}IS}{\dfrac{\neg\, \exists x\, Fx \to \forall y\, \neg\, Fy}{\to (\neg\, \exists x\, Fx) \supset (\forall\, \neg\, Fy)}\,\supset\text{-}IS.}\,\forall\text{-}IS}\,\text{Cut}$$

1.4. Example of an *LK*-derivation (derivation of the 'law of the excluded middle'):

$$\dfrac{\dfrac{\dfrac{\dfrac{\dfrac{A \to A}{\to A, \neg\, A}\,\neg\text{-}IS}{\to A, A \vee \neg\, A}\,\vee\text{-}IS}{\to A \vee \neg\, A, A}\,\text{Interchange}}{\to A \vee \neg\, A, A \vee \neg\, A}\,\vee\text{-}IS}{\to A \vee \neg\, A}\,\text{Contraction.}$$

§ 2. Some remarks concerning the calculi *LJ* and *LK*. The Hauptsatz

(We shall make no further use, in this paper, of remarks 2.1 to 2.3.)
2.1. The schemata are not all mutually independent, i.e., certain schemata could be eliminated with the help of the remaining ones. Yet if they were left out, the '*Hauptsatz*' would no longer be valid.
2.2. In general, we could *simplify* the calculi in various respects if we attached no importance to the *Hauptsatz*. To indicate this briefly: the

inference figures &–IS, ∨–IA, &–IA, ∨–IS, ∀–IA, ∃–IS, ¬–IS, ¬–IA, and ⊃–IA in the calculus LK could be replaced by basic sequents according to the following schemata:

$\mathfrak{A}, \mathfrak{B} \to \mathfrak{A} \& \mathfrak{B}$ $\mathfrak{A} \vee \mathfrak{B} \to \mathfrak{A}, \mathfrak{B}$ $\mathfrak{A} \& \mathfrak{B} \to \mathfrak{A}$ $\mathfrak{A} \& \mathfrak{B} \to \mathfrak{B}$

$\mathfrak{A} \to \mathfrak{A} \vee \mathfrak{B}$ $\mathfrak{B} \to \mathfrak{A} \vee \mathfrak{B}$ $\forall\mathfrak{x}\, \mathfrak{F}\mathfrak{x} \to \mathfrak{F}\mathfrak{a}$ $\mathfrak{F}\mathfrak{a} \to \exists\mathfrak{x}\, \mathfrak{F}\mathfrak{x}$

$\to \mathfrak{A}, \neg \mathfrak{A}$ (law of the excluded middle)

$\neg \mathfrak{A}, \mathfrak{A} \to$ (law of contradiction)

$\mathfrak{A} \supset \mathfrak{B}, \mathfrak{A} \to \mathfrak{B}$.

These basic sequents and our inference figures may easily be shown to be equivalent.

The same possibility exists for the calculus LJ, with the exception of the inference figures ∨–IA and ¬–IS, since LJ-D-sequents may not in fact contain two S-formulae in the succedent (cf. V. § 5).

2.3. The distinction between *intuitionist* and *classical* logic is, externally, of a quite different type in the calculi LJ and LK from that in the calculi NJ and NK. In the case of the latter, the distinction is based on the inclusion or exclusion of the law of the excluded middle, whereas for the calculi LJ and LK the difference is characterized by the restriction on the succedent. (The fact that both distinctions are equivalent will become evident as a result of the equivalence proofs in section V for all calculi discussed in this paper.)

2.4. If ⊃–IS and the ⊃–IA are excluded, the calculus LK is *dual* in the following sense: If we reverse all sequents of an LK-derivation (in which the ⊃-symbol does not occur), i.e., if for $\mathfrak{A}_1, \ldots, \mathfrak{A}_\mu \to \mathfrak{B}_1, \ldots, \mathfrak{B}_\nu$ we put $\mathfrak{B}_\nu, \ldots, \mathfrak{B}_1 \to \mathfrak{A}_\mu, \ldots, \mathfrak{A}_1$, and if we exchange, in inference figures with two upper sequents, the right- and left-hand upper sequents, including their derivations, and also replace every occurrence of & by ∨, ∀ by ∃, ∨ by &, and ∃ by ∀ (in the case of & and ∨ we also have to interchange the respective scopes of the symbols, e.g., for $\mathfrak{B} \vee \mathfrak{A}$ we have to put $\mathfrak{A} \& \mathfrak{B}$), then another LK-derivation results.

This can be seen at once from the schemata. (Special care was taken to arrange them in such a way as to bring out their symmetry.) (Cf. H.-A.'s duality principle, p. 62.)

2.41. In any case, the ⊃-symbol may, in a well-known manner, be eliminated from the calculus NK, by regarding $\mathfrak{A} \supset \mathfrak{B}$ as an abbreviation for $(\neg \mathfrak{A}) \vee \mathfrak{B}$. It may easily be shown that the schemata for the ⊃–IS and the ⊃–IA may then be replaced by the schemata for ∨ and ¬.

The calculus *NJ* has no corresponding property.

2.5. The most important fact for us with regard to the calculi *LJ* and *LK* is the following:

HAUPTSATZ: Every *LJ*- or *LK*-derivation can be transformed into an *LJ*- or *LK*-derivation with the same endsequent and in which the inference figure called a 'cut' does not occur.

The proof follows in § 3.

2.51. In order to give greater clarity to the meaning of the *Hauptsatz*, we shall prove a simple corollary (2.513).

For this purpose we introduce a number of expressions (which will be needed frequently later on) relating to the operational inference figures:

2.511. That *S*-formula which contains the logical symbol in its schema will be called the *principal formula* of an inference figure.

For the &–*IS* and the &–*IA* this is simply the *S*-formula of the form $\mathfrak{A} \& \mathfrak{B}$; for the ∨–*IS* and the ∨–*IA* it is $\mathfrak{A} \vee \mathfrak{B}$; for the ∀–*IS* and the ∀–*IA* it is $\forall \mathfrak{x} \, \mathfrak{F}\mathfrak{x}$; for the ∃–*IS* and the ∃–*IA* it is $\exists \mathfrak{x} \, \mathfrak{F}\mathfrak{x}$; for the ¬–*IS* and the ¬–*IA* it is $\neg \mathfrak{A}$; and for the ⊃–*IS* and the ⊃–*IA* it is $\mathfrak{A} \supset \mathfrak{B}$.

The *S*-formulae designated by $\mathfrak{A}, \mathfrak{B}, \mathfrak{F}\mathfrak{a}$ in the schemata will be called the *side formulae* of the respective inference figures.

They are always subformulae of the principal formula (according to the definition of a subformula in I.2.2).

2.512. We can now easily read off the following facts from the inference figure schemata:

The principal formula occurs always in the lower sequent and the side formulae always in the upper sequents of an operational inference figure.

If a formula occurs as an *S*-formula in an upper sequent of a given inference figure, and if it is here neither a side formula nor the $\mathfrak{D}$ of a cut, then it occurs also as an *S*-formula in the lower sequent.

These two facts entail the following:

If anywhere in an *LJ*- or *LK*-derivation a formula occurs as an *S*-formula, and if we trace the path of the derivation from the formula concerned up to the endsequent, the formula can only vanish from that path if it is the $\mathfrak{D}$ of a cut or the side formula of an operational inference figure. In the latter case, however, there appears, in the next sequent, the *principal formula* of the inference figure of which our side formula is a subformula. To that principal formula we can then, continuing downwards, apply the same consideration, and so on. Thus we obtain the following corollary:

2.513. COROLLARY OF THE HAUPTSATZ (SUBFORMULA PROPERTY): In an *LJ*- or

LK-derivation without cuts, all occurring *D-S*-formulae are *subformulae* of the *S*-formulae that occur in the endsequent.

2.514. Intuitively speaking, these properties of derivations without cuts may be expressed as follows: The *S*-formulae become longer as we descend lower down in the derivation, never shorter. The final result is, as it were, gradually built up from its constituent elements. The proof represented by the derivation is not roundabout in that it contains only concepts which recur in the final result (cf. the synopsis at the beginning of this paper).

Example: The derivation given above (1.3) for $\rightarrow (\neg \exists \mathfrak{x} \mathfrak{F} \mathfrak{x}) \supset (\forall \mathfrak{y} \neg \mathfrak{F} \mathfrak{y})$ may be written without a cut as follows:

$$\cfrac{\cfrac{\cfrac{\cfrac{Fa \rightarrow Fa}{Fa \rightarrow \exists x\, Fx}\ \exists\text{-}IS}{\neg\, \exists x\, Fx, Fa \rightarrow}\ \neg\text{-}IA}{Fa, \neg\, \exists x\, Fx \rightarrow}\ \text{Interchange,}}$$

etc., as above.

§ 3. Proof of the Hauptsatz

The *Hauptsatz* runs as follows:

Every *LJ*- or *LK*-derivation can be transformed into another *LJ*- or *LK*-derivation with the same endsequent, in which no cuts occur.

3.1. *Proof of the Hauptsatz for LK-derivations.*

We introduce a new inference figure (in order to facilitate the proof) which constitutes a modified form of the cut, and which we call a *mix*.

The schema of that figure runs as follows:

$$\cfrac{\Gamma \rightarrow \Theta \quad \Delta \rightarrow \Lambda}{\Gamma, \Delta^* \rightarrow \Theta^*, \Lambda},$$

In order to obtain an inference figure from this schema, Θ and Δ must be replaced by sequences of formulae, separated by commas, in each of which occurs at least once (as a member of the sequence) a formula of the form $\mathfrak{M}$, called the 'mix formula'; and Θ^* and Δ^* must be replaced by the same sequences of formulae, save that all formulae of the form $\mathfrak{M}$ occurring as members of the sequence are omitted. ($\mathfrak{M}$ may be any arbitrary formula.) Γ and Λ must be replaced, as in the other schemata, by arbitrary (possibly empty) sequences of formulae, separated by commas.

Example of a mix:

$$\frac{A \to B, \neg A \quad B \vee C, B, B, D, B \to}{A, B \vee C, D \to \neg A}.$$

B is the mix formula.

We notice at once that every cut may be transformed into a mix by means of a number of thinnings and interchanges. (Conversely, every mix may be transformed into a cut by means of a certain number of preceding interchanges and contractions, though we do not use this fact.)

In the following we shall consider only derivations in which no cuts occur, but which may contain mixes instead.

Since derivations in the old sense may be transformed into derivations of the new kind, it suffices, for the proof of the *Hauptsatz*, to show that a derivation of the new type may be transformed into a derivation with no mix.

Furthermore, the following lemma is already sufficient:

LEMMA: A derivation with a mix for its lowest inference figure, and not containing any other mix, may be transformed into a derivation (with the same endsequent) in which no mix occurs.

From this the theorem as a whole easily follows:

In an arbitrary derivation consider a mix above whose lower sequent no further mix occurs. The derivation for this lower sequent is then of the kind mentioned in the lemma, i.e., it may be transformed in such a way that it no longer contains a mix. In doing so, the rest of the derivation remains unchanged. This operation is then repeated until every mix has systematically been eliminated.

It now remains for us to establish the *proof of the lemma*. (This proof extends into 3.2 incl.)

We have to consider a derivation whose lowest inference figure is a mix and which contains no other mix.

The degree of the mix formula will be called the 'degree of the derivation' (defined in I.2.2).

We shall call the *rank* of the derivation the sum of its rank on the left and its rank on the right. These two terms are defined as follows:

The *left rank* is the largest number of consecutive sequents in a path so that the lowest of these sequents is the *left-hand* upper sequent of the mix and each of the sequents contains the mix formula in the *succedent*.

The *right rank* is (correspondingly) the largest number of consecutive sequents in a path so that the lowest of these sequents is the *right-hand* upper sequent of the mix and each of the sequents contains the mix formula in the *antecedent*.

The lowest possible rank is evidently 2.

To prove the lemma we carry out two complete inductions, one on the degree γ, the other on the rank ρ, of the derivation, i.e., we prove the theorem for a derivation of degree γ, assuming it to hold for derivations of a lower degree (in so far as there are such derivations, i.e., as long as γ is not equal to zero), supposing, therefore, that derivations of lower degree can already be transformed into derivations with no mix. Furthermore, we shall begin by considering the case where the rank ρ of the derivation equals 2 (3.11), and after that the case of $\rho > 2$ (3.12), where we assume that the theorem already holds for derivations of the same degree, but of a lower rank.

In the following German capital letters will generally serve as syntactic variables for *formulae*, and Greek capital letters as syntactic variables for (possibly empty) *sequences of formulae*.

In transforming derivations, we shall occasionally meet 'identical inference figures', i.e., inference figures with identical upper and lower sequents. Since we have not admitted such figures in our calculus, they must be eliminated as soon as they occur; we can do this trivially by omitting one of the two sequents.

The mix formula of the mix that occurs at the end of the derivation is designated by 𝔐. It is of degree γ.

3.10. *Redesignating of free object variables* in preparation for the transformation of derivations.

We wish to obtain a derivation that has the following properties:

3.101. For every $\forall$–*IS* ($\exists$–*IA*) it holds that: Its eigenvariable occurs in the derivation only in sequents *above* the lower sequent of the $\forall$–*IS* ($\exists$–*IA*) and does not occur as an eigenvariable in any other $\forall$–*IS* ($\exists$–*IA*).

3.102. This is achieved by redesignating free object variables in the following way:

We take a $\forall$–*IS* ($\exists$–*IA*) above whose lower sequent either no further inference figures of this kind occur, or if they do, they have already been dealt with in a way still to be described.

In all sequents above the lower sequent of this inference figure we replace the eigenvariable by one and the same free object variable which, so far, has not yet occurred in the derivation. This obviously leaves the $\forall$–*IS*

($\exists$–IA) itself correct, as is easily seen. (The eigenvariable did not in fact occur in its lower sequent.) Furthermore, rest of the derivation remains correct, as is shown by the lemma to follow shortly.

A systematic application of this method to every single $\forall$–IS and $\exists$–IA, thus leaves the derivation correct throughout and the conclusion obviously has the desired property (3.101). Furthermore, as was essential, the degree and rank of the derivation, as well as its endsequent, have remained unaltered.

3.103. Now we give the still outstanding proof of the following lemma. (It is enunciated in a somewhat more general form than is immediately necessary, since we shall have to apply it again later on (3.113.33).)

An *LK*-basic sequent or inference figure becomes a basic sequent or inference figure of the same kind, if we replace a free object variable which is *not the eigenvariable* of the inference figure in all its occurrences in the basic sequent or inference figure, by one and the same free object variable, provided again that this is *not the eigenvariable* of the inference figure.

This holds trivially except for the $\forall$–IS, $\forall$–IA, $\exists$–IS and $\exists$–IA. Even here, however, there is no cause for concern: the restrictions on variables are not violated, since we may neither substitute nor replace the eigenvariable. (This is the reason why both restrictions on variables are necessary.) Furthermore, the formula resulting from $\mathfrak{F}\mathfrak{a}$ is again obtained by substituting $\mathfrak{a}$ for $\mathfrak{x}$ in the formula resulting from $\mathfrak{F}\mathfrak{x}$.

Having prepared the way (3.10), we now proceed to the actual transformation of the derivation which serves to eliminate the mix occurring in it.

As already mentioned, we distinguish the two cases: $\rho = 2$ (3.11) and $\rho > 2$ (3.12).

3.11. Suppose $\rho = 2$.

We distinguish between several individual cases, of which the cases 3.111, 3.112, 3.113.1, 3.113.2 are especially simple in that they allow the mix to be immediately eliminated. The other cases (3.113.3) are the most important since their consideration brings out the basic idea behind the whole transformation. Here we use the induction hypothesis with respect to γ, i.e., we reduce each one of the cases to transformed derivations of a lower degree.

3.111. Suppose the left-hand upper sequent of the mix at the end of the derivation is a basic sequent. The mix then reads:

$$\frac{\mathfrak{M} \to \mathfrak{M} \quad \Delta \to \Lambda}{\mathfrak{M}, \Delta^* \to \Lambda},$$

which is transformed into:

$$\frac{\varDelta \to \varLambda}{\mathfrak{M}, \varDelta^* \to \varLambda}$$ possibly several interchanges and contractions.

That part of the derivation which is above $\varDelta \to \varLambda$ remains the same, and we thus have a derivation without a mix.

3.112. Suppose the right-hand upper sequent of the mix is a basic sequent. The treatment of this case is symmetric to that of the previous one. We have only to regard the two schemata as 'duals' (cf. 2.4).

3.113. Suppose that neither the left- nor the right-hand upper sequent of the mix is a basic sequent. Then both are *lower sequents of inference figures* since $\rho = 2$, and the right and left rank both equal 1, i.e.: In the sequents directly above the *left-hand* upper sequent of the mix, the mix formula $\mathfrak{M}$ does not occur in the *succedent*; in the sequents directly above the *right-hand* upper sequent $\mathfrak{M}$ does not occur in the *antecedent*.

Now the following holds generally: If a formula occurs in the antecedent (succedent) of the lower sequent of an inference figure, it is either a principal formula or the $\mathfrak{D}$ of a thinning, or else it also occurs in the antecedent (succedent) in at least one upper sequent of the inference figure.

This can be seen immediately by looking at the inference figure schemata (1.21, 1.22).

If we now consider the hypotheses in the following three cases, we see at once that they exhaust all the possibilities that exist within case 3.113.

3.113.1. Suppose the left-hand upper sequent of the mix is the lower sequent of a thinning. Then the conclusion of the derivation runs:

$$\frac{\dfrac{\varGamma \to \varTheta}{\varGamma \to \varTheta, \mathfrak{M}} \quad \varDelta \to \varLambda}{\varGamma, \varDelta^* \to \varTheta, \varLambda}.$$

This is transformed into:

$$\frac{\varGamma \to \varTheta}{\varGamma, \varDelta^* \to \varTheta, \varLambda}$$ possibly several thinnings and interchanges.

That part of the derivation which occurs above $\varDelta \to \varLambda$ disappears.

3.113.2. Suppose the right-hand upper sequent of the mix is the lower sequent of a thinning. This case is dealt with symmetrically to the previous one.

3.113.3. The mix formula $\mathfrak{M}$ occurs both in the succedent of the left-hand upper sequent and in the antecedent of the right-hand upper sequent solely as the *principal formula* of one of the operational inference figures.

Depending on whether the terminal symbol of $\mathfrak{M}$ is &, $\vee$, $\forall$, $\exists$, $\neg$, $\supset$, we distinguish the cases 3.113.31 to 3.113.36 (a formula without logical symbols cannot be a principal formula).

3.113.31. Suppose the terminal symbol of $\mathfrak{M}$ is &. In that case the end of the derivation runs:

$$\dfrac{\dfrac{\Gamma_1 \to \Theta_1, \mathfrak{A} \quad \Gamma_1 \to \Theta_1, \mathfrak{B}}{\Gamma_1 \to \Theta_1, \mathfrak{A} \,\&\, \mathfrak{B}}\,\&\text{-}IS \quad \dfrac{\mathfrak{A}, \Gamma_2 \to \Theta_2}{\mathfrak{A} \,\&\, \mathfrak{B}, \Gamma_2 \to \Theta_2}\,\&\text{-}IA}{\Gamma_1, \Gamma_2 \to \Theta_1, \Theta_2}\,\text{mix}$$

(the other form of the &–*IA* is treated analogously).

We transform it into:

$$\dfrac{\dfrac{\Gamma_1 \to \Theta_1, \mathfrak{A} \quad \mathfrak{A}, \Gamma_2 \to \Theta_2}{\Gamma_1, \Gamma_2^* \to \Theta_1^*, \Theta_2}\,\text{mix}}{\Gamma_1, \Gamma_2 \to \Theta_1, \Theta_2}\text{ possibly several thinnings and interchanges.}$$

We can now apply the induction hypothesis with respect to γ to that part of the derivation whose lowest sequent is $\Gamma_1, \Gamma_2^* \to \Theta_1^*, \Theta_2$, because it has a lower degree than γ. ($\mathfrak{A}$ obviously contains fewer logical symbols than $\mathfrak{A}\,\&\,\mathfrak{B}$.) This means that the whole derivation may be transformed into one with no mix.

3.113.32. Suppose the terminal symbol of $\mathfrak{M}$ is $\vee$. This case is dealt with symmetrically to the previous one.

3.113.33. Suppose the terminal symbol of $\mathfrak{M}$ is $\forall$. Then the end of the derivation runs:

$$\dfrac{\dfrac{\Gamma_1 \to \Theta_1, \mathfrak{F}\mathfrak{a}}{\Gamma_1 \to \Theta_1, \forall \mathfrak{x}\mathfrak{F}\mathfrak{x}}\,\forall\text{-}IS \quad \dfrac{\mathfrak{F}\mathfrak{b}, \Gamma_2 \to \Theta_2}{\forall \mathfrak{x}\mathfrak{F}\mathfrak{x}, \Gamma_2 \to \Theta_2}\,\forall\text{-}IA}{\Gamma_1, \Gamma_2 \to \Theta_1, \Theta_2}\,\text{mix.}$$

This is transformed into:

$$\dfrac{\dfrac{\Gamma_1 \to \Theta_1, \mathfrak{F}\mathfrak{b} \quad \mathfrak{F}\mathfrak{b}, \Gamma_2 \to \Theta_2}{\Gamma_1, \Gamma_2^* \to \Theta_1^*, \Theta_2}\,\text{mix}}{\Gamma_1, \Gamma_2 \to \Theta_1, \Theta_2}\text{ possibly several thinnings and interchanges.}$$

Above the left-hand upper sequent of the mix, $\Gamma_1 \to \Theta_1, \mathfrak{F}\mathfrak{b}$, we write the same part of the derivation which previously occurred above $\Gamma_1 \to \Theta_1, \mathfrak{F}\mathfrak{a}$, yet having replaced every occurrence of the free object variable $\mathfrak{a}$ by $\mathfrak{b}$. It now follows from lemma 3.103, together with 3.101, that in performing this operation the part of the derivation *above* $\Gamma_1 \rightarrow \Theta_1, \mathfrak{F}\mathfrak{b}$ has again become a correct part of the derivation. (By virtue of 3.101 neither $\mathfrak{a}$ nor $\mathfrak{b}$ can be the eigenvariable of an inference figure occurring in that part of the derivation.) The same consideration may be applied to that part of the derivation which *includes* the sequent $\Gamma_1 \rightarrow \Theta_1, \mathfrak{F}\mathfrak{b}$, since it too results from $\Gamma_1 \rightarrow \Theta_1, \mathfrak{F}\mathfrak{a}$ by the substitution of $\mathfrak{b}$ for $\mathfrak{a}$. It is now in fact clear that by virtue of the restriction on variables for $\forall$–*IS*, $\mathfrak{a}$ could have occurred neither in Γ_1 and Θ_1, nor in $\mathfrak{F}\mathfrak{x}$. Furthermore, $\mathfrak{F}\mathfrak{a}$ results from $\mathfrak{F}\mathfrak{x}$ by the substitution $\mathfrak{a}$ for $\mathfrak{x}$, and $\mathfrak{F}\mathfrak{b}$ from $\mathfrak{F}\mathfrak{x}$ by the substitution $\mathfrak{b}$ for $\mathfrak{x}$. This is why $\mathfrak{F}\mathfrak{b}$ results from $\mathfrak{F}\mathfrak{a}$ by the substitution $\mathfrak{b}$ for $\mathfrak{a}$.

The mix formula $\mathfrak{F}\mathfrak{b}$ in the new derivation has a lower degree than γ. Therefore, according to the induction hypothesis, the mix may be eliminated.

3.113.34. Suppose the terminal symbol of $\mathfrak{M}$ is $\exists$. This case is dealt with symmetrically to the previous one.

3.113.35. Suppose the terminal symbol of $\mathfrak{M}$ is $\neg$. Then the end of the derivation runs:

$$\dfrac{\dfrac{\mathfrak{A}, \Gamma_1 \rightarrow \Theta_1}{\Gamma_1 \rightarrow \Theta_1, \neg \mathfrak{A}}\;\neg\text{-}IS \qquad \dfrac{\Gamma_2 \rightarrow \Theta_2, \mathfrak{A}}{\neg \mathfrak{A}, \Gamma_2 \rightarrow \Theta_2}\;\neg\text{-}IA}{\Gamma_1, \Gamma_2 \rightarrow \Theta_1, \Theta_2}\;\text{mix.}$$

This is transformed into:

$$\dfrac{\dfrac{\Gamma_2 \rightarrow \Theta_2, \mathfrak{A} \qquad \mathfrak{A}, \Gamma_1 \rightarrow \Theta_1}{\Gamma_2, \Gamma_1^* \rightarrow \Theta_2^*, \Theta_1}\;\text{mix}}{\Gamma_1, \Gamma_2 \rightarrow \Theta_1, \Theta_2}\;\text{possibly several interchanges and thinnings.}$$

The new mix may be eliminated by virtue of the induction hypothesis.

3.113.36. Suppose the terminal symbol of $\mathfrak{M}$ is $\supset$. Then the end of the derivation runs:

$$\dfrac{\dfrac{\mathfrak{A}, \Gamma_1 \rightarrow \Theta_1, \mathfrak{B}}{\Gamma_1 \rightarrow \Theta_1, \mathfrak{A} \supset \mathfrak{B}}\;\supset\text{-}IS \qquad \dfrac{\Gamma \rightarrow \Theta, \mathfrak{A} \qquad \mathfrak{B}, \varDelta \rightarrow \varLambda}{\mathfrak{A} \supset \mathfrak{B}, \Gamma, \varDelta \rightarrow \Theta, \varLambda}\;\supset\text{-}IA}{\Gamma_1, \Gamma, \varDelta \rightarrow \Theta_1, \Theta, \varLambda}\;\text{mix.}$$

This is transformed into:

$$\dfrac{\dfrac{\Gamma \rightarrow \Theta, \mathfrak{A} \qquad \dfrac{\mathfrak{A}, \Gamma_1 \rightarrow \Theta_1, \mathfrak{B} \qquad \mathfrak{B}, \varDelta \rightarrow \varLambda}{\mathfrak{A}, \Gamma_1, \varDelta^* \rightarrow \Theta_1^*, \varLambda}\;\text{mix}}{\Gamma, \Gamma_1^*, \varDelta^{**} \rightarrow \Theta^*, \Theta_1^*, \varLambda}\;\text{mix}}{\Gamma_1, \Gamma, \varDelta \rightarrow \Theta_1, \Theta, \varLambda}\;\text{possibly several interchanges and thinnings.}$$

(The asterisks are, of course, intended as follows: $\varDelta^*$ and $\varTheta^*$ result from $\varDelta$ and $\varTheta_1$ by the omission of all S-formulae of the form $\mathfrak{B}$; $\varGamma^*$, $\varDelta^{**}$ and $\varTheta^*$ result from $\varGamma_1$, $\varDelta^*$ and $\varTheta$ by the omission of all S-formulae of the form $\mathfrak{A}$.)

Now we have two mixes, but both mix formulae are of a lower degree than γ. We first apply the induction hypothesis to the upper mix (i.e., to that part of the derivation whose lowest figure it is). Thus the upper mix may be eliminated. We can then also eliminate the lower mix.

3.12. Suppose $\rho > 2$.

To begin with, we distinguish two main cases: First case: The right rank is greater than 1 (3.121). Second case: The right rank is equal to 1 and the left rank is therefore greater than 1 (3.122).

The second case may essentially be dealt with symmetrically to the first.

3.121. *Suppose the right rank is greater than* 1.

I.e.: The right-hand upper sequent of the mix is the lower sequent of an inference figure, let us call it $\mathfrak{Jf}$, and $\mathfrak{M}$ occurs in the antecedent of at least one upper sequent of $\mathfrak{Jf}$.

The basic idea behind the transformation procedure is the following:

In the case of $\rho = 2$, we generally reduced the derivation to one of a lower *degree*. Now, however, we shall proceed to reduce the derivation to one of the same degree, but of a lower *rank*, in order to be able to use the induction hypothesis with respect to ρ.

The only exception is the first case, 3.121.1, where the mix may be eliminated immediately.

In the remaining cases the reduction to derivations of a lower rank is achieved in the following way: The mix is, as it were, moved up one level within the derivation, beyond the inference figure $\mathfrak{Jf}$. (Case 3.121.231, for example, illustrates this point particularly well.) To speak more precisely, the left-hand upper sequent of the mix (which from now on will be designated by $\varPi \to \varSigma$), at present occurring *beside the lower sequent* of $\mathfrak{Jf}$, is instead written *next to the upper sequents* of $\mathfrak{Jf}$. These now become upper sequents of new mixes. The lower sequents of these mixes are now used as upper sequents of a new inference figure that takes the place of $\mathfrak{Jf}$. This new inference figure takes us back either directly, or after having added further inference figures, to the original endsequent. Each new mix obviously has a rank smaller than ρ, since the left rank remains unchanged and the right rank is diminished by at least 1.

In the strict application of this basic idea special circumstances still arise which make it necessary to distinguish the corresponding cases and to deal with them separately.

3.121.1. Suppose $\mathfrak{M}$ occurs in the antecedent of the left-hand upper sequent of the mix. The end of the derivation runs:

$$\frac{\Pi \to \Sigma \quad \Delta \to \Lambda}{\Pi, \Delta^* \to \Sigma^*, \Lambda}, \text{ thus } \mathfrak{M} \text{ occurs in } \Pi.$$

This is transformed into:

$$\frac{\Delta \to \Lambda}{\Pi, \Delta^* \to \Sigma^*, \Lambda} \text{ possibly several thinnings, contractions and interchanges.}$$

3.121.2. Suppose $\mathfrak{M}$ does not occur in the antecedent of the left-hand upper sequent of the mix. (This hypothesis will be used for the first time in 3.121.222.)

3.121.21. Suppose $\mathfrak{F}$ is a thinning, contraction, or interchange in the *antecedent*. Then the end of the derivation runs:

$$\frac{\Pi \to \Sigma \quad \dfrac{\Psi \to \Theta}{\Xi \to \Theta}\mathfrak{F}}{\Pi, \Xi^* \to \Sigma^*, \Theta} \text{ mix.}$$

This is transformed into:

$$\dfrac{\dfrac{\dfrac{\dfrac{\Pi \to \Sigma \quad \Psi \to \Theta}{\Pi, \Psi^* \to \Sigma^*, \Theta}\text{ mix}}{\Psi^*, \Pi \to \Sigma^*, \Theta} \text{ possibly several interchanges}}{\Xi^*, \Pi \to \Sigma^*, \Theta}\S}{\Pi, \Xi^* \to \Sigma^*, \Theta} \text{ possibly several interchanges.}$$

The inference figure marked § is of the same kind as $\mathfrak{F}$, in so far as the *S*-formulae designated in the schema of $\mathfrak{F}$ (in 1.21) by $\mathfrak{D}$ and $\mathfrak{E}$, were not equal to $\mathfrak{M}$. If $\mathfrak{D}$ or $\mathfrak{E}$ is equal to $\mathfrak{M}$, we have an identical inference figure (Ψ^* equals Ξ^*).

The derivation for the lower sequent of the new mix has the same left rank as the old derivation, whereas its right rank is lower by 1. Thus the mix may be completely eliminated by virtue of the induction hypothesis.

3.121.22. Suppose $\mathfrak{F}$ is an inference figure with *one* upper sequent, but not containing a thinning, contraction, or interchange in the antecedent. Then the end of the derivation runs:

$$\frac{\Pi \to \Sigma \quad \dfrac{\Psi, \Gamma \to \Omega_1}{\Xi, \Gamma \to \Omega_2}\mathfrak{F}}{\Pi, \Xi^*, \Gamma^* \to \Sigma^*, \Omega_2} \text{ mix.}$$

Here we have collected in Γ the same S-formulae that are designated by Γ in the schema of the inference figure (1.21, 1.22). Hence Ψ may be empty or consist of a side formula of the inference figure, and Ξ may be empty or consist of the principal formula of the inference figure.

First of all, the end of the derivation is transformed into:

$$\frac{\dfrac{\dfrac{\Pi \to \Sigma \quad \Psi, \Gamma \to \Omega_1}{\Pi, \Psi^*, \Gamma^* \to \Sigma^*, \Omega_1}\text{ mix}}{\Psi, \Gamma^*, \Pi \to \Sigma^*, \Omega_1}}{\Xi, \Gamma^*, \Pi \to \Sigma^*, \Omega_2}\text{ possibly several interchanges and thinnings.}$$

The lowest inference is obviously an inference figure of the same kind as $\mathfrak{F}$ (taking Γ^*, Π as the Γ of the inference figure and including Σ^* in the Θ of the inference figure).

We must only be careful not to violate the restrictions on variables (if $\mathfrak{F}$ is a $\forall$–IS or $\exists$–IA): Any such violation is precluded by 3.101, which entails that an eigenvariable that may have occurred in $\mathfrak{F}$ cannot have occurred in Π and Σ.

The mix may be eliminated from the new derivation by virtue of the induction hypothesis.

We therefore obtain a derivation with no mix and which is terminated by the following inference figure:

$$\frac{\Psi, \Gamma^*, \Pi \to \Sigma^*, \Omega_1}{\Xi, \Gamma^*, \Pi \to \Sigma^*, \Omega_2}.$$

In general, the endsequent is not yet of the form aimed at. Hence we proceed as follows:

3.121.221. Suppose Ξ does not contain $\mathfrak{M}$.

In that case we perform a number of interchanges, if necessary, and obtain the endsequent of the original derivation.

3.121.222. Suppose Ξ contains $\mathfrak{M}$. Then Ξ is the principal formula of $\mathfrak{F}$ and is identical with $\mathfrak{M}$. We then adjoin:

$$\dfrac{\dfrac{\Pi \to \Sigma \quad \mathfrak{M}, \Gamma^*, \Pi \to \Sigma^*, \Omega_2}{\Pi, \Gamma^*, \Pi^* \to \Sigma^*, \Sigma^*, \Omega_2}\text{ mix}}{\Pi, \Gamma^* \to \Sigma^*, \Omega_2}\text{ possibly several contractions and interchanges.}$$

Once again, this is the endsequent of the original derivation. (Above $\Pi \to \Sigma$ we once more write the derivation associated with it.) Thus we have another mix in the derivation. The left rank of our derivation is the same as that of the original derivation. The right rank is now equal to 1. This is so because directly above the right-hand upper sequent occurs the sequent

$$\Psi, \Gamma^*, \Pi \to \Sigma^*, \Omega_1.$$

𝔐 no longer occurs in its antecedent, for Γ^* does not contain 𝔐, nor does Π, because of 3.121.2; and Ψ contains at most one *side formula* of 𝔍𝔣, which cannot be equal to 𝔐, since the *principal formula* of 𝔍𝔣 is equal to 𝔐.

Hence this mix, too, may be eliminated by virtue of the induction hypothesis.

3.121.23. Suppose 𝔍𝔣 is an inference figure with *two* upper sequents, i.e., a &–*IS*, ∨–*IA*, or a ⊃–*IA*.

(In view of the application to intuitionist logic (3.2) we shall deal with each possibility in greater detail than would be necessary for the classical case.)

3.121.231. Suppose 𝔍𝔣 is a &–*IS*.

Then the end of the derivation runs:

$$\cfrac{\Pi \to \Sigma \quad \cfrac{\Gamma \to \Theta, \mathfrak{A} \quad \Gamma \to \Theta, \mathfrak{B}}{\Gamma \to \Theta, \mathfrak{A} \& \mathfrak{B}} \&\text{–}IS}{\Pi, \Gamma^* \to \Sigma^*, \Theta, \mathfrak{A} \& \mathfrak{B}} \text{mix}.$$

(𝔐 occurs in Γ.) This is transformed into:

$$\cfrac{\cfrac{\Pi \to \Sigma \quad \Gamma \to \Theta, \mathfrak{A}}{\Pi, \Gamma^* \to \Sigma^*, \Theta, \mathfrak{A}} \text{mix} \quad \cfrac{\Pi \to \Sigma \quad \Gamma \to \Theta, \mathfrak{B}}{\Pi, \Gamma^* \to \Sigma^*, \Theta, \mathfrak{B}} \text{mix}}{\Pi, \Gamma^* \to \Sigma^*, \Theta, \mathfrak{A} \& \mathfrak{B}} \&\text{–}IS.$$

Both mixes may be eliminated by virtue of the induction hypothesis.

3.121.232. Suppose 𝔍𝔣 is a ∨–*IA*.

Then the end of the derivation runs:

$$\cfrac{\Pi \to \Sigma \quad \cfrac{\mathfrak{A}, \Gamma \to \Theta \quad \mathfrak{B}, \Gamma \to \Theta}{\mathfrak{A} \vee \mathfrak{B}, \Gamma \to \Theta} \vee\text{–}IA}{\Pi, (\mathfrak{A} \vee \mathfrak{B})^*, \Gamma^* \to \Sigma^*, \Theta} \text{mix}.$$

$((\mathfrak{A} \vee \mathfrak{B})^*$ stands either for $\mathfrak{A} \vee \mathfrak{B}$ or for nothing according as $\mathfrak{A} \vee \mathfrak{B}$ is unequal or equal to 𝔐.)

𝔐 certainly occurs in Γ. (For otherwise 𝔐 would be equal to $\mathfrak{A} \vee \mathfrak{B}$, and the right rank would be equal to 1 contrary to 3.121.)

To begin with, we transform the end of the derivation into:

$$\frac{\dfrac{\Pi \to \Sigma \quad \mathfrak{A}, \Gamma \to \Theta}{\Pi, \mathfrak{A}^*, \Gamma^* \to \Sigma^*, \Theta} \text{ mix}}{\mathfrak{A}, \Pi, \Gamma^* \to \Sigma^*, \Theta} \text{ possibly several interchanges and thinnings}$$

$$\frac{\Pi \to \Sigma \quad \mathfrak{B}, \Gamma' \to \Theta}{\Pi, \mathfrak{B}^*, \Gamma^* \to \Sigma^*, \Theta} \text{ mix}$$
$$\frac{}{\mathfrak{B}, \Pi, \Gamma^* \to \Sigma^*, \Theta} \text{ possibly several interchanges and thinnings}$$
$$\frac{}{\mathfrak{A} \vee \mathfrak{B}, \Pi, \Gamma'^* \to \Sigma^*, \Theta} \vee\text{-}IA.$$

Both mixes may be eliminated by virtue of the induction hypothesis.

From here on the procedure is the same as that in 3.121.221 and 3.121.222, i.e., we distinguish two cases according as $\mathfrak{A} \vee \mathfrak{B}$ is unequal or equal to $\mathfrak{M}$. In the first case we may have to add several interchanges to obtain the endsequent of the original derivation; in the second case we add a mix with $\Pi \to \Sigma$ for its left-hand upper sequent, and thus once again obtain the endsequent of the original derivation by going on to perform a number of contractions and interchanges, if necessary. The mix concerned may be eliminated, since the associated right rank equals 1. (All this as in 3.121.222.)

3.121.233. Suppose $\mathfrak{J}$ is a $\supset$-IA.

Then the end of the derivation runs:

$$\frac{\Pi \to \Sigma \quad \dfrac{\Gamma \to \Theta, \mathfrak{A} \quad \mathfrak{B}, \Delta \to \Lambda}{\mathfrak{A} \supset \mathfrak{B}, \Gamma, \Delta \to \Theta, \Lambda} \supset\text{-}IA}{\Pi, (\mathfrak{A} \supset \mathfrak{B})^*, \Gamma^*, \Delta^* \to \Sigma^*, \Theta, \Lambda} \text{ mix}.$$

3.121.233.1. Suppose $\mathfrak{M}$ occurs in Γ and Δ.

In that case we begin by transforming the derivation into:

$$\frac{\dfrac{\Pi \to \Sigma \quad \Gamma \to \Theta, \mathfrak{A}}{\Pi, \Gamma^* \to \Sigma^*, \Theta, \mathfrak{A}} \text{ mix}}{\mathfrak{A} \supset \mathfrak{B}, \Pi, \Gamma^*, \Pi, \Delta^* \to \Sigma^*, \Theta, \Sigma^*, \Lambda} \begin{array}{l} \dfrac{\Pi \to \Sigma \quad \mathfrak{B}, \Delta \to \Lambda}{\Pi, \mathfrak{B}^*, \Delta^* \to \Sigma^*, \Lambda} \text{ mix} \\ \dfrac{}{\mathfrak{B}, \Pi, \Delta^* \to \Sigma^*, \Lambda} \begin{array}{l}\text{possibly several interchanges and thinnings}\\ \supset\text{-}IA.\end{array} \end{array}$$

Both mixes may be eliminated by virtue of the induction hypothesis. Then we proceed as in 3.121.221 and 3.121.222. (All that may happen in the first case is that beside interchanges a number of contractions become necessary.)

3.121.233.2. Suppose $\mathfrak{M}$ does not occur in both Γ and Δ simultaneously. $\mathfrak{M}$ must occur in either Γ or Δ because of 3.121. Consider the case of $\mathfrak{M}$ occurring in Δ but not in Γ. The second case is treated analogously.

The end of the derivation is transformed into:

$$\frac{\dfrac{\Pi \to \Sigma \qquad \mathfrak{B}, \Delta \to \Lambda}{\Pi, \mathfrak{B}^*, \Delta^* \to \Sigma^*, \Lambda} \text{ mix}}{\underline{\hspace{8cm}} \text{ possibly several interchanges and thinnings}}$$

$$\frac{\Gamma \to \Theta, \mathfrak{A} \qquad \mathfrak{B}, \Pi, \Delta^* \to \Sigma^*, \Lambda}{\mathfrak{A} \supset \mathfrak{B}, \Gamma, \Pi, \Delta^* \to \Theta, \Sigma^*, \Lambda} \supset\text{-}IA.$$

The mix may be eliminated by virtue of the induction hypothesis. We then proceed as in 3.121.221 and 3.121.222. (In the second case, where $\mathfrak{A} \supset \mathfrak{B}$ is equal to $\mathfrak{M}$, the right rank belonging to the new mix equals 1 as always, since $\mathfrak{M}$ does not occur in $\mathfrak{B}, \Pi, \Delta^*$ for the usual reasons, nor does it occur in Γ according to the assumption in the case under consideration.)

3.122. *Suppose the right rank is equal to* 1. *In that case the left rank is greater than* 1.

This case is, in essence, treated dually to 3.121. Special attention is required only for those inference figures with no symmetric counterpart, viz., the $\supset$-*IS* and the $\supset$-*IA*.

The inference figures $\mathfrak{I}\mathfrak{f}$ with *one* upper sequent were incorporated, in 3.121.22, in the general schema:

$$\frac{\Psi, \Gamma \to \Omega_1}{\Xi, \Gamma \to \Omega_2}.$$

The dual schema runs:

$$\frac{\Omega_1 \to \Gamma, \Psi}{\Omega_2 \to \Gamma, \Xi},$$

which also covers the $\supset$-*IS* without any further change. (Γ here represents the formulae designated by Θ in the schemata 1.21, 1.22.)

3.122.1. On the other hand, the case where the inference figure $\mathfrak{I}\mathfrak{f}$ is a $\supset$-*IA*, must be treated separately. Although this treatment will seem very similar to that in 3.121.233, it is not entirely dual.

Thus the end of the derivation runs:

$$\frac{\dfrac{\Gamma \to \Theta, \mathfrak{A} \qquad \mathfrak{B}, \Delta \to \Lambda}{\mathfrak{A} \supset \mathfrak{B}, \Gamma, \Delta \to \Theta, \Lambda} \supset\text{-}IA \qquad \Sigma \to \Pi}{\mathfrak{A} \supset \mathfrak{B}, \Gamma, \Delta, \Sigma^* \to \Theta^*, \Lambda^*, \Pi} \text{ mix.}$$

3.122.11. Suppose $\mathfrak{M}$ occurs both in Θ and Λ. In that case we transform the end of the derivation into:

$$\frac{\dfrac{\Gamma \to \Theta, \mathfrak{A} \quad \Sigma \to \Pi}{\Gamma, \Sigma^* \to \Theta^*, \mathfrak{A}^*, \Pi} \text{ mix}}{\dfrac{\Gamma, \Sigma^* \to \Theta^*, \Pi, \mathfrak{A}^*}{\mathfrak{A} \supset \mathfrak{B}, \Gamma, \Sigma^*, \Delta, \Sigma^* \to \Theta^*, \Pi, \Lambda^*, \Pi}} \begin{matrix}\text{possibly several inter-}\\ \text{changes and thinnings}\end{matrix} \quad \dfrac{\dfrac{\mathfrak{B}, \Delta \to \Lambda \quad \Sigma \to \Pi}{\mathfrak{B}, \Delta, \Sigma^* \to \Lambda^*, \Pi} \text{ mix}}{\supset\text{-}IA}$$

$$\overline{\mathfrak{A} \supset \mathfrak{B}, \Gamma, \Delta, \Sigma^* \to \Theta^*, \Lambda^*, \Pi} \quad \text{possibly several contractions and interchanges.}$$

Both mixes may be eliminated by virtue of the induction hypothesis.

3.122.12. Suppose $\mathfrak{M}$ does not occur in both Θ and Λ simultaneously. It must occur in one of them. We consider the case of $\mathfrak{M}$ occurring in Λ, but not in Θ; the alternative case is completely analogous.

We transform the end of the derivation into:

$$\dfrac{\Gamma \to \Theta, \mathfrak{A} \quad \dfrac{\mathfrak{B}, \Delta \to \Lambda \quad \Sigma \to \Pi}{\mathfrak{B}, \Delta, \Sigma^* \to \Lambda^*, \Pi} \text{ mix}}{\mathfrak{A} \supset \mathfrak{B}, \Gamma, \Delta, \Sigma^* \to \Theta, \Lambda^*, \Pi} \supset\text{-}IA.$$

The mix may be eliminated by virtue of the induction hypothesis.

3.2. *Proof of the Hauptsatz for LJ-derivations.*

In order to transform an *LJ*-derivation into an *LJ*-derivation *without cuts*, we apply *exactly the same procedure* as for *LK*-derivations.

Since an *LJ*-derivation is a special case of an *LK*-derivation, it is clear that the transformation can be carried out. We have only to convince ourselves that with every transformation step an *LJ*-derivation becomes another *LJ*-derivation, i.e., that the *D*-sequents of the transformed derivation do not contain more than one *S*-formula in the succedent, given that this was the case before.

We therefore examine each step of the transformation from that point of view.

3.21. *Replacement of cuts by mixes.* An *LJ*-cut runs:

$$\dfrac{\Gamma \to \mathfrak{D} \quad \mathfrak{D}, \Delta \to \Lambda}{\Gamma, \Delta \to \Lambda},$$

where Λ contains at most one *S*-formula. We transform this cut into:

$$\dfrac{\dfrac{\Gamma \to \mathfrak{D} \quad \mathfrak{D}, \Delta \to \Lambda}{\Gamma, \Delta^* \to \Lambda} \text{ mix}}{\Gamma, \Delta \to \Lambda} \quad \text{possibly several interchanges and thinnings in the antecedent.}$$

This replacement gives us a new *LJ*-derivation.

3.22. By relabelling free object variables (3.10) we trivially get another *LJ*-derivation from a previous one.

3.23. The transformation proper (3.11 and 3.12).

We have to show for each of the cases 3.111 to 3.122.12 that the specified transformations do not introduce any sequents with more than one *S*-formula in the succedent.

3.231. Let us begin with the cases 3.11:

In the cases 3.111, 3.113.1, 3.113.31, 3.113.35 and 3.113.36, only such formulae occur in each succedent of the sequent of a new derivation as had already occurred in the succedent of the sequent of the original derivation.

Essentially the same applies in 3.113.33. The only difference is an additional replacement of free object variables, which does not, of course, alter the number of succedent formulae of a sequent.

Cases 3.112, 3.113.2, 3.113.32, and 3.113.34 were dealt with symmetrically to cases 3.111, 3.113.1, 3.113.31, and 3.113.33, i.e., in order to get one case from another, we read the schemata from right to left instead of from left to right (as well as changing logical symbols, a process which is here of no consequence). Hence in the antecedent of one case we get precisely the same as in the succedent of another. For the antecedents of cases 3.111, 3.113.1, 3.113.31 and 3.113.33, the same applies as for the succedents, viz., in every antecedent of a sequent of the new derivation only such formulae occur as had already occurred in an antecedent of a sequent of the original derivation.

This disposes of all dual cases: 3.112, 3.113.2, 3.113.32 and 3.113.34.

3.232. Now let us look at the cases, 3.12:

3.232.1. For the cases 3.121 it holds generally that Σ^* is empty, since in $\Pi \rightarrow \Sigma$, Σ must contain only *one* formula, and that formula must be equal to $\mathfrak{M}$.

It is now obvious that in every succedent of a sequent only such formulae occur as had already occurred in the succedent of a sequent of the original derivation.

3.232.2. In the cases 3.122 it is somewhat more difficult to see that from an *LJ*-derivation we always get another *LJ*-derivation. We must direct our attention, as was done in our earlier consideration of dual cases, to the *antecedents* in the schemata 3.121.

At this point we distinguish two further subcases:

3.232.21. The case which is dual to 3.121.1 is trivial, since in every antecedent of a sequent of a new derivation (in case 3.121.1) only such formulae occur as had already occurred in an antecedent of a sequent of the original derivation.

3.232.22. In the cases that are dual to 3.121.2, the mix in the end of the derivation runs:

$$\frac{\Omega \to \mathfrak{M} \quad \Sigma \to \Pi}{\Omega, \Sigma^* \to \Pi},$$

where Π contains at most *one* S-formula, and where $\Omega \to \mathfrak{M}$ is the lower sequent of an *LJ*-inference figure in which at least one upper sequent contains $\mathfrak{M}$ as a succedent formula.

If we now look at the inference figure schemata 1.21, 1.22, it becomes easily apparent that such an inference figure can only be a thinning, contraction, or interchange in the antecedent, or a $\vee$–*IA*, a &–*IA*, a $\exists$–*IA*, a $\forall$–*IA*, and a $\supset$–*IA*. Let us disregard for the moment the $\vee$–*IA* and the $\supset$–*IA*. Then all the possibilities enumerated above fall within the case dual to 3.121.22, where both Ψ and Ξ always remain empty. (Γ corresponds to the Θ of the inference figure.) Thus we have the case which is dual to 3.121.221. Furthermore, Γ is equal to $\mathfrak{M}$, i.e., Γ^* is empty, and Π contains at most one formula. Hence in the new derivation there never in fact occurs more than one formula in the succedent of a sequent.

The case of a $\vee$–*IA* is dual to 3.121.231. Again, Γ is equal to $\mathfrak{M}$, Γ^* is empty, and Π contains at most one formula; all is thus in order.

There now remains the case of a $\supset$–*IA*, i.e., 3.122.1. In an *LJ*- $\supset$–*IA*, the Θ of the schema (1.22) is empty. Thus we have the case set out under 3.122.12. Λ^* is also empty, and Π contains at most one formula, which means that here, too, we again obtain an *LJ*-derivation from an *LJ*-derivation.

SECTION IV. SOME APPLICATIONS OF THE HAUPTSATZ

§ 1. Applications of the *Hauptsatz* in propositional logic

1.1. A trivial consequence of the *Hauptsatz* is the already known *consistency* of classical (and intuitionist) *predicate logic* (cf., e.g., D. Hilbert and W. Ackermann, *Grundzüge der theoretischen Logik* (Berlin, 1928, 1st edition), p. 65): the sequent $\to$ (which is derivable from every contradictory sequent $\to \mathfrak{A} \,\&\, \neg\, \mathfrak{A}$, cf. 3.21) cannot be the lower sequent of any inference figure other than of a cut and is therefore not derivable.

1.2. *Solution of the decision problem for intuitionist propositional logic.*

On the basis of the *Hauptsatz* we can state a simple procedure for deciding of a formula of propositional logic – i.e., a formula without object variables –

whether or not it is classically or intuitionistically true. (For classical propositional logic a simple solution has actually been known for some time, cf., e.g., p. 11 of Hilbert-Ackermann.)

First we prove the following *lemma*:

A sequent in whose antecedent one and the same formula does not occur more than three times as an *S*-formula, and in whose succedent, furthermore, one and the same formula occurs no more than three times as an *S*-formula, will be called a 'reduced sequent'. The following lemma now holds:

1.21. Every *LJ*- or *LK*-derivation whose endsequent is reduced, may be transformed into an *LJ*- or *LK*-derivation with the same endsequent, in which all sequents are reduced (and in which no cuts occur if the original derivation did not contain any).

PROOF OF THIS LEMMA: If we eliminate from the antecedent of a sequent, in any places whatever (possibly none), all *S*-formulae occurring more than once, and if we do the same independently in the succedent, so that eventually these formulae occur only once, twice, or three times, we obtain a sequent that will be called a 'reduction instance of the given sequent'.

From a reduction instance of a sequent we may obviously derive all other reduction instances of the same sequent by means of thinnings, contractions, and interchanges such that in the course of these operations only reduced sequents occur.

After these preliminary remarks we now transform the *LJ*- or *LK*-derivation at hand in the following way:

All *basic sequents* as well as the *endsequent* are left intact; they are already reduced sequents.

The *D*-sequents which belong to an *inference figure* are transformed into reduction instances of these sequents in a way about to be indicated. By virtue of our preliminary remark it does not matter if a sequent belonging to two different *D*-inference figures is in each case replaced by a *different reduction instance*, since one sequent is derived very simply from the other by thinnings, contractions, and interchanges so that eventually another complete derivation results. (The same holds for a sequent which, while belonging to an inference figure, is also a basic sequent or an endsequent, since it is of course a reduction instance of itself.)

The transformations of the inference figures are now carried out in the following way:

If a formula occurs more than once within Γ, it is eliminated from Γ, both from the upper sequents and the lower sequent, as many times (from the appropriate places) as is necessary to ensure that finally it occurs in Γ no more than once. The same procedure is used for Δ, Θ, and Λ (i.e., those sequences of formulae that are designated by these letters in the schema III.1.21 and 1.22, of the inference figure concerned).

Having carried out the transformations described, we have now a derivation consisting only of reduced sequents. (An interchange where $\mathfrak{D}$ is identical with $\mathfrak{E}$ may form an exception, yet this figure would be an identical inference figure and could have been avoided.)

The lemma is thus proved.

Given the *Hauptsatz*, together with corollary III. 2.513, and the preceding lemma (1.21), it now holds that:

1.22. For every correctly reduced sequent, both intuitionist and classical, there exists an *LJ*- or *LK*-derivation resp. without cuts consisting only of reduced sequents, and whose *D-S*-formulae are subformulae of the *S*-formula of that sequent.

1.23. Consider now a sequent not containing an object variable. We wish to decide whether or not it is intuitionistically or classically true. We can begin by taking in its place an equivalent reduced sequent $\mathfrak{S}\mathfrak{q}$.

The number of all reduced sequents whose *S*-formulae are subformulae of the *S*-formulae of $\mathfrak{S}\mathfrak{q}$ is obviously finite. The decision procedure may therefore be carried out without further complications in the following way:

We consider the finite system of sequents in question and investigate first of all, which of these sequents are basic sequents. Then we examine each of the remaining sequents to determine whether there occurs an inference figure in which the sequent in question is the lower sequent and in which there occur as upper sequents one or two of those sequents that have already been found to be derivable. If this is the case, the sequent is added to the derivable sequents. (All this is obviously decidable.) We continue in this way until either the sequent $\mathfrak{S}\mathfrak{q}$ itself turns out to be derivable, or until the procedure yields no new derivable sequents. In the latter case the sequent $\mathfrak{S}\mathfrak{q}$ (by virtue of 1.22) is not derivable at all in the calculus under consideration (*LJ* or *LK*). We have therefore succeeded in establishing the validity of that sequent.

1.3. A new proof of the *nonderivability of the law of the excluded middle in intuitionist logic*.

Our decision procedure could have been formulated in a way better suited to the needs of practical application; yet the above presentation (1.2) was intended only to indicate a possibility in principle.

As an example, we shall prove the nonderivability of the law of the excluded middle in intuitionist logic by a method independent of the decision procedure described (although this procedure would have to yield the same result). (This nonderivability has already been proved by Heyting[24] in a completely different way.)

The sequent in question is of the form $\rightarrow A \vee \neg A$. Suppose there exists an *LJ*-derivation for it. According to the *Hauptsatz* there then exists such a derivation without cuts. Its lowest inference figure must be a $\vee$–*IS*, for in all other *LJ*-inference figures either the antecedent of the lower sequent is not empty, or a formula occurs in the succedent whose terminal symbol is not $\vee$; there might still be the case of a thinning in the succedent, but the upper sequent would then be a $\rightarrow$, which, by virtue of 1.1, is not derivable.

Hence either $\rightarrow A$ or $\rightarrow \neg A$ would have to be already derivable (without cuts).

(From the same considerations, incidentally, it follows in general: If $\mathfrak{A} \vee \mathfrak{B}$ is an intuitionistically true formula, then either $\mathfrak{A}$ or $\mathfrak{B}$ is an intuitionistically true formula. In classical logic this does not hold, as the example of $A \vee \neg A$ already shows.)

Now $\rightarrow A$ cannot be the lower sequent of any *LJ*-inference figure whatever (if it is not a cut), unless that figure is another thinning with $\rightarrow$ for its upper sequent. Furthermore, since $\rightarrow A$ is not a basic sequent, it is thus not derivable.

The same considerations show that $\rightarrow \neg A$ is derivable only from $A \rightarrow$ by a $\neg$–*IS* figure, and $A \rightarrow$ is in turn derivable only from $A, A \rightarrow$, since A contains no terminal symbol. Continuing in this way, we always reach only sequents of the type $A, A, \ldots, A \rightarrow$, but never a basic sequent.

Hence $A \vee \neg A$ is not derivable in intuitionist predicate logic.

§ 2. A sharpened form of the *Hauptsatz* for classical predicate logic

2.1. We are here concerned with the following SHARPENING OF THE HAUPTSATZ:

Suppose that we have an *LK*-derivation whose endsequent is of the following kind:

Each *S*-formula of this sequent contains $\forall$ and $\exists$-symbols at most at the beginning, and their scope extends over the whole of the remaining formula.

In that case, the given derivation may be transformed into an *LK*-derivation with the same endsequent and having the following properties:

1. It contains no cuts.

2. It contains a *D*-sequent, let us call it the '*midsequent*', which is such that its derivation (and hence the midsequent itself) contains no $\forall$ and

∃-symbols, and where the only inference figures occurring in the *remaining* part of the derivation, the midsequent included, are ∀–*IS*, ∀–*IA*, ∃–*IS*, ∃–*IA*, and structural inference figures.

2.11. The midsequent divides the derivation, as it were, into an upper part beolnging to propositional logic, and a lower part containing only ∀ and ∃-introductions.

Concerning the *form of the transformed derivation*, the following may still be readily concluded: The lower part, from the midsequent to the endsequent, belongs to only one path since only inference figures with *one* upper sequent occur in it. The *S*-formulae of the midsequent are of the following kind:

Every *S*-formula in the antecedent of the midsequent results from an *S*-formula in the antecedent of the endsequent by the elimination of the ∀ and ∃-symbols (together with the bound object variables beside them), and by the replacement of the bound object variables in the rest of the formula by certain free object variables. The same procedure is followed in the case of succedents.

This follows from the same consideration as in III.2.512.

2.2. Proof of the theorem (2.1)[25].

The transformation of the derivation is carried out in several steps.

2.21. We begin by applying the *Hauptsatz* (III.2.5): The derivation may accordingly be transformed into a derivation without cuts.

2.22. Transformation of basic sequents containing a ∀- or ∃-symbol:

By virtue of the properties of subformulae III.2.513, such sequents can only have the form $\forall \mathfrak{x}\, \mathfrak{F}\mathfrak{x} \to \forall \mathfrak{x}\, \mathfrak{F}\mathfrak{x}$ or $\exists \mathfrak{x}\, \mathfrak{F}\mathfrak{x} \to \exists \mathfrak{x}\, \mathfrak{F}\mathfrak{x}$. They are transformed into (suppose $\mathfrak{a}$ to be a free object variable not yet occurring in the derivation):

$$\frac{\dfrac{\mathfrak{Fa} \to \mathfrak{Fa}}{\forall \mathfrak{x}\, \mathfrak{Fx} \to \mathfrak{Fa}}\ \forall\text{–}IA}{\forall \mathfrak{x}\, \mathfrak{Fx} \to \forall \mathfrak{x}\, \mathfrak{Fx}}\ \forall\text{–}IS \quad \text{or} \quad \frac{\dfrac{\mathfrak{Fa} \to \mathfrak{Fa}}{\mathfrak{Fa} \to \exists \mathfrak{x}\, \mathfrak{Fx}}\ \exists\text{–}IS}{\exists \mathfrak{x}\, \mathfrak{Fx} \to \exists \mathfrak{x}\, \mathfrak{Fx}}\ \exists\text{–}IA.$$

By repeating this procedure sufficiently often we can obviously eliminate all ∀- and ∃-symbols from every basic sequent of the derivation.

2.23. We now perform a complete induction on the 'order' of the derivation, which is defined as follows:

Of the operational inference figures we call those belonging to the symbols &, ∨, ¬, and ⊃ 'propositional inference figures', and the rest, i.e., ∀–*IS*, ∀–*IA*, ∃–*IS*, ∃–*IA*, 'predicate inference figures'. To each predicate inference figure in the derivation we assign the following ordinal number:

We consider that path of the derivation that extends from the lower sequent of the inference figure up to the endsequent of the derivation (including the endsequent) and count the number of lower sequents of the propositional inference figures occurring in it. Their number is the ordinal number.

The sum of the ordinal numbers of all predicate inference figures in the derivation is the *order of the derivation*.

We intend to reduce that order step by step until it becomes zero.

Note that once this has been achieved the rest of the proof of the theorem (2.1) is easily carried out: (The steps involved (2.232) will be such as to preserve the properties that were established in 2.21 and 2.22.)

2.231. In order to do so we assume that the derivation has already been reduced to order zero. From the endsequent we now proceed to the upper sequent of the inference figure above it. We stop as soon as we encounter the lower sequent of a propositional inference figure or a basic sequent; that sequent we call $\mathfrak{S}q$. (It will serve us as 'midsequent', once it has been transformed in a way about to be indicated.)

The derivation of $\mathfrak{S}q$ is now transformed as follows:

We simply omit all D-S-formulae which still contain the symbols $\forall$ and $\exists$. The above derivation remains correct after the described operation since, by virtue of 2.22, its basic sequents are not affected. Furthermore, no principal or side formula of an inference figure has been eliminated, for if such a formula had contained a symbol $\forall$ or $\exists$, the principal formula would certainly have contained that symbol. But no predicate inference figures occur (if they did, the ordinal number of the inference figure would be greater than zero), and by virtue of the subformula property (III.2.513) and the hypothesis of theorem 2.1, the principal formulae of the propositional inference figures cannot contain a $\forall$ or $\exists$. Now every inference figure remains correct if we eliminate, wherever it occurs as an S-formula in the figure, a formula which occurs neither as a principal nor as a side formula. This is easily seen from the schemata III.1.21 and III.1.22. (At worst, an identical inference figure may result, which is then eliminated in the usual manner.)

The sequent $\mathfrak{S}q^*$, which has resulted from $\mathfrak{S}q$ by this transformation, differs from $\mathfrak{S}q$ in that certain S-formulae may possibly have been eliminated. We follow the transformation up with several thinnings and interchanges such that in the end the sequent $\mathfrak{S}q$ reappears, and to it we attach the unaltered lower part of the derivation.

We have now reached our goal: $\mathfrak{S}q^*$ is the 'midsequent', and it obviously satisfies all conditions imposed on the latter by theorem (2.1).

§ 2, A SHAPRENED FORM OF THE *Hauptsatz* FOR CLASSICAL PREDICATE LOGIC

2.232. It now remains for us to carry out the *induction step* of our proof, i.e., the order of the derivation is assumed to be greater than zero, and our task is to diminish it.

2.232.1. We begin by *redesignating the free object variables* in the same way as in III.3.10. As a result of this, the derivation has the following property (III.3.101):

For every $\forall$–*IS* (or $\exists$–*IA*) it holds that the eigenvariable in the derivation occurs only in the sequent above the lower sequent of the $\forall$–*IS* (or $\exists$–*IA*) and does furthermore not occur in any other $\forall$–*IS* or $\exists$–*IA* as an eigenvariable.

The order of the derivation is hereby obviously left unchanged.

2.232.2. We now come to the transformation proper.

To begin with, we observe that in the derivation there occurs a predicate inference figure – let us call it $\mathfrak{F}\mathfrak{f}1$ – with the following property: If we follow that path of the inference figure which extends from the lower sequent to the endsequent, then the first lower sequent of an operational inference figure reached is the lower sequent of a *propositional* inference figure (that inference figure we call $\mathfrak{F}\mathfrak{f}2$). If there were no such instance, the order of the derivation would be equal to zero.

Now our aim is to slide the inference figure $\mathfrak{F}\mathfrak{f}1$ lower down in the derivation beyond $\mathfrak{F}\mathfrak{f}2$. This is easily done by means of the following schemata:

2.232.21. Suppose that $\mathfrak{F}\mathfrak{f}2$ has *one* upper sequent.

2.232.211. Suppose that $\mathfrak{F}\mathfrak{f}1$ is a $\forall$–*IS*. Then that part of the derivation on which the operation is to be carried out runs as follows:

$$\frac{\dfrac{\Gamma \to \Theta, \mathfrak{F}a}{\Gamma \to \Theta, \forall\mathfrak{x}\,\mathfrak{F}\mathfrak{x}}\ \forall\text{–}IS}{\Delta \to \Lambda}\ \mathfrak{F}\mathfrak{f}2, \text{ possibly preceded by structural inference figures.}$$

This we transform into:

$$\frac{\Gamma \to \Theta, \mathfrak{F}a}{\frac{\Gamma \to \mathfrak{F}a, \Theta, \forall\mathfrak{x}\,\mathfrak{F}\mathfrak{x}}{\frac{\Delta \to \mathfrak{F}a, \Lambda}{\frac{\Delta \to \Lambda, \mathfrak{F}a}{\frac{\Delta \to \Lambda, \forall\mathfrak{x}\,\mathfrak{F}\mathfrak{x}}{\Delta \to \Lambda}}}}}$$

possibly several interchanges, as well as a thinning inference figures of exactly the same kind as above, i.e., $\mathfrak{F}\mathfrak{f}2$, possibly preceded by structural inference figures

possibly several interchanges

$\forall$–*IS*

possibly several interchanges and contractions.

The elimination of $\forall\mathfrak{x}\,\mathfrak{F}\mathfrak{x}$ by contraction in the last step of the transformation is made possible by the fact that in Λ, $\forall\mathfrak{x}\,\mathfrak{F}\mathfrak{x}$ must occur as an

S-formula. (For the S-formula $\forall \mathfrak{x} \, \mathfrak{F} \mathfrak{x}$ could not, in the original derivation, have been eliminated from the succedent by means of $\mathfrak{F}2$ and the preceding structural inference figures, since it can obviously not be a side formula of $\mathfrak{F}2$, by virtue of the subformula property III.2.513 and the hypothesis of theorem 2.1.)

The restriction on variables is satisfied by the above $\forall$–IS ($\mathfrak{F}1$) by virtue of 2.232.1.

The order of the derivation has obviously been diminished by 1.

2.232.212. The case where $\mathfrak{F}1$ is a $\exists$–IS is dealt with analogously; all we need do is to replace $\forall$ by $\exists$.

2.232.213. The cases where $\mathfrak{F}1$ is a $\forall$–IA or $\exists$–IA are treated dually to the two preceding cases.

2.232.22. The case where $\mathfrak{F}2$ has *two* upper sequents, i.e., &–IS, ∨–IA, or ⊃–IA, can be dealt with quite correspondingly. At most a number of additional structural inference figures may be required.

2.3. Analogously to theorem 2.1 there are several ways in which the *Hauptsatz* may be further strengthened in the sense that certain restrictions can be placed on the order of occurrence of the operational inference figures in a derivation. For we can permute the inference figures to a large extent by sliding them above and beyond each other as was done above (2.232.2).

We shall not pursue this question further.

§ 3. Application of the *sharpened Hauptsatz* (2.1) to a new[26] consistency proof for arithmetic without complete induction

By *arithmetic* we mean the (elementary, i.e., employing no analytic techniques) theory of the natural numbers. Arithmetic may be formalized by means of our logical calculus *LK* in the following way:

3.1. In arithmetic it is customary to employ 'functions', e.g., x' (equals $x+1$), $x+y$, $x \cdot y$. Since we have not introduced function symbols into our logical formalism, we shall, in order to be able to apply it to arithmetic nevertheless, formalize the propositions of arithmetic in such a way that predicates take the place of functions. In place of the function x', for example, we shall use the predicate $x \text{Pr} y$, which reads: x is the predecessor of y, i.e., $y = x+1$. Furthermore, $[x+y = z]$ will be considered a predicate with three argument places. Thus the symbols $+$ and $=$ have here no independent meaning. A different predicate is $x = y$; the equality symbol here has thus no formal connection at all with the equality symbol in the previous predicate.

§ 3, APPLICATION OF THE *sharpened Hauptsatz* TO A CONSISTENCY PROOF

The number 1, furthermore, will not be written as a symbol for a definite object, since we have only object *variables* in our logical formalism and no symbols for *definite* objects. We shall overcome this difficulty by saying that the predicate 'One x' means informally the same as 'x is the number 1'.

The sentence '$x+1$ is the successor of x', for example, could be rendered thus in our formalism:

$$\forall x \, \forall y \, \forall z \, ((\text{One } y \, \& \, (x+y = z)) \supset x\text{Pr}z).$$

All other natural numbers can be respresented by the predicates One x & xPry; One x & xPry & yPrz, etc.

How are we now to integrate into our calculus the *predicate symbols* just introduced, having admitted only propositional variables? To do so we simply stipulate that the predicate symbols are to be treated in exactly the same way as propositional variables. More precisely: We regard expressions of the form

$$\text{One } \mathfrak{x}, \ \mathfrak{x}\text{Pr}\mathfrak{y}, \ \mathfrak{x} = \mathfrak{y}, \ (\mathfrak{x}+\mathfrak{y} = \mathfrak{z}),$$

where any object variables stand for $\mathfrak{x}, \mathfrak{y}, \mathfrak{z}$, merely as more easily intelligible ways of writing the formulae

$$A\mathfrak{x}, \ B\mathfrak{x}\mathfrak{y}, \ C\mathfrak{x}\mathfrak{y}, \ D\mathfrak{x}\mathfrak{y}\mathfrak{z}.$$

In this sense the axiom formulae that follow are indeed *formulae* in accordance with our definition.

(We cannot, of course, regard the number 1 as a way of writing an object variable, since in our calculus the object variables really function as *variables*, which is not so in the case of propositional variables.)

As '*axiom formulae*' of our arithmetic we shall initially take the following, and shall later, once the consistency proof has been carried out (cf. 3.3), state general criteria for the formation of further admissible axiom formulae:

Equality:

$\forall x \, (x = x)$	(reflexivity)
$\forall x \, \forall y \, (x = y \supset y = x)$	(symmetry)
$\forall x \, \forall y \, \forall z \, ((x = y \, \& \, y = z) \supset x = z)$	(transitivity)

One:

$\exists x \, (\text{One } x)$	(existence of 1)
$\forall x \, \forall y \, ((\text{One } x \, \& \, \text{One } y) \supset x = y)$	(uniqueness of 1)

Predecessor:

$\forall x \, \exists y \, (x\text{Pr}y)$	(existence of successor)

$\forall x \, \forall y \, (x \text{Pr} y \supset \neg \text{ One } y)$ (1 has no predecessor)
$\forall x \, \forall y \, \forall z \, \forall u \, ((x\text{Pr}y \,\&\, z\text{Pr}u \,\&\, x = z) \supset y = u)$ (uniqueness of successor)
$\forall x \, \forall y \, \forall z \, \forall u \, ((x\text{Pr}y \,\&\, z\text{Pr}u \,\&\, y = u) \supset x = u)$ (uniqueness of predecessor).

A formula $\mathfrak{B}$ is called *derivable* in arithmetic without complete induction, if there is an *LK*-derivation for a sequent

$$\mathfrak{A}_1, \ldots, \mathfrak{A}_\mu \to \mathfrak{B}$$

in which $\mathfrak{A}_1, \ldots, \mathfrak{A}_\mu$ are axiom formulae of arithmetic.

The fact that this formal system does actually allow us to represent the types of proof customary in informal arithmetic (as long as they do not use complete induction) cannot be *proved*, since for considerations of an informal character no precisely delimited framework exists. We can merely verify this in the case of individual informal proofs by testing them.

3.2. We shall now prove the *consistency of the formal system just presented*. With the help of the sharpened *Hauptsatz* (2.1) our task is in fact quite simple.

3.21. A 'contradiction' $\mathfrak{A} \,\&\, \neg \mathfrak{A}$ is derivable in our system if and only if there exists an *LK*-derivation for a sequent with an *empty succedent* and with arithmetic axiom formulae in the antecedent, viz.:

From $\Gamma \to \mathfrak{A} \,\&\, \neg \mathfrak{A}$ we obtain $\Gamma \to$ in the following way:

$$\cfrac{\Gamma \to \mathfrak{A} \,\&\, \neg \mathfrak{A} \qquad \cfrac{\cfrac{\cfrac{\cfrac{\cfrac{\mathfrak{A} \to \mathfrak{A}}{\neg \mathfrak{A}, \mathfrak{A} \to} \neg\text{-}IA}{\mathfrak{A} \,\&\, \neg \mathfrak{A}, \mathfrak{A} \to} \&\text{-}IA}{\mathfrak{A}, \mathfrak{A} \,\&\, \neg \mathfrak{A} \to} \text{interchange}}{\mathfrak{A} \,\&\, \neg \mathfrak{A}, \mathfrak{A} \,\&\, \neg \mathfrak{A} \to} \&\text{-}IA}{\mathfrak{A} \,\&\, \neg \mathfrak{A} \to} \text{contraction}}{\Gamma \to} \text{cut.}$$

The converse is obtained by carrying out a thinning in the succedent.

Thus, if our arithmetic is inconsistent, there exists an *LK*-derivation with the endsequent

$$\mathfrak{A}_1, \ldots, \mathfrak{A}_\mu \to,$$

where $\mathfrak{A}_1, \ldots, \mathfrak{A}_\mu$ are arithmetic axiom formulae.

3.22. We now apply the *sharpened Hauptsatz* (2.1). The arithmetic axiom formulae fulfil the requirement laid down for the *S*-formulae of the endsequent. Hence there exists an *LK*-derivation with the same endsequent which has the following properties:

§ 3, APPLICATION OF THE *sharpened Hauptsatz* TO A CONSISTENCY PROOF   113

1. It contains no cuts.
2. It contains a *D*-sequent, the 'midsequent', whose derivation contains no ∀ and ∃-symbols, and whose endsequent results from a number of inference figures ∀–*IA*, ∃–*IA*, thinnings, contractions and interchanges in the antecedent. The midsequent has an empty succedent (2.11).

3.23. We then proceed to redesignate the free object variables as in III.3.10. All mentioned properties remain unchanged, and the following property is added (III.3.101): The eigenvariable of each ∃–*IA* in the derivation occurs only in sequents *above* the lower sequent of the ∃–*IA*.

3.24. Then we replace every occurrence of a free object variable by one and the same natural number in a way to be described presently. In doing so we are left with a figure which we can no longer call an *LK*-derivation. We shall see later to what extent it nevertheless has an informal sense.

The replacement of the free object variables by numbers is carried out in the following order:

3.241. First we replace all free object variables which do not occur as the eigenvariable of a ∃–*IA* by the number 1 throughout. (We could also take another number.)

3.242. Then we take every ∃–*IA* inference figure in the derivation, beginning with the lowest and taking each figure in turn, and replace each eigenvariable (wherever it occurs in the 'derivation') by a number. That number is determined as follows:

The ∃–*IA* can only run:

$$\frac{\text{One } \mathfrak{a}, \Gamma \to \Theta}{\exists x \text{ One } x, \Gamma \to \Theta} \quad \text{or} \quad \frac{v\text{Pr}\mathfrak{a}, \Gamma \to \Theta}{\exists y \, v\text{Pr}y, \Gamma \to \Theta}$$

(by virtue of the subformula property III.2.513; v can be only a number, by virtue of 3.241 and 3.23). In the first case we replace $\mathfrak{a}$ by 1, in the second case by the number that is one greater than v.

3.25. Now we examine the figure which has resulted from the derivation. We are particularly interested in what the (former) midsequent now looks like. We can say this about it:

Its succedent is empty, and each of the antecedent *S*-formulae either has the form One 1 or $v\text{Pr}v'$, where a number stands for v, and where a number one greater than the previous one stands for v'; or it results from an arithmetic axiom formula that has only ∀-symbols at the beginning, by the elimination of the ∀-symbols (and the bound object variables next to them) and the substitution of *numbers* for the bound object variables in the remaining part of the formula. (All this follows from the same consideration as in III.2.512, also cf. 2.11.)

Thus, the *S*-formulae in the antecedent of the midsequent represent *informally true numerical propositions*. It further holds for the '*derivation*' of the midsequent that it has resulted from a derivation containing no ∀- or ∃-symbols, by having all its occurrences of free object variables replaced by numbers. Informally, such a 'derivation' constitutes in effect a *proof in arithmetic using only forms of inference from propositional logic*.

This leads us to the following result:

If our arithmetic is inconsistent, we can derive a contradiction from true numerical propositions through the mere application of inferences from propositional logic.

Here 'true numerical propositions' are propositions of the form One 1, $v\text{Pr}v'$, as well as all numerical special cases of general propositions occurring among the axioms such as, e.g., $3 = 3$, $4 = 5 \supset 5 = 4$, $3\text{Pr}4 \supset \neg$ One 4.

It is almost self-evident that from such propositions no contradictions are derivable by means of propositional logic. A proof for this would hardly be more than a formal paraphrasing of an informally clear situation of fact. Such a proof will therefore not be carried out save for indicating briefly the customary procedure for it:

We determine generally for which numerical values the formulae One μ, $\mu = v$, $\mu\text{Pr}v$, $\mu+v = \rho$, etc., are true and for which values they are false; furthermore, we explain in the customary way (cf., e.g., Hilbert-Ackermann p. 3) the truth or falsity of $\mathfrak{A} \& \mathfrak{B}$, $\mathfrak{A} \vee \mathfrak{B}$, $\neg \mathfrak{A}$, and $\mathfrak{A} \supset \mathfrak{B}$, as functions of the truth or falsity of the subformulae; we then show that all numerical special cases of axiom formulae are 'true'; and finally, that inference figures of propositional logic always lead from true formulae to other true formulae. A contradiction, however, is not a true formula.

3.3. It is easy to see from the remarks made in 3.25 in what way the system of *arithmetic axiom formulae* may be extended without making a contradiction derivable in it: Quite generally, we can allow the introduction of axiom formulae that begin with ∀-symbols spanning the whole formula, which do not contain any ∃-symbols, and of which every numerical special case is informally true. (We could also admit certain formulae containing ∃-symbols, as long as they can be dealt with in the consistency proof in a way analogous to that of the two cases occurring above.)

E.g., the following axiom formulae for addition are admissible:

$$\forall x \, \forall y \, (x\text{Pr}y \supset [x+1 = y])$$

$\forall x \, \forall y \, \forall z \, \forall u \, \forall v \, ((x \mathrm{Pr} y \, \& \, [z+x = u] \, \& \, [z+y = v]) \supset u \mathrm{Pr} v)$
$\forall x \, \forall y \, \forall z \, \forall u \, (([x+y = z] \, \& \, [x+y = u]) \supset z = u)$
$\forall x \, \forall y \, \forall z \, ([x+y = z] \supset [y+x = z])$
etc.

3.4. Arithmetic without complete induction is, however, of little practical significance, since complete induction is constantly required in number theory. Yet the consistency of arithmetic with complete induction has not been conclusively proved to date.

SECTION V. THE EQUIVALENCE OF THE NEW CALCULI *NJ*, *NK*, AND *LJ*, *LK* WITH A CALCULUS MODELLED ON THE FORMALISM OF HILBERT

§ 1. The concept of equivalence

1.1. We shall introduce the following concept of equivalence between *formulae* and *sequents* (which is in harmony with what was said in I.1.1 and I.2.4, concerning the informal sense of the symbol $\wedge$ and of sequents:

Identical formulae are equivalent.

Identical sequents are equivalent.

Two formulae are equivalent if the replacement of every occurrence of the symbol $\wedge$ in one of them by the formula $A \, \& \, \neg A$ yields the other formula.

The sequents $\mathfrak{A}_1, \ldots, \mathfrak{A}_\mu \rightarrow \mathfrak{B}_1, \ldots, \mathfrak{B}_\nu$ is equivalent to the following formula:

If the $\mathfrak{A}$'s and $\mathfrak{B}$'s are not empty:

$$(\mathfrak{A}_1 \, \& \, \ldots \, \& \, \mathfrak{A}_\mu) \supset (\mathfrak{B}_\nu \vee \ldots \vee \mathfrak{B}_1);$$

(this version is more convenient for the equivalence proof than that with $\mathfrak{B}_1 \vee \ldots \vee \mathfrak{B}_\nu$); if the $\mathfrak{A}$'s are empty, but the $\mathfrak{B}$'s are not:

$$\mathfrak{B}_\nu \vee \ldots \vee \mathfrak{B}_1;$$

if the $\mathfrak{B}$'s are empty, but the $\mathfrak{A}$'s are not:

$$(\mathfrak{A}_1 \, \& \, \ldots \, \& \, \mathfrak{A}_\mu) \supset (A \, \& \, \neg A);$$

if the $\mathfrak{A}$'s and the $\mathfrak{B}$'s are empty:

$$A \, \& \, \neg A.$$

The equivalence is transitive.

1.2. (We could of course give a substantially wider definition of equivalence, e.g., two formulae are usually called equivalent if one is derivable from the other. Here we shall content ourselves with the particular definition given, which is adequate for our proofs of equivalence.)

Two *derivations* will be called equivalent if the endformula (endsequent) of one is equivalent to that of the other.

Two *calculi* will be called equivalent if every derivation in one calculus can be transformed into an equivalent derivation in the other calculus.

In § 2 of this section we shall present a calculus (*LHJ* for intuitionist, *LHK* for classical predicate logic) modelled on Hilbert's formalism. In the remaining paragraphs of this section we shall then demonstrate the equivalence of the calculi *LHJ*, *NJ*, and *LJ* (§§ 3–5) as well as the equivalence of the calculi *LHK*, *NK*, and *LK* (§ 6) in the sense just explained. We shall thus successively prove the following:

Every *LHJ*-derivation can be transformed into an equivalent *NJ*-derivation (§ 3); every *NJ*-derivation can be transformed into an equivalent *LJ*-derivation (§ 4); and every *LJ*-derivation can be transformed into an equivalent *LHJ*-derivation (§ 5). This obviously proves the equivalence of all three calculi. The three classical calculi are dealt with analogously in § 6 (6.1–6.3).

§ 2. A logistic calculus according to Hilbert[27] and Glivenko[28]

We shall begin by explaining the *intuitionist* form of the calculus:

An *LHJ*-derivation consists of formulae arranged in tree form, where the initial formulae are basic formulae.

The basic formulae and the inference figures are obtained from the following schemata by the same rule of replacement as in II.2.21, i.e.: For $\mathfrak{A}, \mathfrak{B}, \mathfrak{C}$, put any arbitrary formula; for $\forall \mathfrak{x} \, \mathfrak{F} \mathfrak{x}$ or $\exists \mathfrak{x} \, \mathfrak{F} \mathfrak{x}$ put any arbitrary formula with $\forall$ or $\exists$ for its terminal symbol, where $\mathfrak{x}$ designates the associated bound object variable; for $\mathfrak{F} \mathfrak{a}$ put that formula which results from $\mathfrak{F} \mathfrak{x}$ by the replacement of every occurrence of the bound object variable $\mathfrak{x}$ by the free object variable $\mathfrak{a}$.

Schemata for basic formulae:

2.11. $\mathfrak{A} \supset \mathfrak{A}$

2.12. $\mathfrak{A} \supset (\mathfrak{B} \supset \mathfrak{A})$

2.13. $(\mathfrak{A} \supset (\mathfrak{A} \supset \mathfrak{B})) \supset (\mathfrak{A} \supset \mathfrak{B})$

2.14. $(\mathfrak{A} \supset (\mathfrak{B} \supset \mathfrak{C})) \supset (\mathfrak{B} \supset (\mathfrak{A} \supset \mathfrak{C}))$

2.15. $(\mathfrak{A} \supset \mathfrak{B}) \supset ((\mathfrak{B} \supset \mathfrak{C}) \supset (\mathfrak{A} \supset \mathfrak{C}))$
2.21. $(\mathfrak{A} \& \mathfrak{B}) \supset \mathfrak{A}$
2.22. $(\mathfrak{A} \& \mathfrak{B}) \supset \mathfrak{B}$
2.23. $(\mathfrak{A} \supset \mathfrak{B}) \supset ((\mathfrak{A} \supset \mathfrak{C}) \supset (\mathfrak{A} \supset (\mathfrak{B} \& \mathfrak{C})))$
2.31. $\mathfrak{A} \supset (\mathfrak{A} \vee \mathfrak{B})$
2.32. $\mathfrak{B} \supset (\mathfrak{A} \vee \mathfrak{B})$
2.33. $(\mathfrak{A} \supset \mathfrak{C}) \supset ((\mathfrak{B} \supset \mathfrak{C}) \supset ((\mathfrak{A} \vee \mathfrak{B}) \supset \mathfrak{C}))$
2.41. $(\mathfrak{A} \supset \mathfrak{B}) \supset ((\mathfrak{A} \supset \neg \mathfrak{B}) \supset \neg \mathfrak{A})$
2.42. $(\neg \mathfrak{A}) \supset (\mathfrak{A} \supset \mathfrak{B})$
2.51. $\forall \mathfrak{x} \, \mathfrak{F} \mathfrak{x} \supset \mathfrak{F} \mathfrak{a}$
2.52. $\mathfrak{F} \mathfrak{a} \supset \exists \mathfrak{x} \, \mathfrak{F} \mathfrak{x}$.

(Several of the schemata are dispensable, but independence does not concern us here.)

Schemata for inference figures:

$$\frac{\mathfrak{A} \quad \mathfrak{A} \supset \mathfrak{B}}{\mathfrak{B}} \qquad \frac{\mathfrak{A} \supset \mathfrak{F}\mathfrak{a}}{\mathfrak{A} \supset \forall \mathfrak{x} \, \mathfrak{F}\mathfrak{x}} \qquad \frac{\mathfrak{F}\mathfrak{a} \supset \mathfrak{A}}{(\exists \mathfrak{x} \, \mathfrak{F}\mathfrak{x}) \supset \mathfrak{A}}.$$

Restriction on variables: In the inference figures obtained from the last two schemata, the object variable, designated by $\mathfrak{a}$ in the schema, must not occur in the lower formula (hence not in $\mathfrak{A}$ and $\mathfrak{F}\mathfrak{x}$).

(The calculus *LHJ* is essentially equivalent to that of Heyting[29].)

By including the basic formula schema $\mathfrak{A} \vee \neg \mathfrak{A}$, the *calculus LHK* (classical predicate calculus) results.

(This latter calculus is essentially equivalent to the calculus presented in Hilbert-Ackermann, p. 53.)

§ 3. Transformation of an *LHJ*-derivation into an equivalent *NJ*-derivation

From an *LHJ*-derivation (V.2) we obtain an *NJ*-derivation (II.2) with the same endformula by transforming the *LHJ*-derivation in the following way: (In this transformation all *D*-formulae of this derivation will reappear as *D*-formulae of the *NJ*-derivation, and they will not depend on any assumption formula. Included further will be other *D*-formulae dependent on assumption formulae.)

3.1. The *LHJ-basic formulae* are replaced by *NJ*-derivations according to the following schemata:

$$(2.11) \quad \frac{\overset{1}{\mathfrak{A}}}{\mathfrak{A} \supset \mathfrak{A}} \supset\text{-}I_1 \qquad\qquad (2.12) \quad \frac{\dfrac{\overset{1}{\mathfrak{A}}}{\mathfrak{B} \supset \mathfrak{A}} \supset\text{-}I}{\mathfrak{A} \supset (\mathfrak{B} \supset \mathfrak{A})} \supset\text{-}I_1$$

$$(2.13) \quad \frac{\dfrac{\overset{1}{\mathfrak{A}} \quad \dfrac{\overset{1}{\mathfrak{A}} \quad \overset{2}{\mathfrak{A} \supset (\mathfrak{A} \supset \mathfrak{B})}}{\mathfrak{A} \supset \mathfrak{B}} \supset\text{-}E}{\dfrac{\mathfrak{B}}{\mathfrak{A} \supset \mathfrak{B}} \supset\text{-}I_1}}{(\mathfrak{A} \supset (\mathfrak{A} \supset \mathfrak{B})) \supset (\mathfrak{A} \supset \mathfrak{B})} \supset\text{-}I_2$$

$$(2.14) \quad \frac{\dfrac{\overset{2}{\mathfrak{B}} \quad \dfrac{\overset{1}{\mathfrak{A}} \quad \overset{3}{\mathfrak{A} \supset (\mathfrak{B} \supset \mathfrak{C})}}{\mathfrak{B} \supset \mathfrak{C}} \supset\text{-}E}{\dfrac{\dfrac{\mathfrak{C}}{\mathfrak{A} \supset \mathfrak{C}} \supset\text{-}I_1}{\mathfrak{B} \supset (\mathfrak{A} \supset \mathfrak{C})} \supset\text{-}I_2}}{(\mathfrak{A} \supset (\mathfrak{B} \supset \mathfrak{C})) \supset (\mathfrak{B} \supset (\mathfrak{A} \supset \mathfrak{C}))} \supset\text{-}I_3$$

$$(2.15) \quad \frac{\dfrac{\dfrac{\overset{1}{\mathfrak{A}} \quad \overset{3}{\mathfrak{A} \supset \mathfrak{B}}}{\mathfrak{B}} \supset\text{-}E \quad \overset{2}{\mathfrak{B} \supset \mathfrak{C}}}{\dfrac{\dfrac{\mathfrak{C}}{\mathfrak{A} \supset \mathfrak{C}} \supset\text{-}I_1}{(\mathfrak{B} \supset \mathfrak{C}) \supset (\mathfrak{A} \supset \mathfrak{C})} \supset\text{-}I_2}}{(\mathfrak{A} \supset \mathfrak{B}) \supset ((\mathfrak{B} \supset \mathfrak{C}) \supset (\mathfrak{A} \supset \mathfrak{C}))} \supset\text{-}I_3 \qquad (2.21) \quad \frac{\dfrac{\overset{1}{\mathfrak{A} \& \mathfrak{B}}}{\mathfrak{A}} \&\text{-}E}{(\mathfrak{A} \& \mathfrak{B}) \supset \mathfrak{A}} \supset\text{-}I_1$$

2.22, 2.31, 2.32, 2.51 and 2.52 are dealt with analogously to 2.21.

$$(2.23) \quad \frac{\dfrac{\dfrac{\overset{1}{\mathfrak{A}} \quad \overset{3}{\mathfrak{A} \supset \mathfrak{B}}}{\mathfrak{B}} \supset\text{-}E \quad \dfrac{\overset{1}{\mathfrak{A}} \quad \overset{2}{\mathfrak{A} \supset \mathfrak{C}}}{\mathfrak{C}} \supset\text{-}E}{\dfrac{\dfrac{\mathfrak{B} \& \mathfrak{C}}{\mathfrak{A} \supset (\mathfrak{B} \& \mathfrak{C})} \supset\text{-}I_1}{(\mathfrak{A} \supset \mathfrak{C}) \supset (\mathfrak{A} \supset (\mathfrak{B} \& \mathfrak{C}))} \supset\text{-}I_2}}{(\mathfrak{A} \supset \mathfrak{B}) \supset ((\mathfrak{A} \supset \mathfrak{C}) \supset (\mathfrak{A} \supset (\mathfrak{B} \& \mathfrak{C})))} \supset\text{-}I_3$$

§ 3, TRANSFORMATION OF AN *LHJ*-DERIVATION

$$(2.33) \quad \cfrac{\cfrac{2}{\mathfrak{A} \vee \mathfrak{B}} \quad \cfrac{\overset{1}{\mathfrak{A}} \quad \overset{4}{\mathfrak{A} \supset \mathfrak{C}}}{\mathfrak{C}} \supset\!-E \quad \cfrac{\overset{1}{\mathfrak{B}} \quad \overset{3}{\mathfrak{B} \supset \mathfrak{C}}}{\mathfrak{C}} \supset\!-E}{\cfrac{\cfrac{\mathfrak{C}}{(\mathfrak{A} \vee \mathfrak{B}) \supset \mathfrak{C}} \supset\!-E_2}{\cfrac{(\mathfrak{B} \supset \mathfrak{C}) \supset ((\mathfrak{A} \vee \mathfrak{B}) \supset \mathfrak{C})}{(\mathfrak{A} \supset \mathfrak{C}) \supset ((\mathfrak{B} \supset \mathfrak{C}) \supset ((\mathfrak{A} \vee \mathfrak{B}) \supset \mathfrak{C}))} \supset\!-E_1} \supset\!-E_3} \vee\!-E_1$$

$$(2.41) \quad \cfrac{\cfrac{\overset{1}{\mathfrak{A}} \quad \overset{3}{\mathfrak{A} \supset \mathfrak{B}}}{\mathfrak{B}} \supset\!-E \quad \cfrac{\overset{1}{\mathfrak{A}} \quad \overset{2}{\mathfrak{A} \supset \neg \mathfrak{B}}}{\neg \mathfrak{B}} \supset\!-E}{\cfrac{\cfrac{\wedge}{\neg \mathfrak{A}} \neg\!-I_1}{\cfrac{(\mathfrak{A} \supset \neg \mathfrak{B}) \supset \neg \mathfrak{A}}{(\mathfrak{A} \supset \mathfrak{B}) \supset ((\mathfrak{A} \supset \neg \mathfrak{B}) \supset \neg \mathfrak{A})} \supset\!-I_3} \supset\!-I_2} \neg\!-E$$

$$(2.42) \quad \cfrac{\cfrac{\cfrac{\overset{1}{\mathfrak{A}} \quad \overset{2}{\neg \mathfrak{A}}}{\wedge} \neg\!-E}{\cfrac{\mathfrak{B}}{\mathfrak{A} \supset \mathfrak{B}} \supset\!-I_1}}{(\neg \mathfrak{A}) \supset (\mathfrak{A} \supset \mathfrak{B})} \supset\!-I_2 \,.$$

3.2. The *LHJ-inference figures* are replaced by sections of an *NJ*-derivation according to the following schemata:

$$\cfrac{\mathfrak{A} \quad \mathfrak{A} \supset \mathfrak{B}}{\mathfrak{B}}$$ remains as it is, since it has already the form of a $\supset\!-E$.

$$\cfrac{\mathfrak{A} \supset \mathfrak{F}a}{\mathfrak{A} \supset \forall \mathfrak{x} \, \mathfrak{F}\mathfrak{x}} \quad \text{becomes:} \quad \cfrac{\cfrac{\cfrac{\overset{1}{\mathfrak{A}} \quad \mathfrak{A} \supset \mathfrak{F}a}{\mathfrak{F}a} \supset\!-E}{\cfrac{\forall \mathfrak{x} \, \mathfrak{F}\mathfrak{x}}{\mathfrak{A} \supset \forall \mathfrak{x} \, \mathfrak{F}\mathfrak{x}} \supset\!-I_1} \forall\!-I} \, .$$

$$\frac{\mathfrak{F}a \supset \mathfrak{A}}{(\exists \mathfrak{x}\, \mathfrak{F}\mathfrak{x}) \supset \mathfrak{A}} \quad \text{becomes:} \quad \cfrac{\cfrac{1\ \mathfrak{F}a \quad \overset{2}{\mathfrak{F}a \supset \mathfrak{A}}}{\cfrac{\mathfrak{A}}{(\exists \mathfrak{x}\, \mathfrak{F}\mathfrak{x}) \supset \mathfrak{A}}\supset\text{-}I_1}\exists\text{-}E_2}{}\supset\text{-}E$$

The restriction on variables for $\forall$–I and $\exists$–E is satisfied, as is easily seen, by virtue of the restriction on variables existing for *LHJ*-inference figures.

This completes the transformation of an *LHJ*-derivation into an equivalent *NJ*-derivation.

§ 4. Transformation of an *NJ*-derivation into an equivalent *LJ*-derivation

4.1. We proceed as follows: First we *replace* every *D*-formula of the *NJ*-derivation by the following sequent (cf. III.1.1): In its succedent only the formula itself occurs; in its antecedent occur the assumption formulae upon which the sequent depended, and they occur in the same order from left to right as they did in the *NJ*-derivation. (It is presumably clear what is meant by the order from left to right of the initial formulae of a figure in tree form.)

We then replace every occurrence of the symbol $\wedge$ by $A\ \&\ \neg A$. (The formula resulting from A in this way will be designated by A^*.)

4.2. We thus already have a system of sequents in tree form. The antecedent of the endsequent is empty (II.2.2); it is obviously equivalent to the endsequent of the *NJ*-derivation. The initial sequents all have the form $\mathfrak{D}^* \to \mathfrak{D}^*$ (II.2.2) and are thus already basic sequents of an *LJ*-derivation.

The figures formed from *NJ-inference figures* are transformed into sections of an *LJ*-derivation according to the following schemata:

4.21. The inference figures $\vee$–I, $\forall$–I, and $\exists$–I have become *LJ*-inference figures as a result of the substitution performed. (In the case of a $\forall$–I, the *LJ*-restriction on variables is satisfied by virtue of the *NJ*-restriction on variables.)

4.22. A &–*I* became:

$$\frac{\Gamma \to \mathfrak{A}^* \quad \Delta \to \mathfrak{B}^*}{\Gamma, \Delta \to \mathfrak{A}^*\ \&\ \mathfrak{B}^*}.$$

This is transformed into:

§ 4, TRANSFORMATION OF AN *NJ*-DERIVATION

$$\frac{\dfrac{\Gamma \to \mathfrak{A}^*}{\Gamma, \Delta \to \mathfrak{A}^*} \text{ possibly several inter-changes and contractions} \quad \dfrac{\Delta \to \mathfrak{B}^*}{\Gamma, \Delta \to \mathfrak{B}^*} \text{ possibly several thinnings}}{\Gamma, \Delta \to \mathfrak{A}^* \,\&\, \mathfrak{B}^*} \&\text{-}IS.$$

4.23. A $\supset$-*I* became:

$$\frac{\Gamma_1, \mathfrak{A}^*, \Gamma_2, \ldots, \mathfrak{A}^*, \Gamma_\rho \to \mathfrak{B}^*}{\Gamma_1, \Gamma_2, \ldots, \Gamma_\rho \to \mathfrak{A}^* \supset \mathfrak{B}^*}.$$

This we transform into:

$$\frac{\dfrac{\Gamma_1, \mathfrak{A}^*, \Gamma_2, \ldots, \mathfrak{A}^*, \Gamma_\rho \to \mathfrak{B}^*}{\mathfrak{A}^*, \Gamma_1, \Gamma_2, \ldots, \Gamma_\rho \to \mathfrak{B}^*}}{\Gamma_1, \Gamma_2, \ldots, \Gamma_\rho \to \mathfrak{A}^* \supset \mathfrak{B}^*} \begin{array}{l}\text{possibly several interchanges and}\\ \text{contractions, sometimes a thinning}\\ \supset\text{-}IS.\end{array}$$

4.24. The same procedure applies to a $\neg$-*I*. Finally, we still have to consider the figure

$$\frac{\mathfrak{A}^*, \Gamma \to A \,\&\, \neg A}{\Gamma \to \neg \mathfrak{A}^*}.$$

First we derive $A \,\&\, \neg A \to$ in the calculus *LJ* as follows:

$$\frac{\dfrac{\dfrac{\dfrac{\dfrac{\dfrac{A \to A}{\neg A, A \to} \neg\text{-}IA}{A \,\&\, \neg A, A \to} \&\text{-}IA}{A, A \,\&\, \neg A \to} \text{interchange}}{A \,\&\, \neg A, A \,\&\, \neg A \to} \&\text{-}IA}{A \,\&\, \neg A \to} \text{contraction.}$$

By including this sequent, the figure in question is transformed as follows:

$$\frac{\dfrac{\mathfrak{A}^*, \Gamma \to A \,\&\, \neg A \quad A \,\&\, \neg A \to}{\mathfrak{A}^*, \Gamma \to} \text{cut}}{\Gamma \to \neg \mathfrak{A}^*} \neg\text{-}IS.$$

4.25. By substitution (4.1) the *NJ*-inference figure $\dfrac{\wedge}{\mathfrak{D}}$ became:

$$\frac{\Gamma \to A \,\&\, \neg A}{\Gamma \to \mathfrak{D}^*}.$$

This is transformed into:

$$\frac{\Gamma \to A \,\&\, \neg A \quad A \,\&\, \neg A \to}{\dfrac{\Gamma \to}{\Gamma \to \mathfrak{D}^*}\text{ thinning.}}\text{ cut}$$

The derivation for $A \,\&\, \neg A \to$, as presented in 4.24, should here still be written above that sequent.

4.26. A $\forall$–E became:
$$\frac{\Gamma \to \forall \mathfrak{x}\, \mathfrak{F}^*\mathfrak{x}}{\Gamma \to \mathfrak{F}^*\mathfrak{a}}.$$

This is transformed into:
$$\frac{\Gamma \to \forall \mathfrak{x}\, \mathfrak{F}^*\mathfrak{x} \quad \dfrac{\mathfrak{F}^*\mathfrak{a} \to \mathfrak{F}^*\mathfrak{a}}{\forall \mathfrak{x}\, \mathfrak{F}^*\mathfrak{x} \to \mathfrak{F}^*\mathfrak{a}}\,\forall\text{–}IA}{\Gamma \to \mathfrak{F}^*\mathfrak{a}}\text{ cut.}$$

4.27. The same method is used for $\&$–E.

4.28. A $\supset$–E became:
$$\frac{\Gamma \to \mathfrak{A}^* \quad \Delta \to \mathfrak{A}^* \supset \mathfrak{B}^*}{\Gamma, \Delta \to \mathfrak{B}^*}.$$

This is transformed into:
$$\frac{\Delta \to \mathfrak{A}^* \supset \mathfrak{B}^* \quad \dfrac{\Gamma \to \mathfrak{A}^* \quad \mathfrak{B}^* \to \mathfrak{B}^*}{\mathfrak{A}^* \supset \mathfrak{B}^*, \Gamma \to \mathfrak{B}^*}\,\supset\text{–}IA}{\dfrac{\Delta, \Gamma \to \mathfrak{B}^*}{\Gamma, \Delta \to \mathfrak{B}^*}\text{ possibly several interchanges.}}\text{ cut}$$

4.29. A $\neg$–E became:
$$\frac{\Gamma \to \mathfrak{A}^* \quad \Delta \to \neg\, \mathfrak{A}^*}{\Gamma, \Delta \to A \,\&\, \neg A}.$$

This is transformed into:
$$\frac{\Delta \to \neg\, \mathfrak{A}^* \quad \dfrac{\Gamma \to \mathfrak{A}^*}{\neg\, \mathfrak{A}^*, \Gamma \to}\,\neg\text{–}IA}{\dfrac{\dfrac{\Delta, \Gamma \to}{\Gamma, \Delta \to}\text{ possibly several interchanges}}{\Gamma, \Delta \to A \,\&\, \neg A}\text{ thinning.}}\text{ cut}$$

4.2.10. ∨–*E*. Both right-hand upper sequents are followed up, as in the case of a ⊃–*I* and ¬–*I* (4.23) above, by interchanges, contractions, and thinnings (wherever necessary) so that in each case the result is a sequent in whose antecedent occurs a formula of the form $\mathfrak{A}^*$ or $\mathfrak{B}^*$ at the beginning (whereas the original assumption formulae involved have been absorbed into the rest of the antecedent). Then follows:

$$\frac{\dfrac{\mathfrak{A}^*, \Gamma \to \mathfrak{C}^*}{\mathfrak{A}^*, \Gamma, \Delta \to \mathfrak{C}^*}\text{ possibly several thinnings and interchanges} \quad \dfrac{\mathfrak{B}^*, \Delta \to \mathfrak{C}^*}{\mathfrak{B}^*, \Gamma, \Delta \to \mathfrak{C}^*}\text{ possibly several thinnings and interchanges}}{\dfrac{\Xi \to \mathfrak{A}^* \vee \mathfrak{B}^* \quad \mathfrak{A}^* \vee \mathfrak{B}^*, \Gamma, \Delta \to \mathfrak{C}^*}{\Xi, \Gamma, \Delta \to \mathfrak{C}^*}\text{ cut.}} \; \vee\text{–}IA$$

4.2.11. A ∃–*E* is treated quite similarly: First we move $\mathfrak{F}^*\mathfrak{a}$ in the right-hand upper sequent to the beginning of the antecedent (cf. 4.23); then follows:

$$\frac{\Delta \to \exists \mathfrak{x}\,\mathfrak{F}^*\mathfrak{x} \quad \dfrac{\mathfrak{F}^*\mathfrak{a}, \Gamma \to \mathfrak{C}^*}{\exists \mathfrak{x}\,\mathfrak{F}^*\mathfrak{x}, \Gamma \to \mathfrak{C}^*}\;\exists\text{–}IA}{\Delta, \Gamma \to \mathfrak{C}^*}\text{ cut.}$$

The *LJ*-restriction on variables for ∃–*IA* is satisfied by virtue of the *NJ*-restriction on variables for ∃–*E*.

This completes the transformation of an *NJ*-derivation into an equivalent *LJ*-derivation.

§ 5. Transformation of an *LJ*-derivation into an equivalent *LHJ*-derivation

This transformation is a little more difficult than the two previous ones. We shall carry it out in a number of separate steps.

Preliminary remark: Contractions and interchanges in the succedent do not occur in the calculus *LJ*, since they require the occurrence of at least two *S*-formulae in the succedent.

5.1. We first introduce *new basic sequents* in place of the figures &–*IA*, ∨–*IS*, ∀–*IA*, ∃–*IS*, ¬–*IA*, and ⊃–*IA*; these are to be formed according to the following schemata (rule of replacement as in III.1.2 – the same rule will always apply below; in addition to the letters $\mathfrak{A}$, $\mathfrak{B}$, $\mathfrak{D}$, and $\mathfrak{E}$ we shall also, incidentally, use the letters, $\mathfrak{C}$, $\mathfrak{H}$, and $\mathfrak{J}$):

$\mathfrak{B}\mathfrak{z}1: \mathfrak{A} \& \mathfrak{B} \to \mathfrak{A}$ $\qquad \mathfrak{B}\mathfrak{z}2: \mathfrak{A} \& \mathfrak{B} \to \mathfrak{B}$

$\mathfrak{B}\mathfrak{z}3: \mathfrak{A} \to \mathfrak{A} \vee \mathfrak{B}$ $\qquad \mathfrak{B}\mathfrak{z}4: \mathfrak{B} \to \mathfrak{A} \vee \mathfrak{B}$

$\mathfrak{B}\mathfrak{z}5: \forall \mathfrak{x}\,\mathfrak{F}\mathfrak{x} \to \mathfrak{F}\mathfrak{a}$ $\qquad \mathfrak{B}\mathfrak{z}6: \mathfrak{F}\mathfrak{a} \to \exists \mathfrak{x}\,\mathfrak{F}\mathfrak{x}$

$\mathfrak{B}\mathfrak{z}7: \neg \mathfrak{A}, \mathfrak{A} \to$ $\qquad \mathfrak{B}\mathfrak{z}8: \mathfrak{A} \supset \mathfrak{B}, \mathfrak{A} \to \mathfrak{B}.$

Thus in the *LJ*-derivation to be considered, we transform the inference figures concerned in the following way:

A &–*IA* becomes:

$$\frac{\mathfrak{B}\S1}{\mathfrak{A} \& \mathfrak{B} \to \mathfrak{A} \quad \mathfrak{A}, \Gamma \to \Theta}{\mathfrak{A} \& \mathfrak{B}, \Gamma \to \Theta} \text{ cut.}$$

The other form of the &–*IA* is transformed correspondingly, so is every &–*IA*.

∨–*IS* and ∃–*IS* are dealt with symmetrically.

A ¬–*IA* becomes:

$$\frac{\Gamma \to \mathfrak{A} \quad \dfrac{\mathfrak{B}\S7}{\neg \mathfrak{A}, \mathfrak{A} \to} \text{ interchange}}{\dfrac{\mathfrak{A}, \neg \mathfrak{A} \to}{\Gamma, \neg \mathfrak{A} \to}} \text{ cut}$$
$$\overline{\overline{\neg \mathfrak{A}, \Gamma \to}} \text{ possibly several interchanges.}$$

(The Θ in the schema of ¬–*IA* (III.1.22) must be empty by virtue of the *LJ*-restrictions on succedents; the same holds for the ⊃–*IA*.)

A ⊃–*IA* becomes:

$$\frac{\Gamma \to \mathfrak{A} \quad \dfrac{\mathfrak{B}\S8}{\mathfrak{A} \supset \mathfrak{B}, \mathfrak{A} \to \mathfrak{B}} \text{ interchange}}{\dfrac{\mathfrak{A}, \mathfrak{A} \supset \mathfrak{B} \to \mathfrak{B}}{\Gamma, \mathfrak{A} \supset \mathfrak{B} \to \mathfrak{B}} \text{ cut} \quad \mathfrak{B}, \Delta \to \Lambda}$$
$$\dfrac{\Gamma, \mathfrak{A} \supset \mathfrak{B}, \Delta \to \Lambda}{\overline{\overline{\mathfrak{A} \supset \mathfrak{B}, \Gamma, \Delta \to \Lambda}}} \text{ cut}$$
possibly several interchanges.

5.2. We now write the formula $A \& \neg A$ in the succedent of all *D*-formulae whose *succedent is empty*.

In doing so the basic sequents of the form $\mathfrak{D} \to \mathfrak{D}$, as well as $\mathfrak{B}\S1$ to $\mathfrak{B}\S6$ and $\mathfrak{B}\S8$, also the figures &–*IS*, ∀–*IS*, and ⊃–*IS*, remain unchanged. The other basic sequents and inference figures are transformed into new basic sequents and inference figures according to the following schemata:

$\mathfrak{B}\S9$: $\mathfrak{A}, \neg \mathfrak{A} \to \mathfrak{H}$

$$\mathfrak{F}\mathfrak{f}1: \frac{\Gamma \to \mathfrak{H}}{\mathfrak{D}, \Gamma \to \mathfrak{H}} \qquad \mathfrak{F}\mathfrak{f}2: \frac{\mathfrak{D}, \mathfrak{D}, \Gamma \to \mathfrak{H}}{\mathfrak{D}, \Gamma \to \mathfrak{H}}$$

§ 5, TRANSFORMATION OF AN *LJ*-DERIVATION

$$\mathfrak{F}3: \frac{\varDelta, \mathfrak{D}, \mathfrak{E}, \varGamma \to \mathfrak{H}}{\varDelta, \mathfrak{E}, \mathfrak{D}, \varGamma \to \mathfrak{H}} \qquad \mathfrak{F}4: \frac{\varGamma \to A \,\&\, \neg A}{\varGamma \to \mathfrak{D}}$$

$$\mathfrak{F}5: \frac{\varGamma \to \mathfrak{D} \quad \mathfrak{D}, \varDelta \to \mathfrak{H}}{\varGamma, \varDelta \to \mathfrak{H}} \qquad \mathfrak{F}6: \frac{\mathfrak{A}, \varGamma \to \mathfrak{H} \quad \mathfrak{B}, \varGamma \to \mathfrak{H}}{\mathfrak{A} \vee \mathfrak{B}, \varGamma \to \mathfrak{H}}$$

$$\mathfrak{F}7: \frac{\mathfrak{F}\mathfrak{a}, \varGamma \to \mathfrak{H}}{\exists \mathfrak{x}\, \mathfrak{F}\mathfrak{x}, \varGamma \to \mathfrak{H}} \qquad \mathfrak{F}8: \frac{\mathfrak{A}, \varGamma \to A \,\&\, \neg A}{\varGamma \to \neg \mathfrak{A}}.$$

(For $\mathfrak{F}7$ there exists the following restriction on variables: The free object variable designated by $\mathfrak{a}$ must not occur in the lower sequent.)

5.3. The inference figure $\mathfrak{F}4$ is now replaceable by other figures as follows (this is mainly due to our having kept general the form of the schema $\mathfrak{B}\mathfrak{z}9$):

$$\cfrac{\cfrac{\mathfrak{B}\mathfrak{z}1}{\to A \,\&\, \neg A \quad A \,\&\, \neg A \to A}{\varGamma \to A} \mathfrak{F}5 \quad \cfrac{\cfrac{\mathfrak{B}\mathfrak{z}2}{\varGamma \to A \,\&\, \neg A \quad A \,\&\, \neg A \to \neg A}{\varGamma \to \neg A}\mathfrak{F}5 \quad \cfrac{\mathfrak{B}\mathfrak{z}9}{\neg A, A \to \mathfrak{D}}}{\cfrac{\varGamma, A \to \mathfrak{D}}{A, \varGamma \to \mathfrak{D}} \text{ possibly several } \mathfrak{F}3\text{'s}} \mathfrak{F}5}{\cfrac{\varGamma, \varGamma \to \mathfrak{D}}{\varGamma \to \mathfrak{D}} \text{ possibly several } \mathfrak{F}2\text{'s and } \mathfrak{F}3\text{'s.}} \mathfrak{F}5$$

In a similar way we replace the inference figure $\mathfrak{F}8$ (wherever it occurs in the derivation), only this time we use a new inference figure according to the following schema:

$$\mathfrak{F}9: \frac{\varGamma, \mathfrak{A} \to A \quad \varGamma, \mathfrak{A} \to \neg A}{\varGamma \to \neg \mathfrak{A}}$$

We substitute as follows (in place of $\mathfrak{F}8$):

$$\cfrac{\cfrac{\cfrac{\mathfrak{B}\mathfrak{z}1}{\varGamma \to A \,\&\, \neg A \quad A \,\&\, \neg A \to A}{\mathfrak{A}, \varGamma \to A}\mathfrak{F}5}{\varGamma, \mathfrak{A} \to A} \text{ possibly several } \mathfrak{F}3\text{'s} \quad \cfrac{\cfrac{\mathfrak{B}\mathfrak{z}2}{\mathfrak{A}, \varGamma \to A \,\&\, \neg A \quad A \,\&\, \neg A \to \neg A}{\mathfrak{A}, \varGamma \to \neg A}\mathfrak{F}5}{\varGamma, \mathfrak{A} \to \neg A} \text{ possibly several } \mathfrak{F}3\text{'s}}{\varGamma \to \neg \mathfrak{A}} \mathfrak{F}9.$$

5.4. Now we still introduce two *new inference figures schemata*, viz.:

$$\mathfrak{F}10: \frac{\varGamma, \mathfrak{A} \to \mathfrak{B}}{\varGamma \to \mathfrak{A} \supset \mathfrak{B}}$$

and its converse:

$$\mathfrak{F}11: \frac{\Gamma \to \mathfrak{A} \supset \mathfrak{B}}{\Gamma, \mathfrak{A} \to \mathfrak{B}}.$$

The two types of inference figures are introduced into the derivation in order to enable us to replace a number of other inference figures by more specialized ones (in 5.42 and 5.43).

5.41. To begin with, $\supset$–*IS* inference figures are now replaceable by means of $\mathfrak{F}10$:

A $\supset$–*IS* is transformed into:

$$\frac{\dfrac{\mathfrak{A}, \Gamma \to \mathfrak{B}}{\Gamma, \mathfrak{A} \to \mathfrak{B}} \text{ possibly several } \mathfrak{F}3\text{'s}}{\Gamma \to \mathfrak{A} \supset \mathfrak{B}} \mathfrak{F}10.$$

5.42. The inference figures $\mathfrak{F}1, \mathfrak{F}2, \mathfrak{F}3, \mathfrak{F}5, \mathfrak{F}6$, and $\mathfrak{F}7$ are then transformed in the following way:

As an example we take an $\mathfrak{F}2$, which is transformed into the following figure (suppose Γ equals $\mathfrak{J}_1, \ldots, \mathfrak{J}_\rho$):

$$\frac{\dfrac{\dfrac{\dfrac{\mathfrak{D}, \mathfrak{D}, \mathfrak{J}_1, \ldots, \mathfrak{J}_\rho \to \mathfrak{H}}{\mathfrak{D}, \mathfrak{D}, \mathfrak{J}_1, \ldots, \mathfrak{J}_{\rho-1} \to \mathfrak{J}_\rho \supset \mathfrak{H}} \mathfrak{F}10}{\mathfrak{D}, \mathfrak{D} \to \mathfrak{J}_1 \supset (\mathfrak{J}_2 \supset \ldots (\mathfrak{J}_\rho \supset \mathfrak{H}).)} \text{ several } \mathfrak{F}10\text{'s}}{\dfrac{\mathfrak{D} \to \mathfrak{J}_1 \supset (\mathfrak{J}_2 \supset \ldots (\mathfrak{J}_\rho \supset \mathfrak{H}).)}{\mathfrak{D}, \mathfrak{J}_1, \ldots, \mathfrak{J}_\rho \to \mathfrak{H}} \text{ several } \mathfrak{F}11\text{'s.}}}{} \mathfrak{F}13$$

We proceed quite analogously with all other figures mentioned, i.e., using $\mathfrak{F}10$ and $\mathfrak{F}11$, we replace them by inference figures according to these schemata:

$$\mathfrak{F}12: \frac{\to \mathfrak{C}}{\mathfrak{D} \to \mathfrak{C}} \qquad \mathfrak{F}13: \frac{\mathfrak{D}, \mathfrak{D} \to \mathfrak{C}}{\mathfrak{D} \to \mathfrak{C}} \qquad \mathfrak{F}14: \frac{\Delta, \mathfrak{D}, \mathfrak{E} \to \mathfrak{C}}{\Delta, \mathfrak{E}, \mathfrak{D} \to \mathfrak{C}}$$

$$\mathfrak{F}15: \frac{\Gamma \to \mathfrak{D} \quad \mathfrak{D} \to \mathfrak{C}}{\Gamma \to \mathfrak{C}} \qquad \mathfrak{F}16: \frac{\mathfrak{A} \to \mathfrak{C} \quad \mathfrak{B} \to \mathfrak{C}}{\mathfrak{A} \vee \mathfrak{B} \to \mathfrak{C}} \qquad \mathfrak{F}17: \frac{\mathfrak{F}\mathfrak{a} \to \mathfrak{C}}{\exists \mathfrak{x}\,\mathfrak{F}\mathfrak{x} \to \mathfrak{C}}.$$

(For $\mathfrak{F}17$ there exists a restriction on variables: The free object variable designated by $\mathfrak{a}$ must not occur in the lower sequent.)

5.43. In a similar way we also replace the inference figures $\mathfrak{F}9, \mathfrak{F}13$, and $\mathfrak{F}14$ by the following (using $\mathfrak{F}10$ and $\mathfrak{F}11$):

§ 5, TRANSFORMATION OF AN *LJ*-DERIVATION

$$\mathfrak{F}18: \frac{\Gamma \to \mathfrak{A} \supset A \quad \Gamma \to \mathfrak{A} \supset \neg A}{\Gamma \to \neg \mathfrak{A}} \qquad \mathfrak{F}19: \frac{\to \mathfrak{D} \supset (\mathfrak{D} \supset \mathfrak{C})}{\to \mathfrak{D} \supset \mathfrak{C}}$$

$$\mathfrak{F}20: \frac{\varDelta \to \mathfrak{D} \supset (\mathfrak{E} \supset \mathfrak{C})}{\varDelta \to \mathfrak{E} \supset (\mathfrak{D} \supset \mathfrak{C})}.$$

The basic sequents $\mathfrak{B}\mathfrak{s}8$ and $\mathfrak{B}\mathfrak{s}9$ may be replaced in the same way by: $\mathfrak{A} \supset \mathfrak{B} \to \mathfrak{A} \supset \mathfrak{B}$, this form falls under the schema $\mathfrak{D} \to \mathfrak{D}$; as well as $\mathfrak{B}\mathfrak{s}10: \neg \mathfrak{A} \to \mathfrak{A} \supset \mathfrak{H}$.

5.5. Now comes the *final step*:
Every *D*-sequent
$$\mathfrak{A}_1, \ldots, \mathfrak{A}_\mu \to \mathfrak{B}$$
is replaced by the formula $(\mathfrak{A}_1 \& \ldots \& \mathfrak{A}_\mu) \supset \mathfrak{B}$.

(If the $\mathfrak{A}$'s are empty, we mean $\mathfrak{B}$. An empty succedent no longer occurs, according to 5.2.)

All basic sequents (viz. $\mathfrak{D} \to \mathfrak{D}$, $\mathfrak{B}\mathfrak{s}1$ to $\mathfrak{B}\mathfrak{s}6$, $\mathfrak{B}\mathfrak{s}10$) are thus transformed into *LHJ*-basic sequents.

Of the inference figures, ∀–*IS* and $\mathfrak{F}17$ are also transformed into *LHJ*-inference figures. (∀–*IS*, however, forms an exception if Γ is empty. In that case we first derive (in the *LHJ*-calculus) $(A \supset A) \supset \mathfrak{F}\mathfrak{a}$ from $\mathfrak{F}\mathfrak{a}$ by means of 2.12, and by then applying the *LHJ*-inference figure, we finally obtain $\forall \mathfrak{x} \, \mathfrak{F}\mathfrak{x}$ once again by means of 2.11.)

The figures obtained from the remaining inference figures (which are &–*IS*, $\mathfrak{F}10, 11, 12, 15, 16, 18, 19, 20$) by substitution, are turned into sections of an *LHJ*-derivation in the following way:

An &–*IS* has become (suppose first that Γ is not empty):

$$\frac{\mathfrak{C} \supset \mathfrak{A} \quad \mathfrak{C} \supset \mathfrak{B}}{\mathfrak{C} \supset (\mathfrak{A} \& \mathfrak{B})}.$$

This is transformed into:

$$\frac{\mathfrak{C} \supset \mathfrak{B} \quad \dfrac{\mathfrak{C} \supset \mathfrak{A} \quad (\mathfrak{C} \supset \mathfrak{A}) \supset ((\mathfrak{C} \supset \mathfrak{B}) \supset (\mathfrak{C} \supset (\mathfrak{A} \& \mathfrak{B})))}{(\mathfrak{C} \supset \mathfrak{B}) \supset (\mathfrak{C} \supset (\mathfrak{A} \& \mathfrak{B}))}}{\mathfrak{C} \supset (\mathfrak{A} \& \mathfrak{B})}.$$

If Γ is empty, we proceed as in the case of ∀–*IS*.

The figures obtained from $\mathfrak{F}12, 15, 16,$ and 19 by substitution are dealt with quite analogously using basic formulae according to the schemata 2.12, 2.15, 2.33, and 2.13.

In a similar way $\Im\mathfrak{f}18$ and $\Im\mathfrak{f}20$ are dealt with by means of 2.41 and 2.14 and by the application of 2.15 and 2.14, 2.13.

The only figures now left are those having resulted from $\Im\mathfrak{f}10$ and $\Im\mathfrak{f}11$. Both are trivial for an empty Γ, hence suppose that Γ is not empty. In that case we transform these figures into sections of *LHJ*-derivations as follows:

($\Im\mathfrak{f}10$): From $(\mathfrak{C} \mathbin{\&} \mathfrak{A}) \supset \mathfrak{B}$ we have to derive $\mathfrak{C} \supset (\mathfrak{A} \supset \mathfrak{B})$. Now 2.23 together with 2.11 yields: $(\mathfrak{C} \supset \mathfrak{A}) \supset (\mathfrak{C} \supset (\mathfrak{C} \mathbin{\&} \mathfrak{A}))$. This together with $(\mathfrak{C} \mathbin{\&} \mathfrak{A}) \supset \mathfrak{B}$ and 2.15, 2.14 yields $(\mathfrak{C} \supset \mathfrak{A}) \supset (\mathfrak{C} \supset \mathfrak{B})$, and from this formula together with 2.12, 2.15 yields $\mathfrak{A} \supset (\mathfrak{C} \supset \mathfrak{B})$, and by 2.14 $\mathfrak{C} \supset (\mathfrak{A} \supset \mathfrak{B})$ results.

($\Im\mathfrak{f}11$): From $\mathfrak{C} \supset (\mathfrak{A} \supset \mathfrak{B})$ we derive $(\mathfrak{C} \mathbin{\&} \mathfrak{A}) \supset \mathfrak{B}$ in the *LHJ*-calculus as follows: 2.21 and 2.22 yield $(\mathfrak{C} \mathbin{\&} \mathfrak{A}) \supset \mathfrak{C}$ and $(\mathfrak{C} \mathbin{\&} \mathfrak{A}) \supset \mathfrak{A}$; and from this together with $\mathfrak{C} \supset (\mathfrak{A} \supset \mathfrak{B})$, we obtain $(\mathfrak{C} \mathbin{\&} \mathfrak{A}) \supset \mathfrak{B}$ (by using 2.15, 2.14, 2.15, 2.13).

This completes the transformation of the *LJ*-derivation into an *LHJ*-derivation. Furthermore, the two derivations really are equivalent, since the endsequent of the *LJ*-derivation was affected only by the transformations 5.2 and 5.5, and has thus obviously been transformed into a formula equivalent with it (according to 1.1).

If the results of §§ 3–5 are taken together, the *equivalence of the three calculi LHJ, NJ*, and *LJ* is now fully proved.

§ 6. The equivalence of the calculi *LHK*, *NK*, and *LK*

Now that the equivalence of the different *intuitionist* calculi has been proved, it is fairly easy to deduce that of the *classical* calculi.

6.1. In order to transform an *LHK*-derivation into an equivalent *NK*-derivation we proceed exactly as in § 3. The additional basic formulae according to the schema $\mathfrak{A} \lor \neg \mathfrak{A}$ remain unchanged, and are thus at once basic formulae of the *NK*-derivation.

6.2. In order to transform an *NK*-derivation into an equivalent *LK*-derivation we proceed initially as in § 4. In this way the additional basic formulae according to the schema $\mathfrak{A} \lor \neg \mathfrak{A}$ are transformed into sequents of the form $\rightarrow \mathfrak{A}^* \mathbin{\&} \neg \mathfrak{A}^*$. These we then replace by their *LK*-derivations (according to III.1.4). The transformation of an *NK*-derivation into an equivalent *LK*-derivation is thus complete.

6.3. *Transformation of an LK-derivation into an LHK-derivation.*

We introduce an *auxiliary calculus* differing from the *LK*-calculus in the following respect:

Inference figures may be formed according to the schemata III.1.21, III.1.22, but with the following restrictions: Contractions and interchanges in the succedent are not permissible; in the remaining schemata no substitution may be performed on Θ and Λ; these places thus remain empty.

Furthermore, the following two schemata for inference figures are added (rule of replacement as usual: III.1.2):

$$\mathfrak{F}\mathfrak{f}1: \frac{\Gamma \to \mathfrak{A}, \Theta}{\Gamma, \neg \mathfrak{A} \to \Theta}$$

and its converse:

$$\mathfrak{F}\mathfrak{f}2: \frac{\Gamma, \neg \mathfrak{A} \to \Theta}{\Gamma \to \mathfrak{A}, \Theta}.$$

(Thus, here Θ need not be empty.)

6.31. Transformation of an LK-derivation into a derivation of the auxiliary calculus:

(The procedure is similar to that in 5.4.)

All inference figures, with the exception of contractions and interchanges in the succedent, are transformed according to the following rule: The upper sequents are followed by inference figures $\mathfrak{F}\mathfrak{f}1$, until all formulae of Θ or Λ have been negated and brought into the antecedent (to the right of Γ or Δ). Then follows an inference figure of the same kind as the one just transformed, which is now actually a permissible inference figure in the auxiliary calculus. (The formulae that have been brought into the antecedent are treated as part of Γ or Δ.) Then follow $\mathfrak{F}\mathfrak{f}2$ inference figures, and Θ and Λ are thus brought back into the succedent. (In the case of the $\supset$–IA and the cut, we may first have to carry out interchanges in the antecedent, but these are also permissible inference figures in the auxiliary calculus.)

Now we still have to consider contractions – or interchanges – in the succedent. Here, as in the previous case, the *whole* succedent is negated and brought forward into the antecedent. We then carry out interchanges, a contraction, and further interchanges – or one interchange – in the antecedent, and then the negated formulae are brought back into the succedent (by means of the inference figures $\mathfrak{F}\mathfrak{f}2$).

6.32. Transformation of a derivation of the auxiliary calculus into a derivation of the calculus LJ augmented by the inclusion of the basic sequent schema $\to \mathfrak{A} \vee \neg \mathfrak{A}$:

We begin by transforming all D-sequents as follows:

$\mathfrak{A}_1, \ldots, \mathfrak{A}_\mu \to \mathfrak{B}_1, \ldots, \mathfrak{B}_\nu$ becomes

$\mathfrak{A}_1, \ldots, \mathfrak{A}_\mu \to \mathfrak{B}_\nu \vee \ldots \vee \mathfrak{B}_1$. If the succedent was empty, it remains empty.

Now all basic sequents or inference figures of the auxiliary calculus with the exception of the figures $\mathfrak{I}\mathfrak{f}1$ and $\mathfrak{I}\mathfrak{f}2$, have thus already become basic sequents or inference figures of the calculus *LJ*. This is so since these inference figures have resulted from the schemata III.1.21, III.1.22 (with the exception of the schemata for contraction and interchange in the succedent) by Θ and Λ always having remained empty. At most one formula could therefore occur in the succedent.

Hence we still have to transform the figures which have resulted from the inference figures $\mathfrak{I}\mathfrak{f}1$ and $\mathfrak{I}\mathfrak{f}2$ in the course of the above modification.

6.321. First $\mathfrak{I}\mathfrak{f}1$: If Θ is empty, we replace the inference figure by a $\neg\text{-}IA$, followed by interchanges in the antecedent. Suppose, therefore, that Θ is not empty, where Θ^* designates the formulae belonging to Θ, in reverse order and connected by $\vee$.

After the transformation of the succedents, the inference figure in that case runs as follows:

$$\frac{\Gamma \to \Theta^* \vee \mathfrak{A}}{\Gamma, \neg \mathfrak{A} \to \Theta^*}.$$

This is transformed into the following section of an *LJ*-derivation:

$$\frac{\Gamma \to \Theta^* \vee \mathfrak{A} \qquad \dfrac{\dfrac{\Theta^* \to \Theta^*}{\neg \mathfrak{A}, \Theta^* \to \Theta^*} \text{thinning}}{\Theta^*, \neg \mathfrak{A} \to \Theta^*} \text{interchange} \qquad \dfrac{\dfrac{\dfrac{\dfrac{\mathfrak{A} \to \mathfrak{A}}{\neg \mathfrak{A}, \mathfrak{A} \to} \neg\text{-}IA}{\mathfrak{A}, \neg \mathfrak{A} \to} \text{interchange}}{\mathfrak{A}, \neg \mathfrak{A} \to \Theta^*} \text{thinning}}{\Theta^* \vee \mathfrak{A}, \neg \mathfrak{A} \to \Theta^*} \vee\text{-}IA}{\Gamma, \neg \mathfrak{A} \to \Theta^*} \text{cut.}$$

6.322. After the transformation of its succedents, an inference figure $\mathfrak{I}\mathfrak{f}2$ runs as follows:

$$\frac{\Gamma, \neg \mathfrak{A} \to \Theta^*}{\Gamma \to \Theta^* \vee \mathfrak{A}},$$

where Θ^* has the same meaning as in the previous case. If Θ is empty, assume Θ^* to be empty too, and let $\Theta^* \vee \mathfrak{A}$ mean $\mathfrak{A}$.

It is transformed into the following section of a derivation:

§ 6, THE EQUIVALENCE OF THE CALCULI LHK, NK AND LK

$$\cfrac{\to \mathfrak{A} \vee \neg \mathfrak{A} \quad \cfrac{\cfrac{\cfrac{\mathfrak{A} \to \mathfrak{A}}{\mathfrak{A}, \Gamma \to \mathfrak{A}} \text{possibly several thinnings and interchanges}}{\mathfrak{A}, \Gamma \to \Theta^* \vee \mathfrak{A}} \vee\text{-IS} \quad \cfrac{\cfrac{\cfrac{\Gamma, \neg \mathfrak{A} \to \Theta^*}{\neg \mathfrak{A}, \Gamma \to \Theta^*} \text{possibly several interchanges}}{\neg \mathfrak{A}, \Gamma \to \Theta^* \vee \mathfrak{A}} \vee\text{-IS}}{\mathfrak{A} \vee \neg \mathfrak{A}, \Gamma \to \Theta^* \vee \mathfrak{A}} \vee\text{-IA} \text{ cut.}}{\Gamma \to \Theta^* \vee \mathfrak{A}}}$$

It is easy to see that in the case of an empty Θ all is in order.

6.33. The *LJ*-derivation now obtained, together with the additional basic sequents of the form $\to \mathfrak{A} \vee \neg \mathfrak{A}$, may be transformed, as in § 5, into an *LHJ*-derivation with the inclusion of additional basic formulae of the form $\mathfrak{A} \vee \neg \mathfrak{A}$ (cf. 5.5), i.e., into an *LHK*-derivation. This completes the transformation of the *LK*-derivation into an *LHK*-derivation. At the same time, the endsequent has been transformed (in accordance with 6.32, 5.2, and 5.5) into an equivalent formula (according to 1.1).

By combining the results of 6.1, 6.2, and 6.3, we have now also proved *the equivalence of the three classical calculi of predicate logic*: *LHK*, *NK*, and *LK*.

4. THE CONSISTENCY OF ELEMENTARY NUMBER THEORY

By '*elementary number theory*' I mean the theory of the natural numbers that does not make use of techniques from analysis such as, e.g., irrational numbers or infinite series.

The aim of the present paper is to prove the *consistency* of elementary number theory or, rather, to reduce the question of consistency to certain general fundamental principles.

How such a consistency proof can be carried out at all and for what reasons it is necessary or at least very desirable to do so will be discussed in section I.

The entire paper can be read without any specialized knowledge.

SECTION I. REFLECTIONS ON THE PURPOSE AND POSSIBILITY OF CONSISTENCY PROOFS

In § 1, I shall consider the question why consistency proofs are *necessary* and, in § 2, how such proofs are *possible*[30]. In doing so, I shall briefly restate those aspects of the problem, already familiar to many readers, which are of particular relevance to the rest of this paper.

§ 1. The antinomies of set theory and their significance for mathematics as a whole[31]

1.1. Mathematics is regarded as the most certain of all the sciences. That it could lead to results which contradict one another seems impossible. This faith in the indubitable certainty of mathematical proofs was sadly shaken around 1900 by the discovery of the '*antinomies* (or 'paradoxes') of set theory'. It turned out that in this specialized branch of mathematics contradictions arise without our being able to recognize any specific error in our reasoning.

Particularly instructive is '*Russell's antinomy*', which I shall now discuss in detail.

1.2. A set is a collection of arbitrary objects ('elements of the set'). An 'empty set', which has no elements at all, is also admitted. We now divide the sets into 'sets of the first kind', i.e., sets which contain *themselves* as an element, and 'sets of the second kind', i.e., sets which do *not* contain themselves as an element.

We consider the set $\mathfrak{m}$ which has for its elements the entire collection of the sets of the second kind. Does this set itself belong to the first or the second kind? Both alternatives are absurd: For if the set $\mathfrak{m}$ belongs to the first kind, i.e., if it contains itself as an element, then this contradicts its definition by which all of its elements were supposed to be sets of the second kind. Suppose, therefore, that the set $\mathfrak{m}$ belongs to the second kind, i.e., that it does not contains itself as an element. Since, by definition, it has all sets of the second kind as elements, it must also contain itself as an element and we have thus once again arrived at a contradiction.

1.3. The result is *Russell's antinomy* which shows how easily an obvious contradiction can result from a small number of admittedly somewhat subtle inferences.

What is the actual *significance* of this fact for *mathematics as a whole*? We may be inclined, at first, to dismiss the entire argument as *unmathematical* by claiming that the concept of a 'set of arbitrary objects' is too vague to count as a mathematical concept.

This objection becomes void if we restrict ourselves to quite specific purely *mathematical* objects by making the following stipulation, for example: The only objects admitted as elements of a 'set' are, first: arbitrary natural numbers (1, 2, 3, 4 etc.); second: arbitrary sets consisting of admissible elements.

Example: The following three elements form an admissible set: First, the number 4; second, the set of all natural numbers; third, the set whose two elements are the number 3 and the set of all natural numbers.

Using this purely *mathematical* concept of a set, we can repeat the above (1.2) argument and obtain the *same contradiction*.

1.4. The fact that we happen to have chosen the natural numbers for our initial objects has obviously no bearing on the emergence of the antinomy. It cannot, therefore, be said that a contradiction has been revealed in the domain of the *natural numbers*; the fault must be sought rather in the *logical inferences* employed.

1.5. It is thus natural to go back to look for a *definite error* in the reasoning that has led to the antinomy. We might, for example, argue that the set 𝔪 was defined by referring to the totality of *all* sets (which was indeed subdivided into sets of the first and second kinds, and where 𝔪 was formed with sets of the second kind). The set was then itself regarded as belonging to this totality, which raised the question of whether it belongs to the first or second kind. Such a procedure is *circular*; it is illicit to define an object by means of a totality and to regard it then as belonging to that totality, so that in some sense it contributes to its own definition ('circulus vitiosus').

We might feel that the *correct* interpretation of the set 𝔪 should rather be the following:

If a *definite totality* of sets is given, then this totality may be subdivided into sets of the first and second kinds. Yet if the sets of the second kind (or alternatively, the first kind) are combined into a new set 𝔪, then that set constitutes something *completely new* and cannot itself be regarded as belonging to that totality.

1.6. The impression, at first sight, that the forms of inference leading to the antinomy seem correct derives from the conception of the concept of a 'set' as something '*actualistically*' ((*an sich*)) determined (and the totality of all sets, therefore, constitutes a predetermined closed totality); the critique advanced against this view implies that new sets can be formed only '*constructively*' so that a new set depends in its construction on already existing sets.

1.7. If we were to think that the antinomy has thus been explained away quite satisfactorily, we must at once face up to a new difficulty: The form of reasoning (the circulus vitiosus) which we have just declared to be inadmissible is being used *in analysis* in a quite similar form in the usual proofs of some rather simple theorems, e.g., the theorem: 'A function which is continuous on a closed interval and is of different sign at the endpoints has a zero in the interval.'

The proof of this result is essentially carried out in the following way: The totality of points in the interval is divided into points of the first and second kinds, so that a point is of the first kind if the function has the same sign for all points to the right of it, up to the end of the interval, and it belongs to the second kind if this is not the case. The limit point defined by this subdivision is then the required zero. It belongs itself to the points of the interval. Hence we have the 'circulus vitiosus': The real number concerned is defined by referring to the *totality* of the real numbers (in an interval) and is then itself regarded as belonging to that totality.

This form of inference is nevertheless considered correct in analysis on the following grounds: The number concerned is, after all, not *newly created* by the given definition, it already *actually* exists within the totality of the real numbers and is merely *singled out* from this totality by its definition.

Yet exactly the same could be said about the antinomy mentioned above: The set 𝔪 is already *actually* present in the totality of all sets (defined at 1.3) and is merely singled out by its definition (at 1.2) from this totality.

Considerable *differences* certainly exist between the forms of inference used to derive the antinomy and those customary in proofs in analysis. Yet we must ask ourselves whether these differences are radical enough to justify the further use of these inferences in analysis – since no contradictions have so far arisen – or whether their similarity with the inferences that have led to the antinomies should not prompt us to eliminate these inferences also from analysis. Here the *opinions* of mathematicians concerned with these questions *diverge*.

1.8. We can indeed *challenge* the correctness of *other* forms of inference customary in mathematics because of certain remote analogies that may be drawn between them and inferences leading to the antinomies. Especially radical in this respect are the '*intuitionists*' (Brouwer), who even object to forms of inference customary in *number theory*, not only because these inferences might possibly lead to *contradictions*, but because the theorems to which they lead have no actual *sense* and are therefore worthless. I shall come back to this point later in greater detail (§§ 9–11 and 17.3).

Less radical are the '*logicists*' (Russell). They draw a line between permissible and non-permissible forms of inference, and the antinomies turn out to be a consequence of a nonpermissible circulus vitiosus. At one time the logicists had also disallowed the inference applied in the example from analysis cited above ('ramified theory of types'), but this inference was later readmitted.

1.9. Altogether we are left with the following picture:

The contradictions (antinomies) which had occurred in set theory, a specialized branch of mathematics, had given rise to further doubts about the correctness of certain forms of inference customary in the rest of mathematics. Various attempts to draw a line between permissible and nonpermissible forms of inference have led to different approaches to the subject.

In order to end this unsatisfactory state of affairs, Hilbert drew up the following *programme*:

The *consistency* of the whole of mathematics, in so far as it actually is consistent, is to be *proved* along exact mathematical lines. This proof is to be carried out by means of forms of inference that are completely unimpeachable ('finitist' forms of inference).

How such a consistency proof is conceivable at all will be discussed more fully in § 2.

In the remainder of this paper, I shall then carry out such a consistency proof for *elementary number theory*. Yet even here we shall meet forms of inference whose closer inspection will give us cause for concern. More about this in section III. One point should however be made clear from the outset: *those* forms of inference which might possibly be considered disputable *hardly ever* occur in actual number-theoretical proofs (11.4); we must therefore not be mislead and, because of the great self-evidence of these proofs, consider a consistency proof as superfluous.

§ 2. How are consistency proofs possible?

2.1. *General remarks about consistency proofs.*

2.11. The consistency of geometries is usually proved by appealing to an *arithmetic* model. Here the consistency of *arithmetic* is therefore *presupposed*. In a similar way we can also effect a reduction of some parts of arithmetic to others, e.g., the theory of the complex numbers to that of the real numbers.

What remains to be proved ultimately is the consistency of the theory of the *natural* numbers (elementary number theory) and the theory of the *real* numbers (analysis) of which the former forms a part; and finally the consistency of set theory as far as that theory is consistent.

2.12. This task is *basically different* and more difficult than that of reducing the consistency of one theory to that of another by mapping the objects of the former theory onto the objects of the latter. Let us look more closely at the situation in the case of the *natural* numbers:

These numbers can obviously not be mapped onto a *simpler domain of objects*. Nor are we indeed concerned with the consistency of the *domain of numbers* itself, i.e., with the consistency of the basic relationships between the numbers as determined by the 'axioms' (e.g., the 'Peano axioms' of number theory). To prove the consistency of these axioms without invoking other equivalent assumptions seems inconceivable. We are concerned rather with the consistency of our *logical reasoning* about the natural numbers (starting from their axioms) as it occurs in the *proofs* of number theory. For it is precisely our logical reasoning which in its unrestricted application leads to the antinomy (1.4). We do not of course consider such general constructions as those of arbitrary sets of sets (1.3) as part of *number theory*. Elementary number theory comprises merely *finite* sets (of natural numbers, for example). If *infinite* sets of natural numbers are included, we are already in the domain of the real numbers and hence in *analysis*. This is the fundamental *distinction between elementary number theory and analysis*.

From here we reach *set theory* by extending the concept of a 'set' still further.

How can the *consistency of arithmetic* be actually *proved*?

2.2. *'Proof theory'*.

2.21. The assertion that a mathematical theory is consistent constitutes a proposition about the *proofs* possible in that theory. It says that none of these proofs leads to a contradiction. In order to carry out a consistency proof we must therefore make the possible proofs in the theory themselves *objects* of a new 'metatheory'. The theory that has arbitrary mathematical proofs for its objects is called *'proof theory'* or 'metamathematics'.

2.22. An example of a *theorem in proof theory* is the *'principle of duality'* in projective geometry:

It says roughly that from a theorem about points and straight lines (in the plane) another true theorem results if the word 'point' is replaced by 'straight line' and the word 'straight line' by 'point'. The theorem: 'For any two distinct straight lines there exists exactly one point incident with both straight lines (i.e., lying on them)', for example, has for its dual the theorem: 'For any two distinct points there exists exactly one straight line incident with both points (i.e., passing through them)'.

The principle of duality is *justified* thus: The axioms of projective geometry in the plane are such that the dual of an axiom always yields another axiom. If any theorem has therefore been derived from these axioms, a uniform replacement in the proof of the word 'point' by 'straight line' and of the word 'straight line' by 'point' yields a proof for the *dual* theorem.

This justification is obviously *proof-theoretical* since it is about the 'proof of a theorem'.

(This example also shows that proof theory is capable of advancing mathematics proper.)

2.23. The *'formalization'* of mathematical proofs.

As the objects of our proof theory we take the *proofs* carried out in mathematics proper. These proofs are customarily expressed in the *words* of our language. They have the disadvantage that there are many different ways of expressing the *same* proposition, and that an arbitrariness exists in the order of the words, sometimes even *ambiguity*. In order to make an exact study of proofs possible it is therefore desirable to begin by giving them a uniform uniquely predetermined form. This is achieved by the 'formalization' of the proofs: the words of our language are replaced by definite *symbols*, the logical forms of inference by formal rules for the formation of new formalized propositions from already proved ones.

In section II, I shall carry out such a *formalization* for *elementary number theory*.

The example of the principle of duality (2.22) shows clearly the difficulties that are inherent in proof theory without a formalization: the linguistic expression of the theorem 'for two mutually distinct straight lines there exists exactly one point incident with both straight lines' had to be chosen artificially in such a way that the replacement of 'point' by 'straight line' and vice versa again resulted in another meaningful sentence. Even in carrying out the proof of the principle of duality we are left with the feeling that we have not offered a really rigorous proof. In order to make this proof rigorous, we do in fact require an exact formalization of the propositions and proofs (for the domain of projective geometry).

2.3. *The forms of inference used in the consistency proof; the theorem of Gödel.*

2.31. How can a consistency proof (for elementary number theory, for example) be carried out by means of proof theory?

To begin with, it will have to be made precise what is to be understood by a formalized 'number-theoretical proof'. Then it must be *established* that among all such possible 'proofs' there can exist none which leads to a 'contradiction'. (This is a simple property of 'proofs' which is immediately verifiable for any given 'proof'.)

Such a consistency proof is once again a *mathematical proof* in which certain inferences and derived concepts must be used. Their reliability (especially their consistency) must already be *presupposed*. There can be no '*absolute consistency proof*'. A consistency proof can merely *reduce* the correctness of certain forms of inference to the correctness of other forms of inference.

It is therefore clear that in a consistency proof we can use only forms of inference that count as considerably *more secure* than the forms of inference of the theory whose consistency is to be proved.

2.32. Of the greatest significance at this point is the following proof-theoretical theorem proved by K. Gödel[32]: 'It is not possible to prove the consistency of a formally given (delimited) theory which comprises elementary number theory (nor that of elementary number theory itself) by means of the entire collection of techniques proper to the theory concerned (given that that theory is really consistent).'

From this it follows that the consistency of elementary number theory, for example, cannot be established by means of *part* of the methods of proof used in elementary number theory, nor indeed by *all* of these methods. To what extent, then, is a genuine reinterpretation ((zurückführung)) still possible?

It remains quite conceivable that the consistency of elementary number theory can in fact be verified by means of techniques which, in part, no longer belong to elementary number theory, but which can nevertheless be considered to be *more reliable* than the doubtful components of elementary number theory itself.

2.4. In the following (sections II-IV), I shall carry out a consistency proof for elementary number theory. In doing so, I shall indeed apply techniques of proof which do not belong to elementary number theory (16.2). Several different consistency proofs already exist in the literature[33] all of which reach essentially the same point, viz., the verification of the consistency of elementary number theory with the exclusion of the inference of 'complete induction' which, as is well-known, constitutes a very important and frequently used form of inference in number theory. The inclusion of complete induction in my proof presents certain difficulties (16.2).

SECTION II. THE FORMALIZATION OF ELEMENTARY NUMBER THEORY

As pointed out at 2.23, it is desirable for a proof-theoretical discussion of a mathematical theory to give that theory a precise formally determined structure. In order to prove the consistency of elementary number theory, I shall therefore begin by carrying out such a *formalization of elementary number theory*[34].

This task falls into two parts:
1. The formalization of the *propositions* occurring in elementary number theory (§ 3).
2. The formalization of the *techniques of proof* used in elementary number theory, i.e., forms of inference and derived concepts (§§ 4–6).

§ 3. The formalization of the propositions occurring in elementary number theory

3.1. *Preparatory remarks.*

3.11. A formalization of mathematical propositions represents nothing fundamentally new even outside of proof theory. It is indeed true to say that mathematics has always undergone a successive formalization, i.e., a replacement of language by mathematical symbols. There are, for example, propositions which are written entirely in symbols, e.g.,

$$(a+b) \cdot (a-b) = a^2 - b^2,$$

in words: 'The product of the sum and the difference of the numbers a and b is equal to the difference of the squares of both numbers'.

The proposition 'If $a = b$, then $b = a$', on the other hand, is generally still represented by means of words. Completely formalized, it is written: $a = b \supset b = a$.

3.12. The linguistic expression 'If $\mathfrak{A}$ holds, then $\mathfrak{B}$ holds', formally written as $\mathfrak{A} \supset \mathfrak{B}$, is an example of the *logical connection of propositions* for the purpose of forming a new proposition. Other kinds of *connection* are represented by the symbols &, ∨, ¬, ∀, and ∃, with the following meanings: $\mathfrak{A} \& \mathfrak{B}$ means '$\mathfrak{A}$ holds and $\mathfrak{B}$ holds'; $\mathfrak{A} \vee \mathfrak{B}$: '$\mathfrak{A}$ holds or $\mathfrak{B}$ holds' (i.e., *at least one* of the two propositions holds); ¬$\mathfrak{A}$: '$\mathfrak{A}$ does *not* hold'; $\forall \mathfrak{x}\, \mathfrak{A}(\mathfrak{x})$: '$\mathfrak{A}(\mathfrak{x})$ holds for *all* $\mathfrak{x}$'; $\exists \mathfrak{x}\, \mathfrak{A}(\mathfrak{x})$: 'There is an $\mathfrak{x}$, so that $\mathfrak{A}(\mathfrak{x})$ holds'.

3.13. As an *example*, we shall consider 'Goldbach's conjecture' ('Every even natural number can be represented as the sum of two prime numbers'), which can be formally written as:

$$\forall x \, \{2|x \supset \exists y \, \exists z \, [y+z = x \,\&\, (\text{Prime } y \,\&\, \text{Prime } z)]\}.$$

Here, Prime a stands for 'a is a prime number'; $a|b$, as usual, for 'a is a divisor of b'. All *variables* are taken to refer only to the natural numbers (= positive integers).

3.14. The symbols =, Prime and | are '*predicate symbols*'; once their argument places have been filled by numbers, such symbols constitute a *proposition*. The symbol + is a '*function symbol*'; once its argument places have been filled by numbers, it represents another *number*.

The formal counterpart of a *proposition* is generally called a '*formula*'. (Just as in mathematics, for example, $(a+b) \cdot (a-b) = a^2 - b^2$ is called a 'formula', although in a special sense.)

After these remarks, I shall now give a precise characterization of those formal expressions which are to be admitted into our formalized number theory for the purpose of representing propositions.

3.2. *Precise definition of a formula*[35].

3.21. The following kinds of *symbol* will serve for the formation of formulae:

3.211. Symbols for *definite natural numbers*: 1, 2, 3, 4, 5, 6, 7, 8, 9, 10, 11, 12, ..., briefly called 'numerals'. (No symbols for *other* numbers will be needed.)

3.212. *Variables* for *natural numbers*: These I divide into *free* and *bound* variables (vid. seq.). Any other symbol that has not yet been used may serve as a variable; but it must be stated in each case whether such a symbol is to be a free or a bound variable.

3.213. Symbols for *definite functions*, briefly called 'function symbols': $+$, $\cdot$, and others as needed (cf. 6.1).

3.214. Symbols for *definite predicates*, briefly called 'predicate symbols': $=$, $<$, Prime, | and others as needed (cf. 6.1).

3.215. Symbols for the *logical connectives*: &, $\vee$, $\supset$, $\neg$, $\forall$, $\exists$.

3.22. Definition of *term* (formal expression for a – definite or indefinite – *number*):

3.221. *Numerals* (3.211) and *free variables* (3.212) are terms.

3.222. If $\mathfrak{s}$ and $\mathfrak{t}$ are terms, then so are $\mathfrak{s}+\mathfrak{t}$ and $\mathfrak{s}\cdot\mathfrak{t}$; other terms may be formed analogously by means of further *function symbols* that may have been introduced (3.213).

3.223. *No* expressions *other than* those formed in accordance with 3.221 and 3.222 are terms.

3.224. *Example of a term*: $[(a+2)^3 \cdot b]+4$; where a and b are free variables.

Brackets serve as usual to avoid ambiguities in connection with the grouping of the individual symbols.

3.23. I now define a *formula* (formal counterpart of a number-theoretical *proposition*):

3.231. A *predicate symbol* (3.214) whose 'argument places' are filled by arbitrary *terms* (3.22) is a formula.

Example: $(2+a)\cdot 4 < b$.

3.232. If $\mathfrak{A}$ is a formula, then so is $\neg \mathfrak{A}$. If $\mathfrak{A}$ and $\mathfrak{B}$ are formulae, then so are $\mathfrak{A} \& \mathfrak{B}$, $\mathfrak{A} \vee \mathfrak{B}$, and $\mathfrak{A} \supset \mathfrak{B}$.

3.233. From a given formula we obtain another formula by replacing a free variable occurring in it by a bound variable $\mathfrak{x}$ not yet occurring in the formula and prefixing $\forall \mathfrak{x}$ or $\exists \mathfrak{x}$.

3.234. *No* expressions *other than* those formed in accordance with 3.231, 3.232 and 3.233 are formulae.

3.24. As in the case of terms, *brackets* must be used to display unambiguously the construction of a formula in accordance with 3.232 and 3.233.

Examples of formulae: cf. 3.13, 3.11, 3.231.

The *informal sense* of a formula follows from the remarks in 3.1. It should be observed that a formula with *free variables* constitutes an '*indefinite*' proposition which becomes a '*definite*' proposition only if all free variables in it are replaced by terms without free variables, e.g., numerals[36].

A *minimal term* is a term consisting of one function symbol with numerals in the argument places, e.g.: $1+3$.

A *minimal formula* is a formula consisting of one predicate symbol with numerals in the argument places, e.g.: $4 = 12$.

A *transfinite* formula is a formula containing at least one $\forall$ or $\exists$ symbol.

3.25. *German* and *Greek* letters will be used as 'syntactic variables', i.e., as variables for our proof-theoretical considerations about number theory.

3.3. *The question arises whether our concept of formula is wide enough for the representation of all propositions occurring in elementary number theory.*

Strictly speaking the answer is no. There certainly are propositions in elementary number theory (examples will follow) for which no immediate formal representation exists in terms of the methods formulated. Yet such propositions may safely be disregarded as long as equivalent propositions exist in each case which are representable in our formalism.

Here are a number of important *examples*:

3.31. The only objects of number theory which I have allowed for are the *natural numbers*. Yet the rest of the *integers* as well as, occasionally, the *fractions* are of course also needed in number theory. It is not difficult, however, to *reinterpret* all propositions about integers and fractions as propositions about the natural numbers, by observing that the negative integers *can* be made to *correspond* to pairs of positive integers and the fractions to pairs of integers. (An example: $\frac{a}{b} = \frac{c}{d}$ is interpreted as $a \cdot d = c \cdot b$.) Even in the case where *finite sets* of natural numbers, integers or fractions are included among the objects of number theory (e.g., the 'complete systems of residues') it is still possible to reinterpret all propositions as propositions about the natural numbers, although in this case such interpretations are considerably more complicated. The same holds for propositions in which diophantine equations etc. are taken as objects.

Here I do not intend to discuss these methods of reinterpretation further; they present no fundamental difficulties (especially for the consistency proof) and anyone who concerns himself somewhat more closely with these matters will easily see their feasibility (cf. also 17.2).

If *infinite sets* of natural numbers, integers, or fractions are admitted, such a reinterpretation is in general no longer possible precisely because we are here already dealing with objects from *analysis* (cf. 2.12). It is, after all, customary to define the real numbers themselves as certain infinite sets of rational numbers.

3.32. *Functions* and *predicates* occur in number theory in a variety of forms. In defining a formula I have taken account of this fact by admitting at 3.213 and 3.214 'further symbols as needed'. Further details about the introduction of arbitrary functions and predicates follow in § 6.

3.33. As far as the logical connectives are concerned, finally, the following locutions, for example, are customary:

'The proposition $\mathfrak{A}$ holds *if and only if* the proposition $\mathfrak{B}$ holds.' This *connection of propositions* is of course represented as follows: $(\mathfrak{B} \supset \mathfrak{A}) \& (\mathfrak{A} \supset \mathfrak{B})$.

'There exists *exactly one* number $\mathfrak{x}$, for which the proposition $\mathfrak{A}(\mathfrak{x})$ holds.'

For this we write: $\exists \mathfrak{x} [\mathfrak{A}(\mathfrak{x}) \& \forall \mathfrak{y} (\mathfrak{A}(\mathfrak{y}) \supset \mathfrak{y} = \mathfrak{x})]$, with obviously the same meaning. (Suitable bound variables are to be chosen for $\mathfrak{x}$ and $\mathfrak{y}$; where $\mathfrak{A}(\mathfrak{y})$ is the expression resulting from $\mathfrak{A}(\mathfrak{x})$ by the replacement of $\mathfrak{x}$ by $\mathfrak{y}$.)

'There are *infinitely many* numbers $\mathfrak{x}$ for which the proposition $\mathfrak{A}(\mathfrak{x})$ holds.' This simply means that 'for every number there exists a number greater than the former for which $\mathfrak{A}$ holds', and in this form the proposition is representable in our formalism.

'There are exactly n numbers $\mathfrak{x}$ for which the proposition $\mathfrak{A}(\mathfrak{x})$ holds.'

This proposition – with n left indefinite – can be represented in our formalism only in a considerably paraphrased form, possibly as follows: We include the *finite sets* of natural numbers among the objects of the theory and paraphrase the above proposition thus: 'There exists a set of natural numbers with n elements such that for each one of its elements the proposition $\mathfrak{A}$ holds and every number for which $\mathfrak{A}$ holds belongs to the set.' Here 'number of elements' is a *function*, 'belongs' a *predicate,* and both must be defined in advance. The concept of a *finite set* can once again be *paraphrased* according to 3.31.

There exists of course a variety of other locutions all of which can be reduced to immediately formalizable expressions.

3.34. I shall return to the question of the completeness of the formalism in general after the consistency proof has been carried out (17.1).

§ 4. Example of a proof from elementary number theory

4.1. I now proceed to the formalization of the *techniques of proof* used in elementary number theory. I.e., I shall have to list as completely as possible all *forms of inference* and *methods of deriving concepts* used in proofs of elementary number theory and assign to them a fixed *formalization* which avoids all the ambiguities of their linguistic representation.

Only if a precise formal definition can then be given of what is meant by an elementary number-theoretical '*proof*' can we begin with the *proof theory* of elementary number theory.

I shall begin by giving an *example* of a number-theoretical proof in this paragraph, and shall *classify* the individual *forms of inference* according to definite criteria by means of examples from this proof. In § 5, I shall then give a precise general formulation to these forms of inference.

Finally, in § 6, I shall discuss the *methods of deriving concepts* and the relevant number-theoretical '*axioms*'.

4.2. As an example of a proof from elementary number theory, I shall choose Euclid's well-known proof of the theorem: '*There are infinitely many prime numbers.*'

I shall first carry out the proof *in words* in a version which has been adapted somewhat to the purpose at hand.

In the following (throughout § 4), I shall use the letters $a, b_1, b_2, c, d, l, m, n$, as *free variables*, the letters z, y as *bound variables* (for natural numbers).

The theorem to be proved can be formulated more precisely as follows: '*For every natural number there exists a larger one which is a prime number.*'

Suppose now that a is an arbitrary natural number. We must then show that there exists a prime number which is larger than a. We consider the number $a!+1$. If it is a *prime number* then it satisfies the condition. If it is *not* a prime number then it has a divisor b_1 (excluding 1 and itself). This divisor is larger than a since no number from 2 to a can divide $a!+1$, any such division leaving a remainder of 1. If b_1 is a prime number, it satisfies the condition. If it is not a prime number then it too has a divisor b_2 other than 1 and itself. This number also divides $a!+1$, since it divides b_1. Hence b_2 is also larger than a. By continual repetition of this process we obtain a sequence of numbers: $a!+1, b_1, b_2, \ldots$ whose terms become smaller and smaller. Hence the sequence must terminate at some point, i.e., its last number is a *prime number* which divides $a!+1$ and is larger than a. The existence of a prime number which is larger than a has then been verified. Since a was an *arbitrary* natural number it follows that for *every* natural number there exists a larger one which is a prime number. Q.E.D.

4.3. In the proof I have presupposed various simple theorems as already *known*. These can be *reduced* to still simpler facts by further proofs, although this is unimportant for our present purpose since we are interested, above all, in *the inferences* which occur in the various steps of the above proof.

Here we must keep in mind that through experience we are accustomed to carrying out *entire sequences of proof at once* without being conscious of each individual inference contained in that step. In order to single out the actual *elementary inferences* I shall therefore go through Euclid's proof once again and bring to light *all individual inferences contained in some parts of the proof*. At the same time, I shall *formalize*, in accordance with § 3, the various *propositions* as they occur.

4.4. *Detailed analysis of Euclid's proof.*

The proof contains a somewhat disguised 'complete induction' (cf. the place: 'by continual repetition of this process . . .'). The usual *normal form* of the inference by complete induction is this:

The validity of a proposition is proved for the number 1; then it is shown that if the proposition holds for an arbitrary natural number n it also holds for $n+1$; hence this proposition holds for any natural number.

It will also be convenient to reduce to this normal form the disguised complete induction which here occurs; to do this I shall choose the following proposition as the 'induction proposition', formulated for a number m: 'Either there exists a prime number among the numbers from 1 to m which is greater than a or none of these numbers, except 1, divides $a!+1$. Formally:

$$\{\exists z \, [z \leq m \, \& \, (\text{Prime } z \, \& \, z > a)]\} \vee \forall y \, [(y > 1 \, \& \, y \leq m) \supset \neg \, y|(a!+1)].$$

The proof now runs as follows:

4.41. The induction proposition must first be proved for $m = 1$. Here the second part of the alternative is satisfied automatically since there is obviously no number which is larger than 1 and smaller than or equal to 1. Explicitly: for an arbitrary c it holds that $\neg \, (c > 1 \, \& \, c \leq 1)$; this we assume as given. Then it also holds that $(c > 1 \, \& \, c \leq 1) \supset \neg \, c|(a!+1)$, and, since c was *arbitrary*, $\forall y \, [(y > 1 \, \& \, y \leq 1) \supset \neg \, y|(a!+1)]$. The induction proposition for $m = 1$ follows from this in accordance with the meaning of $\vee$ (3.12), viz.:

$$\{\exists z \, [z \leq 1 \, \& \, (\text{Prime } z \, \& \, z > a)]\} \vee \forall y \, [(y > 1 \, \& \, y \leq 1) \supset \neg \, y|(a!+1)].$$

4.42. Next comes the *'induction step'*, i.e.: we assume that the induction proposition has been proved for an arbitrary number n, so that

$$\{\exists z \, [z \leq n \, \& \, (\text{Prime } z \, \& \, z > a)]\} \vee \forall y \, [(y > 1 \, \& \, y \leq n) \supset \neg \, y|(a!+1)].$$

holds, and that it must now be proved for $n+1$. This is done as follows:

On the basis of the induction hypothesis *two cases* are possible:

1. $\exists z\, [z \leq n\, \&\, (\text{Prime } z\, \&\, z > a)]$;
2. $\forall y\, [(y > 1\, \&\, y \leq n) \supset \neg\, y|(a!+1)]$.

In the first case it follows without difficulty that

$$\exists z\, [z \leq n+1\, \&\, (\text{Prime } z\, \&\, z > a)].$$

I shall not discuss this further. In this case therefore the induction proposition has already been proved for $n+1$, viz.,

$$\{\exists z\, [z \leq n+1\, \&\, (\text{Prime } z\, \&\, z > a)]\}$$
$$\lor\, \forall y\, [(y > 1\, \&\, y \leq n+1) \supset \neg\, y|(a!+1)].$$

Let us now look at the *second* case:

$$\forall y\, [(y > 1\, \&\, y \leq n) \supset \neg\, y|(a!+1)].$$

It holds that $(n+1)|(a!+1) \lor \neg\, (n+1)|(a!+1)$. We can thus distinguish *two subcases*:

First *subcase*: $(n+1)|(a!+1)$. From this it follows that

$$\text{Prime } (n+1)\, \&\, (n+1) > a,$$

which I shall briefly show since the only forms of inference here used are those for which we already have examples in the remaining parts of the proof:

$n+1$ is a *prime number*; for if it had a divisor other than 1 and itself, it would be smaller than $n+1$, and would then also divide $a!+1$, contradicting our assumption that $\forall y\, [(y > 1\, \&\, y \leq n) \supset \neg\, y|(a!+1)]$. Furthermore, $n+1$ is *larger than* a; for the numbers from 2 to a do not divide $a!+1$, such a division always leaving a remainder of 1. Hence it holds in fact that Prime $(n+1)\, \&\, (n+1) > a$; also $n+1 \leq n+1$, hence it holds that $n+1 \leq n+1\, \&\, \text{Prime } (n+1)\, \&\, (n+1) > a$, and consequently also that

$$\exists z\, [z \leq n+1\, \&\, (\text{Prime } z\, \&\, z > a)],$$

and thus

$$\{\exists z\, [z \leq n+1\, \&\, (\text{Prime } z\, \&\, z > a)]\}$$
$$\lor\, \forall y\, [(y > 1\, \&\, y \leq n+1) \supset \neg\, y|(a!+1)].$$

Second *subcase*: $\neg\, (n+1)|(a!+1)$. Suppose that d is an arbitrary number with the property that $d > 1\, \&\, d \leq n+1$. From $d \leq n+1$ follows $d \leq n \lor d = n+1$, which is to be taken as given.

§ 4, EXAMPLE OF A PROOF FROM ELEMENTARY NUMBER THEORY

Suppose first that $d \leq n$; it also holds that

$$\forall y \, [(y > 1 \,\&\, y \leq n) \supset \neg \, y|(a!+1)],$$

hence in particular that $(d > 1 \,\&\, d \leq n) \supset \neg \, d|(a!+1)$. From $d > 1$, together with $d \leq n$, it follows that $d > 1 \,\&\, d \leq n$, and together with the preceding therefore $\neg \, d|(a!+1)$.

If $d = n+1$, however, then, because of $\neg \, (n+1)|(a!+1)$, it also follows that $\neg \, d|(a!+1)$.

Thus it holds in general that $\neg \, d|(a!+1)$, a consequence of the assumption $d > 1 \,\&\, d \leq n+1$. Hence we can write

$$(d > 1 \,\&\, d \leq n+1) \supset \neg \, d|(a!+1)],$$

and further, since d was an arbitrary number,

$$\forall y \, [(y > 1 \,\&\, y \leq n+1) \supset \neg \, y|(a!+1)],$$

and thus once again

$$\{\exists z \, [z \leq n+1 \,\&\, (\text{Prime } z \,\&\, z > a)]\}$$
$$\vee \, \forall y \, [(y > 1 \,\&\, y \leq n+1) \supset \neg \, y|(a!+1)].$$

We have therefore in all cases obtained the induction proposition for $n+1$, and this *completes the induction step*.

4.43. The proof is now quickly *completed*:

From the complete induction follows the validity of the induction proposition for *arbitrary* numbers. We require it only for the number $a!+1$:

$$\{\exists z [z \leq a!+1 \,\&\, (\text{Prime } z \,\&\, z > a)]\}$$
$$\vee \, \forall y \, [(y > 1 \,\&\, y \leq a!+1) \supset \neg \, y|(a!+1)].$$

From the *second* case it follows in particular that

$$(a!+1 > 1 \,\&\, a!+1 \leq a!+1) \supset \neg \, (a!+1)|(a!+1).$$

Yet it holds that $a!+1 > 1 \,\&\, a!+1 \leq a!+1$, which we assume as given; hence it follows that $\neg \, (a!+1)|(a!+1)$. On the other hand it holds of course that $(a!+1)|(a!+1)$; we have thus obtained a *contradiction*, i.e., the second case cannot possibly occur; formally:

$$\neg \, \forall y \, [(y > 1 \,\&\, y \leq a!+1) \supset \neg \, y|(a!+1)].$$

Only the *first* case remains, i.e.: $\exists z \, [z \leq a!+1 \,\&\, \text{Prime } z \,\&\, z > a)]$. Suppose that l is such a number so that $l \leq a!+1$ & (Prime l & $l > a$) holds. Then it holds in particular that Prime l & $l > a$, from which $\exists z$ (Prime z & $z > a$) follows. But a was an *arbitrary* natural number, hence this result holds for *all* natural numbers, i.e., $\forall y \, \exists z$ (Prime z & $z > y$). This is the *conclusion* of Euclid's proof.

4.5. *Classification of the individual forms of inference by reference to examples from Euclid's proof.*

Let us now focus our attention on the individual inferences occurring in the above proof. Here the following *classification* almost suggests itself:

For every logical connective &, ∨, ⊃, ¬, ∀, and ∃ there exist certain *associated* forms of inference. These may be divided into forms of inference by which the connective concerned is *introduced* and other forms of inference by which the same connective is *eliminated* from a proposition. As *examples* I shall cite, in each case, an inference from Euclid's proof:

4.51. A ∀-*introduction* occurs at the end of the proof, viz: after $\exists z$ (Prime z & $z > a$) was proved for the number a, it was inferred that $\forall y \, \exists z$ (Prime z & $z > y$).

A ∀-*elimination* took place at 4.42, subcase 2, where from

$$\forall y \, [(y > 1 \,\&\, y \leq n) \supset \neg y|(a!+1)]$$

it was inferred that $(d > 1 \,\&\, d \leq n) \supset \neg d|(a!+1)$.

4.52. A &-*introduction* (from 4.42, subcase 2): the two propositions $d > 1$ and $d \leq n$ together yielded the proposition $d > 1$ & $d \leq n$.

A &-*elimination* (from 4.43): From $l \leq a!+1$ & (Prime l & $l > a$) it was inferred that Prime l & $l > a$.

4.53. A ∃-*introduction* (from 4.43): From Prime l & $l > a$ it was inferred that $\exists z$ (Prime z & $z > a$).

A ∃-*elimination* (from 4.43): The proposition

$$\exists z \, [z \leq a!+1 \,\&\, (\text{Prime } z \,\&\, z > a)]$$

held. From it it was inferred that $l \leq a!+1$ & (Prime l & $l > a$), where l stood for any one of the numbers which *existed* by virtue of the previous proposition.

4.54. A ∨-*introduction* (from 4.41): From

$$\forall y \, [(y > 1 \,\&\, y \leq 1) \supset \neg y|(a!+1)]$$

it was inferred that

$$\{\exists z \, [z \leq 1 \,\&\, (\text{Prime } z \,\&\, z > a)]\} \vee \forall y \, [(y > 1 \,\&\, y \leq 1) \supset \neg y|(a!+1)].$$

§ 5, THE FORMALIZATION OF FORMS OF INFERENCE

A ∨-*elimination* (from 4.42): the proposition

$$\{\exists z \, [z \leq n \, \& \, (\text{Prime } z \, \& \, z > a)]\} \vee \forall y \, [(y > 1 \, \& \, y \leq n) \supset \neg \, y | (a! + 1)]$$

held. It led to the *distinction of cases*:
  Case 1: $\exists z \, [z \leq n \, \& \, (\text{Prime } z \, \& \, z > a)]$;
  Case 2: $\forall y [(y > 1 \, \& \, y \leq n) \supset \neg \, y|(a!+1)]$.
  This distinction of cases was terminated by the fact that the same proposition

$$\{\exists z [z \leq n+1 \, \& \, (\text{Prime } z \, \& \, z > a)]\}$$
$$\vee \forall y[(y > 1 \, \& \, y \leq n+1) \supset \neg \, y|(a!+1)]$$

could eventually be inferred in both cases.
4.55. A ⊃-*introduction* (from 4.42, subcase 2): Starting with the assumption $d > 1 \, \& \, d \leq n+1$, we reached the result: $\neg \, d|(a!+1)$. Hence

$$(d > 1 \, \& \, d \leq n+1) \supset \neg \, d|(a!+1)$$

held.
  A ⊃-*elimination* (from 4.42, subcase 2): From $d > 1 \, \& \, d \leq n$ and $(d > 1 \, \& \, d \leq n) \supset \neg \, d|(a!+1)$ it was inferred that $\neg \, d|(a!+1)$.
4.56. For *negation* ($\neg$) the situation is not quite as simple; here there exist several distinct forms of inference and these cannot be divided clearly into $\neg$-introductions and $\neg$-eliminations. I shall come back to this later (5.26). Here I shall cite only a single important example from Euclid's proof, viz., a '*reductio ad absurdum*' – inference (from 4.43):

$$\neg \, \forall y \, [(y > 1 \, \& \, y \leq a!+1) \supset \neg \, y|(a!+1)]$$

was inferred from the fact that the *assumption*

$$\forall y[(y > 1 \, \& \, y \leq n+1) \supset \neg \, y|(a!+1)]$$

led to a *contradiction*, viz., to the proposition $\neg \, (a!+1)|(a!+1)$, whereas $(a!+1)|(a!+1)$ is indeed provable.

§ 5. The formalization of the forms of inference occurring in elementary number theory

5.1. *Preliminary remarks.*
  My next task is to formulate in their most *general form* the different kinds

of forms of inference which have been introduced by means of the above examples.

The determination of the individual forms of inference is not entirely *unique*, although the subdivision into *introductions* and *eliminations* of the individual *logical connectives* which I have chosen seems to me especially lucid and natural.

What, then, does the general *pattern* of a form of inference look like?

E.g., as the general form of the *&-elimination* we would be inclined to put simply the following: if a proposition of the form $\mathfrak{A}$ & $\mathfrak{B}$ is proved (where $\mathfrak{A}$ and $\mathfrak{B}$ are arbitrary formulae), then $\mathfrak{A}$ (or $\mathfrak{B}$) also holds.

Yet we must still keep in mind the following: the structure of a mathematical proof does not in general consist merely of a passing from *valid* propositions to other *valid* propositions via inferences. It happens, rather, that a proposition is often *assumed* as valid and further propositions are *deduced* from it whose validity therefore *depends* on the validity of this assumption. Examples from Euclid's proof: The 'reductio' (4.56), the ⊃-introduction (4.55), the induction step in the complete induction (4.42).

In order to describe *completely* the *meaning* of any *proposition* occurring in a proof we must therefore state, in each case, upon which of the *assumptions* that may have been made, the proposition in question *depends*.

I therefore make it a rule that, together with every (formalized) proposition $\mathfrak{B}$ occurring in a formalized proof, the (formalized) *assumptions* $\mathfrak{A}_1, \ldots, \mathfrak{A}_\mu$ upon which the proposition depends must also be *listed* in the following form[37]:

$$\mathfrak{A}_1, \mathfrak{A}_2, \ldots, \mathfrak{A}_\mu \to \mathfrak{B};$$

which reads: From the assumptions $\mathfrak{A}_1, \ldots, \mathfrak{A}_\mu$ follows $\mathfrak{B}$. Such an expression I call a 'sequent'. If there are *no* assumptions, we write $\to \mathfrak{B}$.

An example from Euclid's proof: The proposition $\neg\, d|(a!+1)$ from 4.42, subcase 2, must, in order to display its dependence on assumptions, be represented by the following *sequent*:

$$\forall y\, [(y > 1\, \&\, y \leqq n) \supset \neg\, y|(a!+1)],\, \neg\, (n+1)|(a!+1),$$
$$d > 1\, \&\, d \leqq n+1 \to \neg\, d|(a!+1).$$

Since *every proposition* of the original proof is now represented by a *sequent* in the formalized proof, we can formulate the *forms of inference* directly for *sequents*.

Our earlier example, the &-elimination, would now have to be formulated

§ 5, THE FORMALIZATION OF FORMS OF INFERENCE

thus: 'If the sequent $\mathfrak{G}_1, \ldots, \mathfrak{G}_\mu \to \mathfrak{A} \& \mathfrak{B}$ is proved ($\mu \geq 0$), then $\mathfrak{G}_1, \ldots, \mathfrak{G}_\mu \to \mathfrak{A}$ or $\mathfrak{G}_1, \ldots, \mathfrak{G}_\mu \to \mathfrak{B}$ is also valid.'

In the following, general schemata for the remaining forms of inference will be given in the same way.

5.2. *Precise general formulation of the individual forms of inference.*

5.21. Definition of a *sequent*[38] (formal expression for the meaning of a *proposition in a proof* together with its dependence on possible assumptions):

A *sequent* is an expression of the form:

$$\mathfrak{A}_1, \mathfrak{A}_2, \ldots, \mathfrak{A}_\mu \to \mathfrak{B},$$

where arbitrary *formulae* (3.23) may take the place of $\mathfrak{A}_1, \mathfrak{A}_2, \ldots, \mathfrak{A}_\mu$ and $\mathfrak{B}$. The formulae $\mathfrak{A}_1, \mathfrak{A}_2, \ldots, \mathfrak{A}_\mu$, I call the *antecedent formulae* and $\mathfrak{B}$ the *succedent formula* of the sequent. It is permissible that no *antecedent formulae* occur, then the sequent has the form: $\to \mathfrak{B}$; but there must always be a *succedent formula*.

5.22. Definition of a *derivation* (formal counterpart of a *proof*): A *derivation* consist of a number of consecutive sequents of which each is either a 'basic sequent' or has resulted from certain earlier sequents by a '*structural transformation*' or by the application of a '*rule of inference*'. The definition of the various concepts follows shortly.

The *last sequent* of a derivation contains no antecedent formulae, its succedent formula is called the *endformula* of the derivation. (It represents the *proposition proved* by the proof.)

5.23. Definition of a *basic sequent*:

I distinguish between basic 'logical' and 'mathematical' sequents.

A *basic logical sequent* is a sequent of the form $\mathfrak{D} \to \mathfrak{D}$, where $\mathfrak{D}$ can be any arbitrary formula. (Such a sequent occurs in the formalization of a proof *if and when an assumption $\mathfrak{D}$ is made* in the proof.)

A *basic mathematical sequent* is a sequent of the form $\to \mathfrak{E}$, where the formula $\mathfrak{E}$ represents a 'mathematical axiom'. In § 6, I will explain, in particular, precisely what is to be understood by a *number-theoretical 'axiom'*.

5.24. Definition of a *structural transformation*:

The following kinds of transformation of a sequent are called *structural transformations* (because they affect only the structure of a sequent, independently of the meaning of the individual formulae):

5.241. *Interchange* of two antecedent formulae;

5.242. *Omission* of an antecedent formula identical with another antecedent formula;

5.243. *Adjunction* of an arbitrary formula to the antecedent formulae;

5.244. Replacement of a *bound variable* within a formula throughout the scope of a ∀- or ∃-symbol by another bound variable not yet occurring in the formula.

Transformations according to 5.241, 5.242 and 5.244 obviously leave the meaning of the sequent unchanged, since it makes no difference to the meaning of the sequent in what order the assumptions are listed, or whether one and the same assumption is listed more than once or, finally, what symbol is used for bound variables. All possibilities of transformation mentioned are thus of a purely *formal nature* and informally of no consequence; they must be stated explicitly only because of the special character of our formalization.

A structural transformation according to 5.243 means that to a proposition we may *adjoin* an arbitrary *assumption* upon which, besides other possible assumptions, it is to depend. At first this may seem somewhat strange; yet if a proposition is *true*, for example, we are forced to admit that in that case it also holds on the basis of an *arbitrary assumption*. (If we were to stipulate that this may be asserted only in cases where a 'factual dependence' exists, considerable difficulties would arise because of the possibility of proofs in which only an *apparent* use of an assumption is made).

5.25. Definition of a *rule of inference* (formal counterpart of a form of inference):

Altogether we require thirteen rules of inference.

5.250. The German and Greek *letters* used here have the following *meanings*:

The letters $\mathfrak{A}$, $\mathfrak{B}$, and $\mathfrak{C}$ stand for arbitrary formulae; the expressions $\forall \mathfrak{x}\, \mathfrak{F}(\mathfrak{x})$ and $\exists \mathfrak{x}\, \mathfrak{F}(\mathfrak{x})$ for arbitrary formulae of this form, with $\mathfrak{F}(\mathfrak{a})$ and $\mathfrak{F}(\mathfrak{t})$ denoting the formulae which result if the bound variable $\mathfrak{x}$ is replaced by an arbitrary free variable $\mathfrak{a}$ and an arbitrary term $\mathfrak{t}$, resp.; the letters Γ, Δ, and Θ for arbitrary, possibly empty, sequences of formulae (antecedent formulae of the sequent concerned), separated by commas.

Now *the individual rules of inference*:

5.251. &-*introduction*: from the sequents $\Gamma \to \mathfrak{A}$ and $\Delta \to \mathfrak{B}$ follows the sequent $\Gamma, \Delta \to \mathfrak{A}\,\&\,\mathfrak{B}$.

&-*elimination*: from $\Gamma \to \mathfrak{A}\,\&\,\mathfrak{B}$ follows the $\Gamma \to \mathfrak{A}$ or $\Gamma \to \mathfrak{B}$.

∨-*introduction*: from $\Gamma \to \mathfrak{A}$ follows $\Gamma \to \mathfrak{A} \vee \mathfrak{B}$ or $\Gamma \to \mathfrak{B} \vee \mathfrak{A}$.

∨-*elimination*: from $\Gamma \to \mathfrak{A} \vee \mathfrak{B}$ and $\mathfrak{A}, \Delta \to \mathfrak{C}$ and $\mathfrak{B}, \Theta \to \mathfrak{C}$ follows $\Gamma, \Delta, \Theta \to \mathfrak{C}$.

∀-*introduction*: from $\Gamma \to \mathfrak{F}(\mathfrak{a})$ follows $\Gamma \to \forall \mathfrak{x}\, \mathfrak{F}(\mathfrak{x})$, provided that the free variable $\mathfrak{a}$ does not occur in Γ and $\forall \mathfrak{x}\, \mathfrak{F}(\mathfrak{x})$.

∀-*elimination*: from $\Gamma \to \forall \mathfrak{x}\, \mathfrak{F}(\mathfrak{x})$ follows $\Gamma \to \mathfrak{F}(\mathfrak{t})$.

$\exists$-*introduction*: from $\Gamma \to \mathfrak{F}(t)$ follows $\Gamma \to \exists\mathfrak{x}\,\mathfrak{F}(\mathfrak{x})$.

$\exists$-*elimination*: from $\Gamma \to \exists\mathfrak{x}\,\mathfrak{F}(\mathfrak{x})$ and $\mathfrak{F}(\mathfrak{a}), \Delta \to \mathfrak{C}$ follows $\Gamma, \Delta \to \mathfrak{C}$, provided that the free variable $\mathfrak{a}$ does not occur in Γ, Δ, $\mathfrak{C}$ and $\exists\mathfrak{x}\,\mathfrak{F}(\mathfrak{x})$.

$\supset$-*introduction*: from $\mathfrak{A}, \Gamma \to \mathfrak{B}$ follows $\Gamma \to \mathfrak{A} \supset \mathfrak{B}$.

$\supset$-*elimination*: from $\Gamma \to \mathfrak{A}$ and $\Delta \to \mathfrak{A} \supset \mathfrak{B}$ follows $\Gamma, \Delta \to \mathfrak{B}$.

5.252. '*Reductio*': from $\mathfrak{A}, \Gamma \to \mathfrak{B}$ and $\mathfrak{A}, \Delta \to \neg\mathfrak{B}$ follows $\Gamma, \Delta \to \neg\mathfrak{A}$.

'*Elimination of the double negation*': from $\Gamma \to \neg\neg\mathfrak{A}$ follows $\Gamma \to \mathfrak{A}$.

5.253. '*Complete Induction*': from $\Gamma \to \mathfrak{F}(1)$ and $\mathfrak{F}(\mathfrak{a}), \Delta \to \mathfrak{F}(\mathfrak{a}+1)$ follows $\Gamma, \Delta \to \mathfrak{F}(t)$, provided that the free variable $\mathfrak{a}$ does not occur in $\Gamma, \Delta, \mathfrak{F}(1)$ and $\mathfrak{F}(t)$.

5.26. Some *remarks about the rules of inference*.

In general the formulation of the individual rules of inference should be clear by reference to the appropriate examples of inferences (4.5). Several points should however be explained:

The Γ, Δ and Θ are required since in the most general case we must allow for *arbitrarily many assumptions*.

The formulation using additional hypotheses which seems fairly natural in the cases of the $\supset$-introduction, the 'reductio', and complete induction, may appear rather artificial in the case of the $\vee$ and $\exists$-elimination, if these rules are compared with the corresponding examples of inferences (4.5). However, the formulation is smoothest if in the *distinction of cases* ($\vee$-elimination) the two possibilities that result are simply regarded as *assumptions* which become redundant as soon as the same result ($\mathfrak{C}$) has been obtained from each; in the case of the $\exists$-elimination the situation is similar: the proposition $\mathfrak{F}(\mathfrak{a})$ inferred from $\exists\mathfrak{x}\,\mathfrak{F}(\mathfrak{x})$ is an assumption only in so far as it is *assumed* of the variable $\mathfrak{a}$ occurring in it that it represents any one of the numbers with the property $\mathfrak{F}$ existing by virtue of $\exists\mathfrak{x}\,\mathfrak{F}(\mathfrak{x})$. This assumption *is discharged* as soon as a result ($\mathfrak{C}$) has been deduced from it in which the variable $\mathfrak{a}$ no longer occurs.

This leads me at once to a further point requiring some elaboration: it concerns the *restrictions on free variables* imposed in the case of the $\forall$-introduction, the $\exists$-elimination, and complete induction.

In each case the restriction says that in all formulae involved in the rule of inference (including the assumption formulae) the free variable $\mathfrak{a}$ belonging to the rule of inference may occur only in the formula $\mathfrak{F}(\mathfrak{a})$ or $\mathfrak{F}(\mathfrak{a}+1)$. It is easily seen by means of examples that this requirement is necessary in general and actually quite obvious; in the case of mathematical proofs it is fulfilled automatically. (By its very purpose, the variable $\mathfrak{a}$ is naturally out of place in the remaining formulae.)

The following must be said about the *rules of inference* for *negation*: as already mentioned at 4.56, the choice of elementary forms of inference is here *more arbitrary* than in the case of the other logical connectives. I should like to *mention* the following simple *alternative rules of inference* that might have been adopted:

From $\mathfrak{A}, \Gamma \to \mathfrak{B}$ and $\neg \mathfrak{A}, \Delta \to \mathfrak{B}$ follows $\Gamma, \Delta \to \mathfrak{B}$.
From $\Gamma \to \mathfrak{A} \vee \mathfrak{B}$ and $\Delta \to \neg \mathfrak{B}$ follows $\Gamma, \Delta \to \mathfrak{A}$ (example at 4.43).
From $\Gamma \to \neg \mathfrak{B}$ and $\mathfrak{A}, \Delta \to \mathfrak{B}$ follows $\Gamma, \Delta \to \neg \mathfrak{A}$.
From $\Gamma \to \neg \mathfrak{A}$ follows $\Gamma \to \mathfrak{A} \supset \mathfrak{B}$ (example at 4.41).
From $\Gamma \to \mathfrak{A}$ and $\Delta \to \neg \mathfrak{A}$ follows $\Gamma, \Delta \to \mathfrak{B}$.

As *basic logical sequents* for the $\neg$-connective we could also have taken the following: $\to \mathfrak{A} \vee \neg \mathfrak{A}$ 'law of the excluded middle' (example at 4.42); $\to \neg (\mathfrak{A} \& \neg \mathfrak{A})$, 'law of contradiction'.

However, the two rules of inference which I have chosen (5.252) are *sufficient*; the remaining rules and the basic sequents listed here are already contained in them (if the rules of inference for the other logical connectives are included); this may be verified without any essential difficulties.

5.3. Are our rules of inference actually sufficient for the representation of all inferences occurring in elementary number theory?

5.31. The completeness of the purely logical rules of inference, i.e., the rules belonging to the connectives $\&, \vee, \supset, \neg, \forall, \exists$, has already been proved elsewhere[39] (completeness here means that all correct inferences of the same type are representable by the stated rules).

To these forms of inference we must now, for the purpose of *elementary number theory*, add 'complete induction'. Here the question of the completeness of the rules of inference becomes a rather difficult problem; I shall return to it after the consistency proof (17.1) has been carried out. At this point I should merely like to observe the following: It may be considered as fairly certain that all inferences occurring in the usual number-theoretical proofs are representable in our system as long as no use is made of techniques from analysis. The same may also be said of the frequently used 'intuitive' inferences, even if this is not immediately obvious from looking at them.

In order to verify this in general each individual proof would of course have to be examined separately and this would be extremely laborious.

5.32. I shall content myself with a number of particularly important *examples*:

Complete induction occurs frequently in certain modified forms, which are reducible to our normal form as follows:

5.321. First the 'method of infinite descent', which runs as follows:

From $\Gamma \to \mathfrak{F}(t)$ and $\mathfrak{F}(\mathfrak{a}+1), \Delta \to \mathfrak{F}(\mathfrak{a})$ follows $\Gamma, \Delta \to \mathfrak{F}(1)$. Again $\mathfrak{a}$ must not occur in $\Gamma, \Delta, \mathfrak{F}(1)$ and $\mathfrak{F}(t)$. This sequent is transformed thus: $\neg \mathfrak{F}(\mathfrak{a}) \to \neg \mathfrak{F}(\mathfrak{a})$ is a basic sequent. From it follows (5.243)

$$\mathfrak{F}(\mathfrak{a}+1), \neg \mathfrak{F}(\mathfrak{a}) \to \neg \mathfrak{F}(\mathfrak{a}),$$

and this, together with $\mathfrak{F}(\mathfrak{a}+1), \Delta \to \mathfrak{F}(\mathfrak{a})$, by 'reductio' (5.252) yields $\Delta, \neg \mathfrak{F}(\mathfrak{a}) \to \neg \mathfrak{F}(\mathfrak{a}+1)$, and finally (5.241) $\neg \mathfrak{F}(\mathfrak{a}), \Delta \to \neg \mathfrak{F}(\mathfrak{a}+1)$. If we then include the basic sequent $\neg \mathfrak{F}(1) \to \neg \mathfrak{F}(1)$, we can apply the rule of complete induction in its earlier form (5.253), with $\neg \mathfrak{F}$ as the induction proposition, and obtain $\neg \mathfrak{F}(1), \Delta \to \neg \mathfrak{F}(t)$. By including $\Gamma \to \mathfrak{F}(t)$, and thus also obtaining (5.243) $\neg \mathfrak{F}(1), \Gamma \to \mathfrak{F}(t)$ as valid, we deduce $\Gamma, \Delta \to \neg\neg \mathfrak{F}(1)$ by 'reductio', and from it, by 'elimination of the double negation' (5.252): $\Gamma, \Delta \to \mathfrak{F}(1)$.

5.322. A further example consists of the following modified complete induction:

From $\Gamma \to \mathfrak{F}(1)$ and $\forall \mathfrak{x} \, [\mathfrak{x} \leq \mathfrak{a} \supset \mathfrak{F}(\mathfrak{x})], \Delta \to \mathfrak{F}(\mathfrak{a}+1)$ follows $\Gamma, \Delta \to \mathfrak{F}(t)$. Again $\mathfrak{a}$ must not occur in $\Gamma, \Delta, \mathfrak{F}(1)$ and $\mathfrak{F}(t)$; $\mathfrak{x}$ designates a bound variable not occurring in $\mathfrak{F}(1)$.

This induction is easily turned into a normal complete induction (5.253) with the following induction proposition (stated for an arbitrary number m): $\forall \mathfrak{x} \, [\mathfrak{x} \leq m \supset \mathfrak{F}(\mathfrak{x})]$, in words possibly: 'For all numbers from 1 to m, $\mathfrak{F}$ holds'.

5.323. The corresponding 'infinite descent' form runs: From $\Gamma \to \mathfrak{F}(t)$ and $\mathfrak{F}(\mathfrak{a}+1), \Delta \to \exists \mathfrak{x} \, [\mathfrak{x} \leq \mathfrak{a} \,\&\, \mathfrak{F}(\mathfrak{x})]$ follows $\Gamma, \Delta \to \mathfrak{F}(1)$. This form can be reduced to the normal form of a complete induction in the same way as the two previous examples.

The induction in Euclid's proof was originally of *this* kind (4.2) and was then reduced to its normal form (4.4).

§ 6. Derived concepts and axioms in elementary number theory

6.1. In a proof there may also occur '*derived concepts*' in addition to the actual *inferences*; these are introductions of new objects, functions, or predicates.

What kinds of derived concepts are in practice used in number theory?

The introduction of new *objects* such as negative numbers etc. has already been discussed at 3.31, and it was pointed out that these objects are basically dispensable.

The introduction of a new *function* or a *predicate* usually takes the form of a verbal 'definition' of these concepts.

Examples:

The function a^b is defined as 'the number a, taken b times as factor'.

The function $a!$ is defined as 'the product of the numbers from 1 to a'.

The number (a, b) is defined as 'the greatest common divisor of a and b'.

The predicate 'a is a perfect number' means the same as 'the number a is equal to the sum of its proper divisors'.

The predicate $a \neq b$ means the same as $\neg (a = b)$.

The predicate $a|b$ means the same as $\exists z (a \cdot z = b)$.

The function $\left(\dfrac{a}{b}\right)$, the 'Legendre symbol', is defined for the case where b is an odd prime number as follows: $\left(\dfrac{a}{b}\right) = 0$ if $b|a$ holds; if $\neg b|a$ holds, then $\left(\dfrac{a}{b}\right) = 1$ if the number a is a quadratic residue mod b, and $\left(\dfrac{a}{b}\right) = -1$ if a is not a quadratic residue mod b.

The function $ak(a, b, c)$, the 'Ackermann function', a function significant for certain questions of proof theory, may be defined thus[40] ('recursively'):

$$ak(a, b, 0) = a+b,$$

$$ak(a, b, 1) = a \cdot b,$$

$$ak(a, b, 2) = a^b,$$

and further for $c \geq 2$:

$$ak(a, 0, c+1) = a,$$

$$ak(a, b+1, c+1) = ak(a, ak(a, b, c+1), c).$$

I shall *not set up general formal schemata* for these and other methods of forming concepts. It will turn out that even without such schemata these concepts may be incorporated wholesale in the consistency proof. The same holds for the 'axioms', about which I shall now say a few words.

6.2. In number-theoretical proofs we start from certain simple, immediately obvious propositions for which no further proof is offered. These are the 'axioms'. They are closely related to the *derived concepts* in so far as these axioms state basic facts about the predicates and functions occurring in them. Actually, a new concept may be formally introduced by merely stating a number of axioms about it ('implicit definition'). An example: The function (a, b) is completely characterizable by the axioms:

$\forall x \, \forall y \, [(x, y)|x \, \& \, (x, y)|y]$ and $\forall x \, \forall y \, \neg \exists z \, [(z|x \, \& \, z|y) \, \& \, z > (x, y)]$.

The choice of the axioms is *not uniquely determined*. Our aim might be to make do with as few simple axioms as possible[41]. For the purpose of carrying out number-theoretical proofs *in practice*, however, a larger number of axioms is usually stipulated without concern for redundancy, independence, etc. For my *consistency proof* it is fairly immaterial which axioms are chosen. As in the case of the derived concepts, I shall content myself, for the time being, with the statement of several *examples* from which it can be seen what kinds of proposition qualify as axioms:

Some axioms for the predicate = and the function +, formalized:

$\forall x \, (x = x)$ $\qquad\qquad\qquad \forall x \, \neg \, (x+1 = x)$

$\forall x \, \forall y \, (x = y \supset y = x)$ $\qquad\qquad \forall x \, \forall y \, (x+y = y+x)$

$\forall x \, \forall y \, \forall z \, [(x = y \, \& \, y = z) \supset x = z]$ $\quad \forall x \, \forall y \, \forall z \, [(x+y)+z = x+(y+z)]$.

6.3. *The concept of 'the ... such that'.*

The following special kind of construction is also worth mentioning: If a proposition of the form

$\forall \mathfrak{x}_1 \, \forall \mathfrak{x}_2 \ldots \forall \mathfrak{x}_\nu \, \exists \mathfrak{y} \, \{\mathfrak{F}(\mathfrak{x}_1, \mathfrak{x}_2, \ldots, \mathfrak{x}_\nu, \mathfrak{y})$
$\qquad\qquad\qquad \& \, \forall \mathfrak{z} \, [\mathfrak{F}(\mathfrak{x}_1, \mathfrak{x}_2, \ldots, \mathfrak{x}_\nu, \mathfrak{z}) \supset \mathfrak{z} = \mathfrak{y}]\}$,

in words possibly: 'For every combination of numbers $\mathfrak{x}_1, \ldots, \mathfrak{x}_\nu$ there exists one and only one number $\mathfrak{y}$ such that $\mathfrak{F}(\mathfrak{x}_1, \ldots, \mathfrak{x}_\nu, \mathfrak{y})$ holds', *has been proved*, then *a function may be introduced* which *represents* precisely this value ($\mathfrak{y}$) in its dependence on the combination of numbers ($\mathfrak{x}_1, \ldots, \mathfrak{x}_\nu$) ('the ... such that'). Formally: For this function one might use the expression (written for the arguments $a_1, \ldots, a_\nu$): $\iota_\mathfrak{y} \mathfrak{F}(a_1, \ldots, a_\nu, \mathfrak{y})$; for this expression the following then holds:

$$\forall \mathfrak{x}_1 \ldots \forall \mathfrak{x}_\nu \, \mathfrak{F}(\mathfrak{x}_1, \ldots, \mathfrak{x}_\nu, \iota_\mathfrak{y} \mathfrak{F}(\mathfrak{x}_1, \ldots, \mathfrak{x}_\nu, \mathfrak{y})).$$

The $\mathfrak{x}$'s may also be *empty*, in which case the ι-symbol represents a single *number*.

Such derived concepts which are *not* generally *needed* in practical elementary number theory, or which can be replaced by 'definitions' of the kind mentioned above (6.1), are immaterial for the question of *consistency* since they may always be *eliminated* from a derivation[42].

SECTION III. DISPUTABLE AND INDISPUTABLE FORMS OF INFERENCE IN ELEMENTARY NUMBER THEORY[43]

The task of the consistency proof will be (2.31) to justify the *disputable* forms of inference (including derived concepts and axioms) on the basis of *indisputable* inferences. For a proper understanding of my consistency proof for elementary number theory, which follows in section IV, we shall therefore have to examine precisely what forms of inference and other techniques of proof from elementary number theory are indeed *disputable*, and which others can be accepted as undoubtedly correct. An *unequivocal* delimitation is not possible (cf. 1.8); but we can certainly produce arguments which will make the admissibility of some methods of proof very *plausible*, whereas a corresponding justification fails for other methods in cases where there exists a remote analogy to the fallacies arising in the *antinomies* of set theory, and which make these techniques appear *disputable*.

We shall now develop such arguments by first considering the mathematical theory with a *finite* domain of objects (§ 7) and by then discussing the peculiarities and difficulties arising from the generalization to an *infinite* domain of objects (§ 8–11).

§ 7. Mathematics over finite domains of objects

7.1. The mathematical treatment of a *finite* domain of objects proceeds as follows:

The *objects* of the domain are *enumerated*; in doing so, each object receives a definite *designation* referring to no other object.

A *function* or a *predicate* is defined thus: Suppose the number of argument places is v. For every ordered v-tuple of objects, it is determined which object is the associated functional value or, in the case of predicates, whether the predicate does or does not hold for this combination of objects.

We could also permit functions and predicates to remain undefined for some combinations of objects; this constitutes an unimportant complication.

Since there are always only *finitely many* ordered v-tuples of objects, every function and every predicate may be *completely* described by such a 'definition table'.

7.2. *For every definite proposition* (3.24) which has been constructed in accordance with 3.22, 3.23, from the given objects, functions, and predicates together with the logical connectives, it can furthermore be *'calculated'* according to the following formal rule whether the proposition is *true* or *false*:

The proposition is represented by a formula without free variables. If it contains the symbol $\forall$, then the term $\forall \mathfrak{x}\, \mathfrak{F}(\mathfrak{x})$ concerned is replaced by $[\ldots [\mathfrak{F}(\mathfrak{g}_1)\, \&\, \mathfrak{F}(\mathfrak{g}_2)]\, \&\, \mathfrak{F}(\mathfrak{g}_3)]\, \&\, \ldots]\, \&\, \mathfrak{F}(\mathfrak{g}_\rho)]$, where $\mathfrak{g}_1, \ldots, \mathfrak{g}_\rho$ represents the entire collection of objects of the domain. The same is done for *every* $\forall$ that occurs, and each $\exists$ is replaced by a corresponding expression with $\vee$ instead of $\&$.

Then every *term* that occurs is 'evaluated' on the basis of the definition tables for the functions occurring in it, i.e., the term is replaced by the object symbol which represents its 'value'. If several function symbols are nested, then the calculation is carried out step by step working from the inside to the outside.

For each occurring *minimal formula* (3.24) we then determine on the basis of the definition table of the predicate concerned whether it represents a true or a false proposition. Then follows the determination of the truth or falsity of the subformulae built up by the various logical connectives; this is done step by step from the inside out according to the following instructions:

$\mathfrak{A}\, \&\, \mathfrak{B}$ is true if $\mathfrak{A}$ and $\mathfrak{B}$ are both true, otherwise false.

$\mathfrak{A} \vee \mathfrak{B}$ is true if $\mathfrak{A}$ is true and also if $\mathfrak{B}$ is true; it is false only if $\mathfrak{A}$ and $\mathfrak{B}$ are both false.

$\mathfrak{A} \supset \mathfrak{B}$ is false if $\mathfrak{A}$ is true and $\mathfrak{B}$ is false; in every other case $\mathfrak{A} \supset \mathfrak{B}$ is true.

$\neg\, \mathfrak{A}$ is true if $\mathfrak{A}$ is false, but false if $\mathfrak{A}$ is true.

The entire procedure follows at once from the actual *sense* which we associate with the formal symbols. For us it is important only to realize that in a theory with a *finite* domain of objects every well-defined proposition is *decidable*, i.e., that it can be determined by a definite procedure in finitely many steps whether the proposition is true or false.

7.3. It is easily proved that the *logical rules of inference* (5.2), applied to this theory, are correct in the sense that their application to 'true' basic mathematical sequents leads to 'true' derivable sequents. Here the concept of the *'truth' of a sequent* is to be determined formally in agreement with its informal sense as follows: a sequent without free variables is *false* if all antecedent formulae are true and the succedent formula is false; in every other case it is *true*. A sequent with *free variables* is *true* if *every arbitrary* replacement of object symbols yields a true sequent.

A verification of this statement would mean no more than a *confirmation* of the fact that we have indeed chosen our formal rules of inference in such a way that they are in *harmony* with the informal sense of the logical connectives.

7.4. It should still be noted that *in practice* the above method of introducing objects, functions, and predicates and of 'evaluating' the propositions is rarely used in mathematical theories with *finite* domains of objects; for a large number of objects this would become far too lengthy. In such cases the methods used are rather like those applied in the case of an *infinite* domain of objects described below.

§ 8. Decidable concepts and propositions over an infinite domain of objects

8.1. What becomes different if we wish to develop the theory with an *infinite* domain of objects such as the *natural numbers*, for example?

8.11. It is then no longer possible *to enumerate the objects explicitly* since there are *infinitely many* of them.

The place of such an enumeration is taken by a *construction rule* of the following kind: 1 designates a natural number. So does $1+1$, $1+1+1$, generally: From an expression representing a natural number an expression for another natural number is obtained by adjoining $+1$. (The symbols 2, 3, 4, etc. may be introduced afterwards as abbreviations for $1+1$, $1+1+1$, $1+1+1+1$, etc.; this is of secondary importance.)

This rule, which must be expressed in *finitely* many words, generates the *infinite* number sequence because it contains the *possibility* of continuing this constructive process through a repetitive *procedure*. ('Potential infinity'.)

8.12. Nor can *functions* and *predicates*, as in the case of a finite domain, be defined by an enumeration of all individual values. If we wanted to give a definition table for a number-theoretical function with one argument, for example, we would have to state successively its value for the arguments 1, 2, 3, 4, etc., hence for *infinitely many* values. This is impossible. Instead, we specify a *calculation rule*; e.g., for the function $2 \cdot a : 2 \cdot 1$ is 2; $2 \cdot (b+1)$ is equal to $(2 \cdot b)+2$. This rule *makes it possible* to calculate the associated functional values uniquely one by one for each natural number.

Generally, a function or a predicate is considered to be *decidably defined* if a *decision procedure* is given for it, i.e.: for every given enumeration of natural numbers it must be possible to *calculate* uniquely the associated *functional value* by means of this procedure or, in the case of predicates, it must be *decidable* uniquely whether the *predicate concerned holds or does not hold* for this collection of numbers.

For all examples of definitions of functions and predicates given at 6.1 such decision procedures can be stated. In the case of derived concepts formed according to 6.3 this may at times no longer be possible. By eliminating these derived concepts we have transferred the doubts associated with them to the *logical forms of inference*; these will be further discussed below (§§ 9–11).

8.2. Let us now consider the propositions in the theory over the infinite domain of natural numbers.

Of every given definite proposition in which the connectives '*all*' and '*there is*' do *not* occur it can be *decided*, as in the case of a finite domain, whether it is true or false. The procedure is the same as at 7.2. Instead of being determined by a *definition table*, the values of the terms of a proposition as well as the truth or falsity of the minimal formulae are now determined by the appropriate *decision rule* for the functions or predicates concerned.

The application of the *logical rules of inference* to propositions of this kind can also be shown to be admissible in the same way as in the case of a finite domain.

It should still be mentioned that a corresponding result also holds for propositions in which the connectives 'all' and 'there is' *refer only to finitely many numbers*. Such propositions can be decided in the way described, ∀ and ∃ must be replaced by & and ∨ as at 7.2, and the appropriate forms of inference, i.e., the ∀- and ∃-forms of inference (5.251) as well as complete induction (5.253) can also be shown to be admissible in the same way, as long as the domain of the – free and bound – variables that occur is limited to the numbers from 1 to a fixed number $\mathfrak{n}$.

§ 9. The 'actualist' interpretation of transfinite propositions[44]

9.1. Let us now turn to the essentially *transfinite propositions*, i.e., propositions in which the connectives 'all' or 'there is' refer to the totality of all natural numbers. Here we are confronted with a *fundamentally new state of affairs*.

First we must note that the *decision rule* which is applicable in the case of a finite domain (7.2, 8.2) does not carry over to such transfinite propositions.

In the case of a proposition about *all* natural numbers, for example, we would have to test *infinitely many* individual cases, which is impossible. No decision rule for arbitrary transfinite propositions is known and it is doubtful whether such a rule can ever be given. If there were such a rule, we could then decide, for example, by *calculation* whether 'Fermat's last theorem' (as well as Goldbach's conjecture, etc.) is true or false.

What *sense* then can be ascribed to a proposition whose truth cannot be *verified*?

9.2. The traditional view is this: it is *'actually'* predetermined whether a transfinite proposition such as Fermat's last theorem is 'true' or 'false' independently of whether we know or shall ever know which of the two is the case. Every transfinite proposition is thought of as having a definite *actual sense*; in particular, the sense of a ∀-proposition is thought to be this: 'For *every* single one of the infinitely many natural numbers the proposition concerned holds'; the sense of a ∃-proposition: 'In the infinite totality of the natural numbers there *somewhere* exists a number for which the proposition concerned holds'.

From this interpretation it is inferred further that for *transfinite* propositions the same *logical forms of inference* are valid as for the finite case, since the 'actualist' sense of the logical connectives in transfinite propositions corresponds exactly to that in the finite case.

9.3. At this point there now exists ample cause for *criticism*, as long as we have decided to draw the *utmost consequences* from the insights gained in considering the *antinomies* of set theory. This I will now do and shall, as a result of a critical examination of Russell's antinomy (1.6), lay down the following *principle*:

An infinite totality must not be regarded as actually existing and closed (actual infinity), but only as something becoming which can be extended constructively further and further from something finite (potential infinity).

9.4. The constructive methods for the *introduction of objects, functions, and predicates* stated in § 8 are in line with this principle. They were explicitly based on the idea of a gradual *progression* in the number sequence, starting at the beginning, and not on the idea of a *completed* totality of all natural numbers. The same holds true for the propositions discussed at 8.2, since they also refer to only *finitely* many objects and not yet to an *infinite* totality.

9.5. The 'actualist' interpretation of *transfinite propositions* described at 9.2, however, is *no longer compatible* with this principle, for it is based on the idea of the closed infinite number sequence.

At the same time, the view that the logical *forms of inference* can simply be transferred from finite to infinite domains of objects must be *rejected*.

I remind the reader of a *similar* although more trivial case of an inadmissible generalization from the finite to the infinite, viz., the well-known fallacy: 'Every (finite) set of natural numbers contains a largest number; hence the (infinite) set of all natural numbers contains a largest number.' This argument leads to contradictions since it does not in fact hold true.

9.6. Having rejected the actualist interpretation of transfinite propositions, we are still left with the possibility of *ascribing a 'finitist' sense to such* *propositions, i.e., of interpreting* them in each case as expressions for definite *finitely* characterizable states of affairs.

Once this view has been adopted, the relevant *logical forms of inference* must be examined for their compatibility with this interpretation of the *propositions*.

Such an examination will be carried out in § 10 below for an extensive portion of the transfinite propositions and their associated forms of inference. In § 11, I shall discuss the remaining propositional forms and their forms of inference; there our method will meet with difficulties and the significance of the *intuitionist* (1.8) *delimitation* between permissible and nonpermissible forms of inference within number theory will become apparent; another still *stricter delimitation* will also turn out to be defensible.

§ 10. Finitist interpretation of the connectives ∀, &, ∃ and ∨ in transfinite propositions

We start with a number theory whose propositions refer to only *finitely* many numbers. Then we adjoin step by step certain types of *transfinite* propositions.

10.1. The ∀-*connective*.

10.11. We shall begin with the simplest form of a transfinite proposition: $\forall \mathfrak{x}\, \mathfrak{F}(\mathfrak{x})$, where $\mathfrak{F}$ shall not yet contain a ∀ or ∃, so that the truth of $\mathfrak{F}(\mathfrak{x})$ is *verifiable* for each individual number substituted for $\mathfrak{x}$ (8.2).

True propositions of this form are, for example:

$$\forall x\, (2|x \vee \neg\, 2|x); \qquad \forall x\, (x = x).$$

Such propositions will undoubtedly be regarded as significant ((*sinnvoll*)) and true. After all, we need not associate the idea of a *closed* infinite number of individual propositions with this ∀, but can, rather, *interpret* its sense '*finitistically*' as follows: 'If, starting with 1, we substitute for $\mathfrak{x}$ successive natural numbers then, however far we may progress in the formation of numbers, a true proposition results in each case.'

10.12. This interpretation may be generalized to the case where $\mathfrak{F}$ is an arbitrary proposition to which a finitist sense has already been ascribed: $\forall \mathfrak{x}\, \mathfrak{F}(\mathfrak{x})$ may be significantly asserted if $\mathfrak{F}(\mathfrak{x})$ represents a significant and true proposition for arbitrary successive replacements of $\mathfrak{x}$ by numbers.

10.13. The *forms of inference* associated with the ∀-connective, the ∀-introduction, and the ∀-elimination (5.251), are in harmony with this interpretation: A ∀ is *introduced* if a proof is available that on the basis of certain assumptions (Γ) – *transfinite* assumptions are totally without sense at this point and ruled out for the time being – $\mathfrak{F}(\mathfrak{a})$ is true, and from this is inferred that on the basis of the same assumptions $\forall \mathfrak{x}\, \mathfrak{F}(\mathfrak{x})$ holds. This is in order, for if an arbitrary number $\mathfrak{n}$ is given, then it may be substituted for $\mathfrak{a}$ – in the whole proof – and a proof for $\mathfrak{F}(\mathfrak{n})$ results (under the same assumptions Γ which, by virtue of the restriction on variables for the $\forall$-introduction, do not contain $\mathfrak{a}$ and have thus obviously remained unaffected by this substitution). In the case of the $\forall$-elimination, $\Gamma \to \mathfrak{F}(\mathfrak{t})$ is deduced from $\Gamma \to \forall \mathfrak{x}\, \mathfrak{F}(\mathfrak{x})$. Once possible occurrences of free variables have been replaced by numbers, the term $\mathfrak{t}$ represents a definite number $\mathfrak{n}$; in keeping with its finitist sense the proposition $\forall \mathfrak{x}\, \mathfrak{F}(\mathfrak{x})$ also guarantees that $\mathfrak{F}(\mathfrak{n})$ holds; hence this form of inference is also acceptable.

10.14. The usual *number-theoretical axioms* may be formulated in such a way that they follow from propositions without $\forall$ or $\exists$ by a number of $\forall$-inferences ranging over the entire proposition (cf. 6.2). The conclusion that, in terms of the finitist interpretation of the $\forall$, and on the basis of the decidable definitions of the functions and predicates occurring in them, these axioms are *true* is of such *self-evidence* that it requires no further investigation.

It seems hardly possible that this conclusion could be reduced to something basically simpler.

10.2. The *&-connective*.

A transfinite proposition of the form $\mathfrak{A}\, \&\, \mathfrak{B}$ is significant and may be asserted if $\mathfrak{A}$ and $\mathfrak{B}$ have already been recognized as significant and valid propositions. The rules for the &-introduction and &-elimination are obviously in harmony with this interpretation. Here, as above, transfinite assumptions (Γ, Δ) are excluded for the time being.

10.3. The *$\exists$-connective*.

The reader may so far have the impression that the 'finitist interpretation' attributes to transfinite propositions really only *the same* sense as that usually associated with such propositions. That this is not the case emerges from the following discussion of the $\exists$ and $\vee$ (cf. 10.6).

What sense should we concede to a proposition of the form $\exists \mathfrak{x}\, \mathfrak{F}(\mathfrak{x})$? The *actualist* interpretation that *somewhere* in the infinite number sequence there exists a number with the property $\mathfrak{F}$ is for us without sense. If, on the other hand, the proposition $\mathfrak{F}(\mathfrak{n})$ has been recognized as significant and valid for a definite number $\mathfrak{n}$, we wish to be able to conclude ($\exists$-introduction): $\exists \mathfrak{x}\, \mathfrak{F}(\mathfrak{x})$. There are no objections to this; the proposition $\exists \mathfrak{x}\, \mathfrak{F}(\mathfrak{x})$ now constitutes only a *weakening* of the proposition $\mathfrak{F}(\mathfrak{n})$ ('Partialaussage'

§ 10, FINIST INTERPRETATION OF ∀, &, ∃ AND ∨ IN TRANSFINITE PROPOSITIONS

for Hilbert, 'Urteilsabstrakt' for Weyl) in that it now attests merely that we have found a number $\mathfrak{n}$ with the property $\mathfrak{F}$, although this number itself is no longer mentioned. Thus, $\exists \mathfrak{x}\, \mathfrak{F}(\mathfrak{x})$ acquires in this way a finitist sense.

If, instead of being introduced in $\mathfrak{F}(\mathfrak{n})$, the ∃ is introduced in a proposition $\mathfrak{F}(\mathfrak{t})$ containing an *arbitrary term* $\mathfrak{t}$, then nothing has essentially changed. For if the occurring free variables are replaced by definite numbers (which is after all what free variables stand for) then, by virtue of the decidable definitions of functions, $\mathfrak{t}$ becomes a definite calculable number $\mathfrak{n}$. If a ∃-introduction is accompanied by the occurrence of nontransfinite assumptions (Γ) the situation has not essentially changed.

How can we infer further propositions from a proved proposition of the form $\exists \mathfrak{x}\, \mathfrak{F}(\mathfrak{x})$ by the *elimination of* ∃ on the basis of the *finitist sense* of that proposition? In contrast with the situation in the case of ∀ and &, it is obviously not possible, *to reclaim* the proposition $\mathfrak{F}(\mathfrak{n})$ from $\exists \mathfrak{x}\, \mathfrak{F}(\mathfrak{x})$, which had provided the justification for the assertion of $\exists \mathfrak{x}\, \mathfrak{F}(\mathfrak{x})$, precisely because the value of $\mathfrak{n}$ is no longer apparent from $\exists \mathfrak{x}\, \mathfrak{F}(\mathfrak{x})$. We can nevertheless proceed as follows: we conclude $\mathfrak{F}(\mathfrak{a})$, where $\mathfrak{a}$ is a free variable taking the place of the number $\mathfrak{n}$ whose value need not be known at this time. If we then succeed in deducing from $\mathfrak{F}(\mathfrak{a})$ a certain proposition $\mathfrak{C}$ no longer containing $\mathfrak{a}$, then this proposition holds. We have thus a ∃-elimination in accordance with 5.251.

This is the first rule, so far, in which an *associated assumption*, viz., $\mathfrak{F}(\mathfrak{a})$, occurs. This assumption may be *transfinite*. Although we have previously not granted a sense to transfinite propositions as *assumptions* but only as *proved* propositions, we can here say: the fact that $\exists \mathfrak{x}\, (\mathfrak{F}\mathfrak{x})$ has been proved and is significant, means that a number $\mathfrak{n}$ must have been known and is reconstructible on the basis of the proof of $\exists \mathfrak{x}\, (\mathfrak{F}\mathfrak{x})$, so that $\mathfrak{F}(\mathfrak{n})$ also represents a significant true proposition. Here the assumption $\mathfrak{F}(\mathfrak{a})$ is not *regarded* as an arbitrary assumption but as the true proposition $\mathfrak{F}(\mathfrak{n})$, where $\mathfrak{a}$ merely denotes the number $\mathfrak{n}$. The proof of $\mathfrak{C}$ from the assumption $\mathfrak{F}(\mathfrak{a})$ thus no longer appears as hypothetical, but as an ordinary *direct* proof; and precisely this is its sense.

10.4. The ∨-connective can easily be handled analogously to ∃, just as & was handled analogously to ∀: a transfinite proposition of the form $\mathfrak{A} \vee \mathfrak{B}$ is significant and may be asserted if one of the propositions $\mathfrak{A}$ or $\mathfrak{B}$ *has been recognized* as significant and valid. The rule of the ∨-*introduction* corresponds completely to this interpretation. A ∨-*elimination* is carried out thus: if $\mathfrak{A} \vee \mathfrak{B}$ is given and if the same proposition $\mathfrak{C}$ follows from the assumption $\mathfrak{A}$ as well as from the assumption $\mathfrak{B}$, then $\mathfrak{C}$ holds. This is in order since

$\mathfrak{A} \vee \mathfrak{B}$ entails that either $\mathfrak{A}$ or $\mathfrak{B}$ has at some point been recognized as valid. In this way a proof for $\mathfrak{C}$ from $\mathfrak{A}$, or a proof for $\mathfrak{C}$ from $\mathfrak{B}$, can be made independent of the assumption $\mathfrak{A}$, or $\mathfrak{B}$, as was done in the case of the ∃-elimination, and we obtain a direct proof. The second proof becomes redundant and it is thus immaterial whether it has a *sense* or not.

10.5. At this point it should be explained briefly how the rule of *complete induction* is immediately compatible with the finitist interpretation: Suppose that $\mathfrak{F}(1)$ is a significant valid proposition. The term $\mathfrak{t}$ in the conclusion $\mathfrak{F}(\mathfrak{t})$ represents a definite number $\mathfrak{n}$, once possible occurrences of free variables have been replaced by numbers. By replacing $\mathfrak{a}$ successively by the numbers 1, 2, 3, up to $\mathfrak{n}-1$, in the proof of $\mathfrak{F}(\mathfrak{a}+1)$ from $\mathfrak{F}(\mathfrak{a})$, we have formed a *direct proof*, starting from the valid proposition $\mathfrak{F}(1)$ via $\mathfrak{F}(2)$, $\mathfrak{F}(3)$, etc. up to $\mathfrak{F}(\mathfrak{n})$, so that finally, $\mathfrak{F}(\mathfrak{n})$ is now a valid, significant proposition.

This may sound trivial; what is essential is that the assumption $\mathfrak{F}(\mathfrak{a})$, *which may have been without sense* (if it was transfinite) *has been afforded a sense* by the possibility of transforming the relevant portion of the proof into a direct proof in which $\mathfrak{F}(\mathfrak{a})$ *no longer functions as an assumption*.

10.6. The finitist interpretation given to the connectives ∨ and ∃ *differs* from the actualist interpretation not only *conceptually* but also *in its practical consequences,* as the following *examples* show:

According to the actualist interpretation the proposition 'Fermat's last theorem is either true or not true' is *true*. According to the finitist interpretation of ∨, however, this proposition *cannot* be asserted. For, this would require that *one* of the two propositions has already been established as valid. But up to now this has not been done.

A corresponding example containing a ∃ is the proposition

$$\exists x \{[\forall y \, \forall z \, \forall u \, \forall v \, (v > 2 \supset y^v + z^v \neq u^v)]$$
$$\vee \, [\exists y \, \exists z \, \exists u \, (x > 2 \, \& \, y^x + z^x = u^x)]\},$$

in words, possibly: 'There exists a number x so that either Fermat's theorem is true or there exists a counterexample with the exponent x'. According to the actualist interpretation this proposition is true, but according to the finitist interpretation of the ∃, it may not be asserted since at present no such number is known.

Consequently, neither of these two propositions is *provable* by the *forms of inference* discussed *so far,* since it was possible to ascribe a finitist sense to these forms of inference; the additional forms of inference relating to ¬ are needed for this purpose (cf. 11.2).

10.7. The finitist interpretation of transfinite propositions containing the connectives ∀, &, ∃ and ∨, which has been attempted in these paragraphs, and the justification of the associated forms of inference is in many respects *incomplete*; the meaning of propositions in which a number of such connectives occur in *nested* form, in particular, still needs to be discussed in greater detail. I shall not do this, since I am here concerned only with examining *fundamentals*.

A purely formal *consistency proof* for this part of number theory could be developed later on the basis of these considerations. Such a proof would be of little value, however, since it itself would have to *make use* of transfinite propositions and the same associated forms of inference which it is intended to *'justify'*. Such a proof would therefore not represent *an appeal to more elementary facts*, although it would of course *confirm* the finitist character of the formalized rules of inference. We would, however, have to have a clear idea *beforehand* of *what* can be considered finitist (in order to be able to carry out the consistency proof proper with finitist methods of proof).

§ 11. The connectives ⊃ and ¬ in transfinite propositions: the intuitionist view

11.1. The ⊃-*connective*.

We now intend to include transfinite propositions containing the connective ⊃.

What does $\mathfrak{A} \supset \mathfrak{B}$ mean? Suppose, for example, that there exists a *proof* in which the proposition $\mathfrak{B}$ is proved on the basis of the assumption $\mathfrak{A}$ by means of inferences that have already been recognized as permissible. From this we infer, by ⊃-introduction: $\mathfrak{A} \supset \mathfrak{B}$. This proposition is merely intended to express the fact that *a proof is available* which permits a proof of the proposition $\mathfrak{B}$ from the proposition $\mathfrak{A}$, once the proposition $\mathfrak{A}$ is proved. The ⊃-*elimination* is in harmony with this interpretation: here $\mathfrak{B}$ is inferred from $\mathfrak{A}$ and $\mathfrak{A} \supset \mathfrak{B}$; this is in order, since $\mathfrak{A} \supset \mathfrak{B}$ indicates precisely the existence of a proof for $\mathfrak{B}$ in the case where $\mathfrak{A}$ is already proved.

In interpreting $\mathfrak{A} \supset \mathfrak{B}$ in this way, I have presupposed that the available proof of $\mathfrak{B}$ from the assumption $\mathfrak{A}$ contains merely inferences *already recognized as permissible*. On the other hand, such a proof may itself contain other ⊃-*inferences* and then our interpretation *breaks down*. For, it is *circular* to justify the ⊃-inferences on the basis of a ⊃-interpretation which itself already involves the presupposition of the admissibility of *the same* form of inference. The ⊃-inferences *which occur in the proof* would in that case have to be justified *beforehand*; but this has its difficulties, especially if the assumption $\mathfrak{A}$ has itself the form $\mathfrak{C} \supset \mathfrak{D}$; if this happens, we have actually no proof for $\mathfrak{D}$ from $\mathfrak{C}$ on the basis of which we could ascribe a *sense* to $\mathfrak{C} \supset \mathfrak{D}$.

In order to cope with this difficulty, we would really have to formulate a more complicated rule of interpretation. This represents one of the *principal objectives* of the *consistency proof* which follows in section IV.

11.2. *The $\neg$-connective* presents even greater obstacles to a finitist interpretation than the $\supset$. Transfinite propositions were actually always interpreted in such a way that they could in each case be regarded as something that had previously been recognized as valid. In its actualist interpretation, $\neg \mathfrak{A}$ does not however express the fact that something *holds*, but rather purely negatively that something, viz., the proposition $\mathfrak{A}$, *does not hold*.

The following *positive* interpretation seems nonetheless possible: $\neg \mathfrak{A}$ is to be regarded as significant and true if a proof exists to the effect that from the assumption of the validity of $\mathfrak{A}$ a falsehood is certain to follow. And here the $\neg$-connective is reinterpreted in terms of the $\supset$-connective, since $\neg \mathfrak{A}$ can certainly be defined as equivalent with $\mathfrak{A} \supset 1 = 2$. The '*reductio*' is in *harmony* with this interpretation, as may be shown quite formally: From $\mathfrak{A}, \Gamma \to \mathfrak{B}$ and $\mathfrak{A}, \Delta \to \mathfrak{B} \supset 1 = 2$ we wish to derive $\Gamma, \Delta \to \mathfrak{A} \supset 1 = 2$. This is done as follows: By $\supset$-elimination we obtain $\mathfrak{A}, \Gamma, \mathfrak{A}, \Delta \to 1 = 2$, hence (5.242) $\mathfrak{A}, \Gamma, \Delta \to 1 = 2$, and from this, by $\supset$-introduction, $\Gamma, \Delta \to \mathfrak{A} \supset 1 = 2$. This completes the reduction of the '*reductio*' to the $\supset$-forms of inference.

It should be noted that in this reinterpretation of the $\neg$ in terms of the $\supset$, all *doubts* associated with the $\supset$ naturally carry over to the $\neg$-connective to a corresponding degree.

Now there actually arises a further difficulty: The '*elimination of the double negation*' cannot at all be shown to agree with the given $\neg$-interpretation. There is no compelling reason why the validity of

$$(\mathfrak{A} \supset 1 = 2) \supset 1 = 2$$

should follow from the validity of $\mathfrak{A}$.

This form of inference *conflicts* in fact quite *categorically* with the remaining forms of inference. In the case of the logical connectives $\forall, \&, \exists, \vee$ and $\supset$ we had in each case an *introduction* and an *elimination inference* corresponding to each other in a certain way. (Cf. the discussion in § 10 and in 11.1.) In the case of the $\neg$-connective, the '*reductio*' can be regarded both as an introduction (of $\neg$ in $\neg \mathfrak{A}$) and an elimination (of $\neg$ in $\neg \mathfrak{B}$);

the 'elimination of the double negation', however, represents an *additional* ¬-elimination which does not correspond to the ¬-introduction by 'reductio'. Double negation renders possible *indirect proofs of positive propositions* ($\mathfrak{A}$) from their denials by means of contradiction, in cases where a positive proof of the same proposition may be completely inaccessible. In this way we can, for example, prove the two propositions containing ∨ and ∃ given as examples at 10.6, whereas under the finitist interpretation of ∨ and ∃ these propositions may not even be asserted.

From this it follows that there is no way at all of including the 'elimination of the double negation' in a finitist interpretation of the kind given for ∨ and ∃.

11.3. *Here the intuitionists draw the line in number theory* by disallowing the 'elimination of the double negation' in the case of *transfinite* propositions $\mathfrak{A}$. This delimitation is often also effected by disallowing the 'law of the excluded middle', $\mathfrak{A} \vee \neg \mathfrak{A}$, in the case of a transfinite $\mathfrak{A}$; this comes to the same thing[45].

The *'finitist'* interpretation' of the logical connectives ∀, &, ∃ and ∨ in transfinite propositions described in § 10 agrees essentially with the interpretation of the *intuitionists*. Yet they allow *a more general use* of the ⊃-connective; the ¬-connective is interpreted as at 11.2 by reducing it to ⊃, and to this corresponds the expression '$\mathfrak{A}$ is *absurd*' in place of '$\mathfrak{A}$ does not hold' for ¬ $\mathfrak{A}$.

The 'elimination of the double negation' undoubtedly stands in definite contrast to the remaining forms of inference to such a degree that it might quite reasonably be disallowed. In fact, I consider a still *more radical critique*, especially of the general use of the ⊃ (11.1), as equally well justified.

A theorem by Gödel about the equivalence of intuitionist number theory and elementary number theory as a whole.

As was first proved by K. Gödel[46], it is possible to *eliminate* the *'elimination of the double negation'* with a *transfinite* $\mathfrak{A}$ from any given elementary number-theoretical proof by a special interpretation of transfinite propositions, so that every proof of this kind becomes intuitionistically acceptable.

In this way, *the whole of* actualist number theory becomes reduced to *intuitionist* number theory. In particular, the former is *consistent* if the latter is.

The *interpretation involved* takes the following form: the logical connectives &, ∀, ⊃ and ¬ are ascribed their *intuitionist* sense. Not so for ∨ and ∃;

$\mathfrak{A} \vee \mathfrak{B}$ is interpreted as $\neg((\neg \mathfrak{A}) \& \neg \mathfrak{B})$, $\exists \mathfrak{x} \mathfrak{F}(\mathfrak{x})$ as $\neg \forall \mathfrak{x} \neg \mathfrak{F}(\mathfrak{x})$. The reason for this interpretation is that the $\vee$ and $\exists$ cannot here be given their intuitionist meaning, since the examples of propositions stated at 10.6 are provable in actualist number theory, but not in intuitionist number theory. If $\vee$ and $\exists$ in these examples are replaced by $\&$, $\forall$ and $\neg$ in the way described, then propositions result which are also *intuitionistically* provable.

In *my consistency proof* the 'elimination of the double negation' actually presents no essential difficulties (13.93).

11.4. The forms of inference which we have not been able to justify so far by means of a finitist interpretation, and which are therefore disputable for the time being, *occur very rarely in proofs carried out in practical number theory*. It follows from our discussion that such inferences are principally the 'elimination of the double negation' (and the 'law of the excluded middle') applied to transfinite propositions, as well as the use of transfinite propositions containing nested $\supset$- and $\neg$-connectives.

Transfinite propositions of a more complicated structure hardly ever occur in practice. In *Euclid's proof* presented in § 4, for example, the only essentially transfinite propositions are the two propositions occurring at the end: $\exists z$ (Prime $z \& z > a$) and $\forall y \exists z$ (Prime $z \& z > y$). The whole proof is entirely finitist. The other transfinite propositions which occur in it, i.e., those containing $\forall$ or $\exists$, are such that their bound variables range only over a *finite* segment of the number sequence.

As an example of a more difficult proof I have looked through Rev. Zeller's proof of the 'law of quadratic reciprocity'[47], and here I have also been unable to find a 'disputable inference'.

We are indeed justified in having the *impression of an unquestionable correctness* in the case of this and similar proofs. In these proofs we tend automatically to look more for a *finitist* than an *actualist* interpretation of transfinite propositions.

The task of the *consistency proof* for *elementary number theory* is thus more that of justifying *theoretically possible* rather than actually *occurring* inferences.

SECTION IV. THE CONSISTENCY PROOF

I shall now *prove* the consistency of elementary number theory as a whole as formalized in section II.

In carrying out this consistency proof we must make certain, as was pointed out in 2.31, that the inferences and derived concepts used in the proof *itself* are *indisputable* or at least considerably more reliable than the doubtful forms of inference of elementary number theory. It follows from our discussion in section III that this requirement can be regarded as met if the techniques of proof used are *'finitist'* (in the sense of §§ 9–11). The extent of our success in this direction will be examined more closely in section V (16.1).

§§ 13–15 contain the *core* of the consistency proof, whereas § 12 is concerned with some relatively simple *preliminaries*.

§ 12. The elimination of the symbols ∨, ∃, and ⊃ from a given derivation

Suppose a number-theoretical derivation (5.22) as given. It is to be shown that it is consistent, i.e., that its endformula cannot have the form $\mathfrak{A} \,\&\, \neg \mathfrak{A}$.

We begin by stating a rule for the *transformation of the given derivation*. As a result of this transformation, the connectives ∨, ∃ and ⊃ *will no longer occur* in the derivation.

12.1. In actualist logic, which we are in effect dealing with in unrestricted number theory, the different *logical connectives* can be *represented by other connectives* in various ways. By means of *three* connectives, viz., ¬, any one of the three connectives &, ∨ and ⊃, as well as any one of the two connectives ∀ and ∃, all others may be expressed. I shall make use of this fact to facilitate the consistency proof and shall retain the symbols &, ∀ and ¬ and express ∨, ∃ and ⊃ in terms of these.

This does not mean that the ambiguities (11.1) associated with the ⊃ are thus conjured away, they stay with us in an equivalent form in the ¬.

The replacement takes the form:

For $\mathfrak{A} \vee \mathfrak{B}$ we put $\neg\,((\neg\,\mathfrak{A}) \,\&\, \neg\,\mathfrak{B})$.
For $\mathfrak{A} \supset \mathfrak{B}$ we put $\neg\,(\mathfrak{A} \,\&\, \neg\,\mathfrak{B})$.
For $\exists \mathfrak{x}\, \mathfrak{F}(\mathfrak{x})$ we put $\neg\,\forall \mathfrak{x}\, \neg\,\mathfrak{F}(\mathfrak{x})$.

All ∨-, ∃- and ⊃-symbols occurring in the derivation are replaced in this way. The *order* in which this is done is obviously immaterial.

12.2. We must now examine to what extent the given *derivation* has *remained correct* after these replacements and, where this is not the case, *modify* the derivation accordingly. That such a modification is possible is very plausible since the new formulations for the ∨, ∃ and ⊃ are indeed *equivalent* to the original ones in the actualist interpretation. The precise formal verification is therefore not difficult:

Basic logical sequents (5.23) have become other basic logical sequents.

The same holds true for *basic mathematical sequents*, as long as we presuppose that a mathematical axiom in which the $\vee$-, $\exists$- and $\supset$-connectives occur, becomes another mathematical axiom after the replacement of these connectives by $\neg$, & and $\forall$. This requirement is easily met: the simplest way is to formulate all axioms from the very beginning without $\vee$, $\exists$, and $\supset$.

Structural transformations (5.24) and application instances of the *rules of inference* (5.25) have obviously remained correct, as long as we are not dealing with one of the *rules involving the connectives* $\vee$, $\exists$ and $\supset$. The latter rules must be replaced by applications of *other* rules of inference in accordance with the following instructions:

A $\vee$-*introduction*: 'From $\Gamma \to \mathfrak{A}$ follows $\Gamma \to \mathfrak{A} \vee \mathfrak{B}$', after the replacement takes the form: 'From $\Gamma^* \to \mathfrak{A}^*$ follows $\Gamma^* \to \neg((\neg \mathfrak{A}^*) \& \neg \mathfrak{B}^*)$'. $\mathfrak{A}^*$ designates the formula which has resulted from $\mathfrak{A}$ by the replacement; $\mathfrak{B}^*$ and Γ^* are to be understood in the same way.

In words, the new version by means of the forms of inference for & and $\neg$ reads as follows: $\mathfrak{A}^*$ holds on the assumptions Γ^*. If $(\neg \mathfrak{A}^*) \& \neg \mathfrak{B}^*$ were to hold, then so would $\neg \mathfrak{A}^*$, in particular, and this cannot be the case since it contradicts $\mathfrak{A}^*$, i.e., $\neg((\neg \mathfrak{A}^*) \& \neg \mathfrak{B}^*)$ holds on the assumptions Γ^*.

To this corresponds the following *formal* instruction: the appropriate place in the derivation is to be transformed thus:

$$(\neg \mathfrak{A}^*) \& \neg \mathfrak{B}^* \to (\neg \mathfrak{A}^*) \& \neg \mathfrak{B}^*$$

is a basic sequent; by &-elimination we obtain $(\neg \mathfrak{A}^*) \& \neg \mathfrak{B}^* \to \neg \mathfrak{A}^*$, this together with the sequent $(\neg \mathfrak{A}^*) \& \neg \mathfrak{B}^*, \Gamma^* \to \mathfrak{A}^*$, obtained from $\Gamma^* \to \mathfrak{A}^*$ by means of 5.243, by 'reductio' yields $\Gamma^* \to \neg((\neg \mathfrak{A}^*) \& \neg \mathfrak{B}^*)$.

The other form of the $\vee$-introduction is dealt with in the same way.

A $\vee$-*elimination* has the following form after the replacement:

'From $\Gamma^* \to \neg((\neg \mathfrak{A}^*) \& \neg \mathfrak{B}^*)$ and $\mathfrak{A}^*, \Delta^* \to \mathfrak{C}^*$ and $\mathfrak{B}^*, \Theta^* \to \mathfrak{C}^*$ follows $\Gamma^*, \Delta^*, \Theta^* \to \mathfrak{C}^*$'.

This is transformed thus: $\neg \mathfrak{C}^* \to \neg \mathfrak{C}^*$ yields $\mathfrak{A}^*, \neg \mathfrak{C}^* \to \neg \mathfrak{C}^*$, this together with $\mathfrak{A}^*, \Delta^* \to \mathfrak{C}^*$, by 'reductio' yields $\Delta^*, \neg \mathfrak{C}^* \to \neg \mathfrak{A}^*$; similarly $\mathfrak{B}^*, \neg \mathfrak{C}^* \to \neg \mathfrak{C}^*$, together with $\mathfrak{B}^*, \Theta^* \to \mathfrak{C}^*$, yields the sequent $\Theta^*, \neg \mathfrak{C}^* \to \neg \mathfrak{B}^*$; taking both results together, we obtain $\Delta^*, \neg \mathfrak{C}^*, \Theta^*, \neg \mathfrak{C}^* \to (\neg \mathfrak{A}^*) \& \neg \mathfrak{B}^*$ by &-introduction, hence (5.242, 5.241) $\neg \mathfrak{C}^*, \Delta^*, \Theta^* \to (\neg \mathfrak{A}^*) \& \neg \mathfrak{B}^*$; from $\Gamma^* \to \neg((\neg \mathfrak{A}^*) \& \neg \mathfrak{B}^*)$ follows $\neg \mathfrak{C}^*, \Gamma^* \to \neg((\neg \mathfrak{A}^*) \& \neg \mathfrak{B}^*)$, thus, by 'reductio', we obtain

$\varDelta^*, \Theta^*, \varGamma^* \to \neg \neg \mathfrak{C}^*$, and finally, by 'elimination of the double negation', $\varDelta^*, \Theta^*, \varGamma^* \to \mathfrak{C}^*$, hence (5.241) $\varGamma^*, \varDelta^*, \Theta^* \to \mathfrak{C}^*$.

A ∃-*introduction* or ∃-*elimination* is dealt with analogously to the ∨-introduction or ∨-elimination; a ∀-elimination takes the place of a &-elimination, and a ∀-introduction the place of a &-introduction in the appropriate place of the derivation. The details are straightforward.

A ⊃-*introduction*, after the replacement, takes the form:

'From $\mathfrak{A}^*, \varGamma^* \to \mathfrak{B}^*$ follows $\varGamma^* \to \neg (\mathfrak{A}^* \& \neg \mathfrak{B}^*)$'. This is transformed thus: $\mathfrak{A}^* \& \mathfrak{B}^* \to \mathfrak{A}^* \& \neg \mathfrak{B}^*$ yields $\mathfrak{A}^* \& \neg \mathfrak{B}^* \to \mathfrak{A}^*$ as well as $\mathfrak{A}^* \& \neg \mathfrak{B}^* \to \neg \mathfrak{B}^*$, hence also $\mathfrak{A}^*, \mathfrak{A}^* \& \neg \mathfrak{B}^* \to \neg \mathfrak{B}^*$; this, together with $\mathfrak{A}^*, \varGamma^* \to \mathfrak{B}^*$, yields $\varGamma^*, \mathfrak{A}^* \& \neg \mathfrak{B}^* \to \neg \mathfrak{A}^*$, hence $\mathfrak{A}^* \& \neg \mathfrak{B}^*, \varGamma^* \to \neg \mathfrak{A}^*$. By including $\mathfrak{A}^* \& \neg \mathfrak{B}^* \to \mathfrak{A}^*$, we obtain $\varGamma^* \to \neg (\mathfrak{A}^* \& \neg \mathfrak{B}^*)$.

A ⊃-*elimination*, after the replacement, takes the form:

'From $\varGamma^* \to \mathfrak{A}^*$ and $\varDelta^* \to \neg (\mathfrak{A}^* \& \neg \mathfrak{B}^*)$ follows $\varGamma^*, \varDelta^* \to \mathfrak{B}^*$'.

This is transformed thus: $\varGamma^* \to \mathfrak{A}^*$ and $\neg \mathfrak{B}^* \to \neg \mathfrak{B}^*$ yield $\varGamma^*, \neg \mathfrak{B}^* \to \mathfrak{A}^* \& \neg \mathfrak{B}^*$, hence $\neg \mathfrak{B}^*, \varGamma^* \to \mathfrak{A}^* \& \neg \mathfrak{B}^*$; by including $\neg \mathfrak{B}^*, \varDelta^* \to \neg (\mathfrak{A}^* \& \neg \mathfrak{B}^*)$, we obtain $\varGamma^*, \varDelta^* \to \neg \neg \mathfrak{B}^*$ and from this $\varGamma^*, \varDelta^* \to \mathfrak{B}^*$.

12.3. We have thus succeeded in transforming the given derivation into a derivation in which the symbols ∨, ∃ and ⊃ *no longer occur*. It should be observed that the *endformula* of the derivation has undergone a change only if it contained a ∨, ∃ or ⊃.

12.4. It is worth noting that according to what was said at 11.3, the given derivation is now already essentially an *intuitionistically* admissible number-theoretical derivation; for wherever the 'elimination of the double negation' is still used it could be replaced by other rules of inference.

§ 13. The reduction of sequents

The concept of the '*stability of a reduction rule*' for a sequent, to be defined below, will serve as the formal replacement of the informal *concept of truth*; it provides us with a *special finitist interpretation* of propositions and takes the place of their *actualist interpretation* (cf. §§ 9–11).

In a sequent in which the connectives ∨, ∃ and ⊃ no longer occur, an individual *reduction step* can be carried out in the following way (13.11 to 13.53):

13.11. Suppose that the sequent contains at least one free *variable*. In that case we replace every occurrence of this free variable by one and the same arbitrarily chosen *numeral*.

13.12. Suppose that the sequent contains no free variables and that somewhere in one of its formulae a *minimal term* (3.24) occurs (e.g., as part of a longer term). In that case we replace the minimal term by its *associated 'functional value'*, i.e., by that number which, by virtue of the definition of the function concerned (cf. 8.12), represents the value of the term for the given numbers taken as arguments.

Thus I am now assuming of the *functions* that they are *decidably defined* in the sense of 8.12.

13.21. Suppose that the sequent contains no free variables and no minimal terms and that its *succedent formula* (5.21) has the form $\forall \mathfrak{x}\, \mathfrak{F}(\mathfrak{x})$. In that case we replace it by a formula $\mathfrak{F}(\mathfrak{n})$, i.e., by a formula which results from $\mathfrak{F}(\mathfrak{x})$ by the substitution of an arbitrarily *chosen numeral* $\mathfrak{n}$ for the variable $\mathfrak{x}$.

13.22. Suppose that the sequent contains no free variables and no minimal terms and that its succedent formula has the form $\mathfrak{A}\,\&\,\mathfrak{B}$. In that case we replace it by the formula $\mathfrak{A}$ or by the formula $\mathfrak{B}$, *as we please*.

13.23. Suppose that the sequent contains no free variables and no minimal terms and that its succedent formula has the form $\neg\,\mathfrak{A}$. In that case we replace it by the formula $1 = 2^{48}$ and, at the same time, adjoin the formula $\mathfrak{A}$ (in the last place) to the antecedent formulae of the sequent (cf. 11.2).

13.3. If none of the possibilities listed above applies, the *succedent formula* of the sequent must be a *minimal formula* (3.24).

I am now assuming of *predicates*, as was done for functions above, that they are *decidably defined* in the sense of 8.12.

We can consequently decide of a given minimal formula on the basis of the definition of the predicate concerned, whether it represents *a true or false* proposition.

13.4. Suppose that the sequent contains no free variables and no minimal terms and that its succedent formula is a *true minimal formula*; or: that the succedent formula is a *false minimal formula* (e.g., $1 = 2$) and that one of its antecedent formulae is also a *false minimal formula*.

For such an obviously *true* sequent (cf. 7.3.) *no reduction step is defined*.

13.5. Suppose that the sequent contains no free variables and no minimal terms, that its succedent formula is a false minimal formula, and that none of its antecedent formulae are false minimal formulae. In that case the following three different kinds of reduction step are permissible (counterpart to 13.2):

13.51. Suppose that an *antecedent formula* has the form $\forall \mathfrak{x}\, \mathfrak{F}(\mathfrak{x})$. To it we adjoin an antecedent formula $\mathfrak{F}(\mathfrak{n})$, i.e., a formula which results from $\mathfrak{F}(\mathfrak{x})$ by the substitution of a numeral $\mathfrak{n}$ for the variable $\mathfrak{x}$. In doing so we may either retain or omit the formula $\forall \mathfrak{x}\, \mathfrak{F}(\mathfrak{x})$.

13.52. Suppose that an antecedent formula has the form $\mathfrak{A} \mathbin{\&} \mathfrak{B}$. In that case we adjoin to it either the formula $\mathfrak{A}$ or the formula $\mathfrak{B}$. In doing so we may omit or retain the formula $\mathfrak{A} \mathbin{\&} \mathfrak{B}$.

13.53. Suppose that an antecedent formula has the form $\neg \mathfrak{A}$. We replace it by the succedent formula $\mathfrak{A}$. In doing so we may either omit or retain the formula $\neg \mathfrak{A}$.

13.6. A *reduction rule* for a sequent in which the connectives $\vee$, $\exists$ and $\supset$ do not occur is a rule which renders possible in each case the 'reduction' of a sequent in *finitely many individual reduction steps* (in accordance with 13.11 to 13.53) to one of the correct *reduced forms* (13.4) regardless of how we may *choose* the numeral $\mathfrak{n}$, or which of the two formulae $\mathfrak{A}$ and $\mathfrak{B}$ (in the case of 13.22) we may *choose* when carrying out a reduction step in which there exists an 'option', i.e., one of the steps described at 13.11, 13.21 and 13.22.

13.7. Wherever several possibilities are open to us in any of the *other* reduction steps (e.g., in the case of 13.5), no actual option exists since we require it to be *determined* by the reduction rule which kind of reduction step is to take place; also, e.g., what numeral $\mathfrak{n}$ is to be used in the adjunction of an antecedent formula $\mathfrak{F}(\mathfrak{n})$, and whether or not the formula $\forall \mathfrak{x}\, \mathfrak{F}(\mathfrak{x})$ is to be omitted in the process.

13.8. *Comments concerning the reduction process.*

13.81. *The reduction of true sequents containing no variables.*

In order to illustrate the reduction concept, I shall first show that for sequents without variables and without the symbols $\vee$, $\exists$ and $\supset$, the concept of the stability of a *reduction rule* coincides with the concept of truth in the sense of a *calculation procedure* (7.2, 7.3):

Such a 'true' sequent is to be brought to its reduced form according to the following *rule*: First, all *terms* that may occur are to be replaced by their 'numerical values' (13.12). If the reduced form (13.4) has not yet been reached, a reduction step is to be carried out by which the sequent is transformed into another 'true' sequent in which *fewer logical connectives* occur then before. This is always possible. After all, reductions according to 13.22 and 13.23 certainly meet this requirement. In the case of 13.5, the following reduction step, among the various possibilities, is to be applied:

If a false antecedent formula of the form $\mathfrak{A} \mathbin{\&} \mathfrak{B}$ occurs, then either $\mathfrak{A}$ or $\mathfrak{B}$ must be false; in that case the formula $\mathfrak{A} \mathbin{\&} \mathfrak{B}$ is replaced by $\mathfrak{A}$ or $\mathfrak{B}$, resp. If a false antecedent formula of the form $\neg \mathfrak{A}$ occurs, it is omitted and the succedent formula is replaced by $\mathfrak{A}$.

Each one of the given reduction steps obviously leads to another *true* sequent, furthermore to one with *fewer logical connectives* than before. The continuation of this process obviously leads to the reduced form of the sequent in finitely many steps.

The fact that, conversely, every sequent without variables for which a *reduction rule* is available is *true* follows from the fact that a *false* sequent, as is easily verified, would be transformed into another false sequent by every permissible reduction step or that, in the case of a reduction step according to 13.22, the *choice* of $\mathfrak{A}$ or $\mathfrak{B}$ could be made in such a way that this is the case.

13.82. These considerations can be extended without difficulty to the case of sequents containing $\forall$-symbols ranging over only *finitely* many numbers. The reduction of the $\forall$ then proceeds analogously to that of the &.

13.83. If we proceed to the *infinite* domain of objects of all natural numbers, the statement of a reduction rule for an arbitrary derivable sequent is in general no longer as simple. Since it is here no longer true that all formulae are *decidable* we may, at times, be forced to make use of the option *to retain* the transformed antecedent formula in reduction steps according to 13.51 to 13.53, whereas it was always possible to *omit* this formula in the case of a finite domain (13.81, 13.82).

As an example, I shall give a reduction rule for the proposition 'Fermat's last theorem is either true or not true', stated at 10.6 which, according to its finitist interpretation at that point, is not a true proposition; after the replacement of the $\vee$ and written as a sequent, this proposition has the form:

$$\rightarrow \neg \{[\neg \forall x \forall y \forall z \forall u \neg (u > 2 \,\&\, x^u + y^u = z^u)] \\ \&\, [\neg \neg \forall x \forall y \forall z \forall u \neg (u > 2 \,\&\, x^u + y^u = z^u)]\}.$$

This is reduced as follows: First we obtain (13.23)

$$[\neg \forall x \forall y \forall z \forall u \neg (u > 2 \,\&\, x^u + y^u = z^u)] \\ \&\, [\neg \neg \forall x \forall y \forall z \forall u \neg (u > 2 \,\&\, x^u + y^u = z^u)] \rightarrow 1 = 2.$$

By two reductions according to 13.52 we obtain

$$\neg \forall x \forall y \forall z \forall u \neg (u > 2 \,\&\, x^u + y^u = z^u),\\ \neg \neg \forall x \forall y \forall z \forall u \neg (u > 2 \,\&\, x^u + y^u = z^u) \rightarrow 1 = 2;$$

Further (13.53):

$$\neg \forall x \forall y \forall z \forall u \neg (u > 2 \,\&\, x^u + y^u = z^u) \\ \rightarrow \neg \forall x \forall y \forall z \forall u \neg (u > 2 \,\&\, x^u + y^u = z^u).$$

The reduction of this *basic logical sequent* must now be completed along the lines described *in general* at 13.92.

13.90. In the following I shall prove that *reduction rules* can be stated for all sequents occurring in an arbitrarily given derivation, once the derivation has been transformed according to § 12.

From this the *consistency* will then follow at once:

For if a sequent of the form $\rightarrow \mathfrak{A} \,\&\, \neg\, \mathfrak{A}$ were derivable, then $\rightarrow 1 = 2$, for example, would also be derivable. This is so since $\rightarrow \mathfrak{A}$ as well as $\rightarrow \neg\, \mathfrak{A}$ follow from $\rightarrow \mathfrak{A} \,\&\, \neg\, \mathfrak{A}$ by &-elimination, hence also (5.243) $\neg\, 1 = 2 \rightarrow \mathfrak{A}$ and $\neg\, 1 = 2 \rightarrow \neg\, \mathfrak{A}$; by 'reductio' we obtain $\rightarrow \neg\,\neg\, 1 = 2$, and by 'elimination of the double negation' $\rightarrow 1 = 2$. (In the same way *any arbitrary proposition* can be derived from a contradiction.) No reduction rule can however be stated for the sequent $\rightarrow 1 = 2$, since there is no reduction step that might possibly be applied to it, nor is it in reduced form (13.4), since $1 = 2$ is false.

13.91. In the case of *basic mathematical sequents*, I am assuming that the given reduction rules have been formulated in such a way that they do not make use of the possibility, which exists for reduction steps carried out according to 13.5, *of retaining* the transformed antecedent formula.

For all customary number-theoretical axioms such rules are easily stated. Let us, for instance, consider the examples given at 6.2: these must first be written as sequents and the $\supset$ replaced by & and $\neg$; the resulting sequent can then be *reduced* by first eliminating the ∀-symbols according to 13.1 and by replacing their associated variables by arbitrary numerals and then proceeding as described at 13.81. The justification is that the formulae which result at each step are indeed 'true'.

13.92. *Basic logical sequents* are to be reduced according to the following simple rule:

Suppose that a sequent of the form $\mathfrak{A} \rightarrow \mathfrak{A}$ is given. We first replace the free variables by arbitrary numerals (13.11), then the minimal terms by *those* numerals that represent their values (13.12). The latter procedure must be repeated until no further minimal terms occur – for it can certainly happen that new minimal terms arise during the computation. The sequent finally has the form $\mathfrak{A}^* \rightarrow \mathfrak{A}^*$.

The succedent formula $\mathfrak{A}^*$ is then reduced by means of reduction steps according to 13.21, 13.22 and, if necessary, 13.12 until it has the form $\neg\, \mathfrak{C}$ or is a minimal formula. In the case of reductions according to 13.21 or 13.22 the replacement numerals or formulae may be *chosen arbitrarily*.

If the succedent formula has become a *true minimal formula*, then the reduction procedure terminates (13.4).

If the succedent formula has become a *false minimal formula*, then further reduction steps must be carried out according to 13.51, 13.52 and 13.12 in such a way that the *antecedent formula* $\mathfrak{A}^*$ *undergoes precisely the same transformations, in the same order, as the succedent* formula $\mathfrak{A}^*$ did earlier. If the antecedent formula has taken the form $\forall \mathfrak{x}\, \mathfrak{F}(\mathfrak{x})$, for example, it must be replaced by a formula $\mathfrak{F}(\mathfrak{n})$, and for the replacement numeral $\mathfrak{n}$ *the same numeral must be taken* that was *chosen* in the corresponding reduction of the succedent formula. Reduction steps according to 13.52 are dealt with correspondingly. Thus the antecedent formula eventually becomes identical with the succedent formula and the procedure once again *terminates*, since the reduced form (13.4) has been reached.

If the succedent formula has taken the form $\neg\, \mathfrak{C}$, a reduction according to 13.23 must first be carried out. The sequent then runs: $\mathfrak{A}^*, \mathfrak{C} \to 1 = 2$. As in the previous case, this sequent is reduced in such a way that the antecedent formula $\mathfrak{A}^*$ is transformed *in the same way* as was the succedent formula $\mathfrak{A}^*$, so that finally $\neg\, \mathfrak{C}$ appears in its place. Then the sequent runs $\neg\, \mathfrak{C}, \mathfrak{C} \to 1 = 2$. By means of 13.53 it is reduced to $\mathfrak{C} \to \mathfrak{C}$. This is another *basic logical sequent*; the formula $\mathfrak{C}$ contains at least one *fewer* logical connective than $\mathfrak{A}^*$, and this procedure must consequently end after finitely many steps. A reduction rule has thus been given for arbitrary basic logical sequents.

13.93. In a similar way arbitrary sequents of the form $\mathfrak{A}\,\&\,\mathfrak{B} \to \mathfrak{A}$, $\mathfrak{A}\,\&\,\mathfrak{B} \to \mathfrak{B}$, $\mathfrak{A}, \mathfrak{B} \to \mathfrak{A}\,\&\,\mathfrak{B}$, $\forall \mathfrak{x}\, \mathfrak{F}(\mathfrak{x}) \to \mathfrak{F}(\mathfrak{t})$, $\mathfrak{A}, \neg\, \mathfrak{A} \to 1 = 2$, or $\neg\,\neg\, \mathfrak{A} \to \mathfrak{A}$ may be reduced, a fact which will be used later.

Here, too, the free variables and minimal terms are first replaced according to 13.11 and 13.12. The sequent $\mathfrak{A}^*\,\&\,\mathfrak{B}^* \to \mathfrak{A}^*$ then has a form which also occurred in the reduction of the logical *basic sequent* $\mathfrak{A}^*\,\&\,\mathfrak{B}^* \to \mathfrak{A}^*\,\&\,\mathfrak{B}^*$ according to 13.92; hence the reduction of the sequent in question can be completed in the same way as that of the latter. The same holds true for $\mathfrak{A}^*\,\&\,\mathfrak{B}^* \to \mathfrak{B}^*$ and, correspondingly, for $(\forall \mathfrak{x}\, \mathfrak{F}(\mathfrak{x}))^* \to (\mathfrak{F}(\mathfrak{t}))^*$; here the basic sequent $(\forall \mathfrak{x}\, \mathfrak{F}(\mathfrak{x}))^* \to (\forall \mathfrak{x}\, \mathfrak{F}(\mathfrak{x}))^*$ must be used. In the case of $\mathfrak{A}^*, \mathfrak{B}^* \to \mathfrak{A}^*\,\&\,\mathfrak{B}^*$, a reduction step according to 13.22 must be carried out; from it either $\mathfrak{A}^*, \mathfrak{B}^* \to \mathfrak{A}^*$ or $\mathfrak{A}^*, \mathfrak{B}^* \to \mathfrak{B}^*$ follows, whichever we wish. The reduction is then continued in exactly the same way as that of the basic sequent $\mathfrak{A}^* \to \mathfrak{A}^*$ or $\mathfrak{B}^* \to \mathfrak{B}^*$; the additional antecedent formula is disregarded and presents no problem. In the case of $\mathfrak{A}^*, \neg\, \mathfrak{A}^* \to 1 = 2$, a reduction step according to 13.53 yields $\mathfrak{A}^* \to \mathfrak{A}^*$, hence another basic sequent.

In the case of $\neg\neg \mathfrak{A}^* \to \mathfrak{A}^*$, reduction steps are carried out on the succedent formula according to 13.21, 13.22 and 13.12, until it has the form $\neg \mathfrak{C}$ or is a minimal formula. If it has become a true minimal formula, then the reduction is at an end. If it has assumed the form $\neg \mathfrak{C}$, it is reduced according to 13.23 to $\neg\neg \mathfrak{A}^*, \mathfrak{C} \to 1 = 2$, further, by 13.53, to $\mathfrak{C} \to \neg \mathfrak{A}^*$ then by (13.23) to $\mathfrak{C}, \mathfrak{A}^* \to 1 = 2$. The same procedure is followed in the case where the succedent formula has become a false minimal formula; in that case, $\to \mathfrak{A}^*$ is obtained first, and then $\mathfrak{A}^* \to 1 = 2$.

In both cases we have obtained a sequent which had *also* occurred in the reduction of the *basic logical sequent* $\mathfrak{A}^* \to \mathfrak{A}^*$ according to the procedure stated at 13.92 (or by a procedure that is not *essentially* different). Once again we need only follow the procedure stated *there* in order to complete the reduction of the sequent at hand.

It should be noted that in any reduction steps according to 13.5 in the reduction procedures at 13.92 and 13.93, the antecedent formula involved was never *retained*.

§ 14. Reduction steps on derivations [49]

In order to reduce arbitrary derived sequents we shall state a procedure by which certain reduction steps are carried out on the entire *derivation* of the sequent concerned. For this purpose I shall *modify* somewhat the concept of a derivation used so far (14.1), and shall then explain how an individual *reduction step* is to be carried out on such a *derivation* (14.2).

14.1. *Modification of the concept of a derivation.*

The new concept of a derivation results from the old one (5.2) as follows: 5.22 continues to apply, even though the 'endsequent' of the derivation may now also contain *antecedent formulae* (so that we can speak of a 'derivation for a sequent'). The symbols ∨, ∃ and ⊃ must not occur in the derivation. No sequent of the derivation may be used to obtain *more than one* further sequent (by the application of a rule of inference).

It is easily seen that a derivation in the old sense can be transformed into a derivation with the same endsequent which also satisfies this condition. We need merely work backwards from the endsequent and write down correspondingly often those sequents which have been used more than once, together with the sequents used for their derivation.

Basic mathematical sequents must fulfil the requirement 13.91; together with these all their 'reduction instances', i.e., all sequents which may arise in the course of a given reduction procedure, are also admitted as basic mathematical sequents.

As *basic logical sequents* we may take arbitrary sequents of the form $\mathfrak{A} \to \mathfrak{A}$, $\mathfrak{A} \& \mathfrak{B} \to \mathfrak{A}$, $\mathfrak{A} \& \mathfrak{B} \to \mathfrak{B}$, $\mathfrak{A}, \mathfrak{B} \to \mathfrak{A} \& \mathfrak{B}$, $\forall \mathfrak{x} \, \mathfrak{F}(\mathfrak{x}) \to \mathfrak{F}(\mathfrak{t})$, $\mathfrak{A}, \neg \mathfrak{A} \to 1 = 2$, or $\neg \neg \mathfrak{A} \to \mathfrak{A}$, as well as all sequents which may occur in the reduction of one of these sequents according to 13.92, 13.93.

Structural transformations in their old form are no longer allowed. From the rules of inference we retain the rule of ∀-introduction and 'complete induction' with the following modification: a ∀-introduction or a 'complete induction' in whose sequents no free variables other than $\mathfrak{a}$ occur, remains permissible if in all associated sequents not containing the variable $\mathfrak{a}$, the minimal terms which occur are replaced by their 'numerical values', until all minimal terms have been eliminated (for motivation cf. 14.22).

The following new '¬-introduction' rule is added: From $\Gamma, \mathfrak{A} \to 1 = 2$ results $\Gamma \to \neg \mathfrak{A}$.

We shall adopt one additional rule of inference – the *'chain rule'* – : From a sequence of sequents (at least one) of a given form, a sequent of the following kind results: for its *succedent formula* we take the succedent formula of any one of the sequents of the sequence. If this formula is a false minimal formula, any other may be taken in its place. For its *antecedent formulae* we write down, in arbitrary order, all antecedent formulae of the sequent concerned, together with all antecedent formulae of *earlier* sequents in the sequence. In carrying out this inference, we may *omit* formulae for which the following holds: the same formula *occurs* already among the formulae written down (i.e. those not omitted); or: the formula is the same as the *succedent formula* of a sequent *occurring earlier* in the sequence than the sequent from whose antecedent formulae it is taken. *Other antecedent formulae may be inserted* among the formulae already written down. Finally, the completed sequent may be modified by one or more applications of the substitution rule for bound variables as in 5.244.

The 'chain rule' has thus been formulated *flexibly* enough to allow for the transformation of a derivation in the old sense, which we assume to be already freed of the symbols ∨, ∃ and ⊃ by the method described in § 12 (and which we also suppose to fulfil the conditions for functions, predicates and axioms in 13.12, 13.3, 13.91), into a derivation in the new sense without any change in its endsequent.

Reason: All *structural transformations* are special cases of the 'chain rule'. The omitted *rules of inference* may be replaced by the new basic sequents that have taken their place, together with the 'chain rule', e.g., the &-introduction: $\Gamma \to \mathfrak{A}$ and $\varDelta \to \mathfrak{B}$ and $\mathfrak{A}, \mathfrak{B} \to \mathfrak{A} \& \mathfrak{B}$ by the 'chain rule' yields $\Gamma, \varDelta \to \mathfrak{A} \& \mathfrak{B}$. The ∀-elimination: $\Gamma \to \forall \mathfrak{x} \, \mathfrak{F}(\mathfrak{x})$ and $\forall \mathfrak{x} \, \mathfrak{F}(\mathfrak{x}) \to \mathfrak{F}(\mathfrak{t})$ by the 'chain rule' yields $\Gamma \to \mathfrak{F}(\mathfrak{t})$. The &-elimination and the 'elimination of the double negation' are replaced correspondingly. Finally the 'reductio': from $\mathfrak{A}, \Gamma \to \mathfrak{B}$ and $\mathfrak{A}, \varDelta \to \neg \mathfrak{B}$ and $\mathfrak{B}, \neg \mathfrak{B} \to 1 = 2$ we obtain by the 'chain rule' $\Gamma, \varDelta, \mathfrak{A} \to 1 = 2$, and by $\neg$-introduction finally $\Gamma, \varDelta \to \neg \mathfrak{A}$.

The *new* concept of a derivation is thus not narrower than the *old one*, and for the purpose of stating a reduction rule for any one of the sequents occurring in a derivation we can, without loss of generality, assume as given a derivation in the *new* sense for the sequent concerned. In the following, I shall designate as 'premisses' those sequents from which a new sequent, the 'conclusion', results in the course of the application of a rule of inference.

It is fairly obvious that in view of its informal meaning the 'chain rule' constitutes a 'correct' inference. It can actually be shown to be replaceable by the old rules of inference together with structural transformations.

In formulating the 'chain rule', we have allowed for the case in which no real use is made of some of the premisses; this turns out to be of practical value for the reduction procedure. The extensive replacement of the rules of inference by combinations of basic sequents and the 'chain rule' is also motivated by convenience; it has the virtue of changing the original *vertical* arrangement of inferences into a *horizontal* arrangement.

Finally, I shall also presuppose that it has been *stated* for each sequent of a given derivation whether it is a *basic sequent* and of what *kind* or from *what* preceding sequents and by *what* rules of inference it has been obtained; I shall assume, in general, that it has been stated how the individual sequents, formulae, etc., involved in an application of a rule of inference, *correspond* to the designations used in the associated general schema: in this way the need for resolving possible *ambiguities* does not arise.

14.2. *Reduction steps on derivations.*

I shall now define the concept of a reduction step on a derivation (14.1) and at the same time prove the following: in such a step the derivation concerned is transformed into another *derivation* and its *endsequent* is modified in the following way:

The possible occurrences of free variables are replaced by arbitrarily chosen numerals; any minimal terms that may be present are then replaced by their 'numerical values' until all minimal terms have been eliminated; and, furthermore, *at most one* reduction step according to 13.2 or 13.5 is carried out on the sequent. (It *may* thus happen that an endsequent without free variables or terms remains entirely unchanged.)

The reduction step on derivations is *unambiguous*, except in the cases in which the endsequent undergoes one or more transformations according to a reduction step on sequents involving a choice (13.11, 13.21, 13.22); here the choice may be made arbitrarily; if this has been done, the reduction step is *then* also unambiguous.

If the endsequent of the derivation is in reduced form according to 13.4, *no* reduction step is defined for this derivation. In other cases we carry out a reduction step now to be defined (recursively). In the following we therefore assume that the endsequent is *not* in reduced form.

14.21. If the endsequent of the derivation is a *basic sequent* then the reduction step on it is carried out according to the reduction rules 13.91–13.93, which clearly also cover all basic sequents in their present sense: a replacement of all possible occurrences of free variables and terms must here take place, followed by precisely *one* step according to 13.2 or 13.5 (or none at all, if the reduced form has already been reached). Thus the assertions about the reduction step on derivations made above have obviously been established.

14.22. We now consider the case where the endsequent is the result of the application of a *rule of inference* and we presuppose that, for the derivations of the premisses, the concept of a reduction step has already been defined and the validity of the associated assertions demonstrated.

The reduction step on the entire derivation begins with the following *preliminary* (replacement of free variables and minimal terms):

We begin by replacing all occurrences of free variables in the endsequent by arbitrarily chosen numerals. Then we replace *the same* variables (i.e. the variables that were replaced in the endsequent) in the entire derivation by the same numerals, and replace the *remaining* free variables by 1, with one important exception: the free variable occurring in a $\forall$-introduction or 'complete induction' and designated by $\mathfrak{a}$ at 5.25 must *not* be replaced in the premisses $\Gamma \to \mathfrak{F}(\mathfrak{a})$ or $\mathfrak{F}(\mathfrak{a}), \Delta \to \mathfrak{F}(\mathfrak{a}+1)$, nor in *any* sequent belonging to the *derivation* of that sequent.

Next we replace all *minimal terms* occurring in the derivation one by one by their 'numerical values', with one important exception: *no* replacement takes place in the premisses of a $\forall$-introduction or 'complete induction' containing $\mathfrak{a}$, nor in any sequent belonging to the derivation of that sequent.

Both replacement procedures obviously leave the derivation correct. Essential to this in the replacement of free variables is, first, the restriction on the variable $\mathfrak{a}$ in the case of a ∀-introduction and a 'complete induction' as formulated at 5.25, and the requirement (14.1) that every derivational sequent serves as a premiss for at most *one* application of a rule of inference. These two facts make it actually possible to separate completely from the rest of the variables the variables to be replaced, so that by this distinction no error is introduced into any application of a rule of inference.

In the case of a term replacement the special requirement formulated at 14.1 for the ∀-introduction and the 'complete induction' is important (which is why it was introduced); for, the original normal form of these rules of inference (5.25) may be destroyed by the replacement.

After this 'preliminary' comes the actual reduction step, according to the following rules. If the endsequent is already in *reduced form*, the reduction step terminates at this point.

14.23. Suppose that the endsequent is the result of a ∀-*introduction* or a ¬-*introduction*. It is then eliminated and its premiss taken for the new endsequent, where, in the case of a ∀-introduction, every occurrence of the free variable $\mathfrak{a}$ must be replaced throughout the derivation of this premiss by an arbitrarily chosen numeral and every minimal term by its 'numerical value', subject to the same restrictions as at 14.22; *not to be replaced*, however, are terms in which the variable $\mathfrak{a}$ occurred earlier.

The derivation has obviously remained correct, and the endsequent has become a reduced endsequent in the sense of 13.21 or 13.23.

14.24. Suppose that the endsequent is the result of a '*complete induction*'. The numerical value of the term $\mathfrak{t}$ will be denoted by the numeral $\mathfrak{n}$; $\mathfrak{m}$ shall be the numeral for the number smaller by 1 (if $\mathfrak{n}$ is not equal to 1). The free variable $\mathfrak{a}$ in the derivation of the premiss $\mathfrak{F}(\mathfrak{a}), \Delta \to \mathfrak{F}(\mathfrak{a}+1)$ is then replaced successively by the numerals 1, 2, 3, etc. up to $\mathfrak{m}$, subject to the same restriction as at 14.22, and all *minimal terms* that may have resulted are then replaced by their 'numerical values', also subject to the same restriction as at 14.22. The derivation as a whole is then completed by the application of the '*chain rule*', which makes it possible to derive the endsequent $\Gamma, \Delta \to (\mathfrak{F}(\mathfrak{n}))^*$ once again from $\Gamma \to (\mathfrak{F}(1))^*$ and the newly derived sequents $(\mathfrak{F}(1))^*, \Delta \to (\mathfrak{F}(2))^*$ and $(\mathfrak{F}(2))^*, \Delta \to (\mathfrak{F}(3))^*$ etc. up to $(\mathfrak{F}(\mathfrak{m}))^*, \Delta \to (\mathfrak{F}(\mathfrak{n}))^*$. The asterisk denotes in each case the changes that have occurred through the replacement of minimal terms. By virtue of the preparatory replacement of terms (14.22) and the further replacement of the terms carried just out, *all occurrences* of minimal terms have finally been eliminated, so that the related $\mathfrak{F}^*$-expressions have indeed become *identical* with one another, even if they had not been identical before. If $\mathfrak{n}$ equals 1, then we merely put 1 for $\mathfrak{a}$, and by the 'chain rule' the endsequent $\Gamma, \Delta \to (\mathfrak{F}(1))^*$ results from $\Gamma \to (\mathfrak{F}(1))^*$ and $(\mathfrak{F}(1))^*, \Delta \to (\mathfrak{F}(2))^*$.

14.25. The last case to be considered is that in which the endsequent is the conclusion of a 'chain rule' inference. This is the most difficult reduction, since the chain rule in some sense amasses the difficulties of all inferences.

The premiss whose succedent formula provides the succedent formula of the endsequent, I shall call the *'major premiss'*. If the succedent formula of the endsequent is a false minimal formula, we choose as major premiss the *first* premiss (in the given order) whose succedent formula is *also* a false minimal formula. This does not change the correctness of the 'chain rule', even if a later premiss was the major premiss before; it may merely happen that certain antecedent formulae of the endsequent can no longer be regarded as taken from the *premisses*, but rather as newly *adjoined*.

From these preliminaries it follows that the major premiss can in no case be in reduced form (13.4), for otherwise the endsequent would obviously also have to be in reduced form, and this was assumed not to be the case. Hence a *reduction step* can be carried out on the derivation of the major premiss. In this respect I shall distinguish four cases to be dealt with in turn (14.251–14.254).

14.251. Suppose that the major premiss undergoes a change according to 13.2 in the reduction step on its derivation. In that case the endsequent is subjected to the appropriate reduction step for sequents according to 13.2; any *choice* that arises is to be made arbitrarily. The reduction step for derivations is then carried out on the derivation of the major premiss and, wherever a choice exists, the *same* choice is to be made as before. The succedent formulae of both sequents are now the same once again (up to possible redesignations of bound variables) and the 'chain rule' is once again correct. Thus, the reduction step for the entire derivation is completed.

14.252. Suppose that the major premiss undergoes a change according to 13.5 in the reduction step on its derivation, and that the affected antecedent formula is one of the formulae that has been *included* among the antecedent formulae of the endsequent (when the latter was formed by the 'chain rule') *or* that it was omitted because an identical formula *had* already *occurred* among the antecedent formulae. In that case the reduction step is carried out on the derivation of the major premiss and, so that the 'chain rule' becomes again correct, the endsequent is modified according to the *corresponding* reduction step on sequents (13.5). I.e., if the affected antecedent formula was *itself* absorbed into the endsequent, then the same reduction step is here carried out on *that* formula; but if it was omitted because it was identical with an already *existing* formula, then the reduction step is carried out on *the latter* formula and it is *retained*, regardless of whether the corresponding formula in the reduction of the *premiss* is omitted or retained.

14.253. (Principal case.) Suppose that the major premiss, say $\varDelta \to \mathfrak{C}$, undergoes a change according to 13.5 in the reduction step on its derivation and that the affected antecedent formula ($\mathfrak{B}$) is a formula that was not included among the antecedent formulae of the endsequent because it agreed with the succedent formula of an earlier premiss; suppose further that *this* premiss, call it $\varGamma \to \mathfrak{B}$, undergoes a *change* during the reduction step on its derivation which, in that case, must necessarily be a change according to 13.2. (Since $\mathfrak{B}$ cannot be a minimal formula.) – Suppose that the endsequent of the whole derivation has the form $\varTheta \to \mathfrak{D}$. I shall distinguish *three subcases* depending on whether $\mathfrak{B}$ has the form $\forall \mathfrak{x}\, \mathfrak{F}(\mathfrak{x})$, $\mathfrak{A} \,\&\, \mathfrak{B}$ or $\neg\, \mathfrak{A}$. The treatment of the three cases is not essentially different.

Suppose first that $\mathfrak{B}$ has the form $\forall \mathfrak{x}\, \mathfrak{F}(\mathfrak{x})$. In that case an antecedent formula $\mathfrak{F}(\mathfrak{n})$ is adjoined in the reduction step according to 13.51 on $\varDelta \to \mathfrak{C}$, and $\forall \mathfrak{x}\, \mathfrak{F}(\mathfrak{x})$ is either retained or omitted; in the reduction step on $\varGamma \to \forall \mathfrak{x}\, \mathfrak{F}(\mathfrak{x})$, which must be carried out according to 13.21, the *same* symbol $\mathfrak{n}$ may be chosen for the numeral to be substituted, so that $\varGamma \to \mathfrak{F}(\mathfrak{n})$ results. We now form three 'chain-rule' inferences: the premisses of the first are those of the original 'chain-rule' inference, but with $\varGamma \to \mathfrak{F}(\mathfrak{n})$ in place of $\varGamma \to \forall \mathfrak{x}\, \mathfrak{F}(\mathfrak{x})$; its conclusion: $\varTheta \to \mathfrak{F}(\mathfrak{n})$. A correct result. The premisses of the second are those of the original 'chain-rule' inference, except that $\varDelta \to \mathfrak{C}$ is replaced by the sequent that was reduced according to 13.51; its conclusion: $\varTheta, \mathfrak{F}(\mathfrak{n}) \to \mathfrak{D}$. This is also a *correct* 'chain-rule' inference. The third 'chain-rule' inference again yields the endsequent $\varTheta \to \mathfrak{D}$ from $\varTheta \to \mathfrak{F}(\mathfrak{n})$ and $\varTheta, \mathfrak{F}(\mathfrak{n}) \to \mathfrak{D}$.– Together with each one of the sequents used we must of course write down the complete *derivation* of each sequent so that altogether we now have another correct derivation.

If $\mathfrak{B}$ has the form $\mathfrak{A} \,\&\, \mathfrak{B}$, then we adjoin antecedent formula $\mathfrak{A}$ or $\mathfrak{B}$ in carrying out a reduction step on $\varDelta \to \mathfrak{C}$ according to 13.52. $\varGamma \to \mathfrak{A} \,\&\, \mathfrak{B}$ becomes either $\varGamma \to \mathfrak{A}$ or $\varGamma \to \mathfrak{B}$, as desired; the choice should be made so that the *same* formula occurs as in $\varDelta \to \mathfrak{C}$. The procedure is then continued exactly as in the previous case.

If $\mathfrak{B}$ has the form $\neg\, \mathfrak{A}$, then $\varDelta \to \mathfrak{C}$ is reduced to $\varDelta^* \to \mathfrak{A}$, and $\varGamma \to \neg\, \mathfrak{A}$ to $\varGamma, \mathfrak{A} \to 1 = 2$. We now form, as before, two 'chain-rule' inferences with the conclusions $\varTheta, \mathfrak{A} \to 1 = 2$ and $\varTheta \to \mathfrak{A}$. With their order interchanged, these two inferences again yield $\Theta \to \mathfrak{D}$ by a third 'chain-rule' inference. This is so since $\mathfrak{D}$, like $\mathfrak{C}$ and $1 = 2$, is a false minimal formula.

14.254. We are still left with the following possibilities: the major premiss remains *unchanged* in the reduction step on its derivation; or: its change is of the kind assumed at 14.253, and the premiss $\Gamma \to \mathfrak{B}$ remains *unchanged* in the reduction step on its derivation. – In both cases we carry out the reduction step on the derivation of the premisses that have remained unchanged, and this completes the reduction. However, in the case of a reduction step according to 14.253 on the *derivation of the premisses* (which after all leaves the endsequent, i.e., the premisses *unchanged*) we proceed somewhat *differently*, viz.: we carry out this reduction, but without completing the prescribed third 'chain-rule' inference; instead, we take the two premisses, rather than the conclusion, of the 'chain-rule' inference and adjoin them to the premisses of *that* 'chain-rule' inference which concludes the derivation as a whole. This obviously leaves the 'chain-rule' inference correct. The endsequent is not changed.

The definition of a *reduction step on a derivation is thus complete.*

§ 15. Ordinal numbers and proof of finiteness

It remains to be shown that a successive application of a *reduction step* on a given derivation always leads to the reduced form (of the endsequent) *in finitely many steps*, regardless of the choices made in those cases in which a choice exists. In doing so, we shall at the same time have given a reduction rule (13.6) for arbitrary derived *sequents*, since the reduction of the *derivation* of the sequent (according to § 14) automatically involves the reduction of the *sequent* (according to § 13).

In order to prove the finiteness of the procedure we shall have to show that each reduction step in a definite sense '*simplifies*' a derivation. For this purpose I shall correlate with each derivation an '*ordinal number*' representing a *measure* for the '*complexity*' of the derivation (15.1, 15.2). It can then be shown that with every reduction step on a derivation the ordinal number of that derivation (in general) *diminishes* (15.3). However, the *finiteness* of the reduction procedure is hereby not immediately guaranteed; for the ordering of the derivations (corresponding to the well-ordering of their ordinal numbers) is of a special kind, since it may happen that in terms of its complexity a derivation ranks *above infinitely many* other derivations. E.g., a derivation whose endsequent has taken the form $\to \forall \mathfrak{x}\, \mathfrak{F}(\mathfrak{x})$, as a result of a 'complete induction' and a $\forall$-introduction, must be regarded as more complex than any one of the infinitely many special instances obtained by substituting definite numerals for $\mathfrak{x}$ and decomposing the 'complete induction' (14.23, 14.24). The situation may be complicated still further by a multiple nesting of such instances. Thus the 'ordinal numbers' here have the nature of 'transfinite ordinal numbers' (cf. footnote 21) and the inductive comprehension of their totality is not possible by ordinary complete induction, but only by *transfinite induction* whose validity requires a special verification (15.4).

15.1. *Definition of ordinal numbers* (recursive).

As 'ordinal numbers' I shall use certain positive *finite decimal fractions* formed according to the following rule:

Ordinal numbers with the characteristic 0 are precisely the following numbers: 0.1, 0.11, 0.111, 0.1111, ... i.e., in general: any number with the characteristic 0, whose mantissa consists of finitely many 1's; also the number 0.2.

Neither here nor below shall we permit the adjunction of zeros at the end of an ordinal number; the representations of the numbers thus become unique. – I shall call one mantissa *smaller* than another mantissa if this relationship holds between the *numbers* resulting from these mantissas if the latter are prefixed by '0'.

The mantissa of an ordinal number with the characteristic $\rho+1$ ($\rho \geqq 0$) is obtained by taking several mutually distinct ordinal numbers (at least one) with the characteristic ρ, ordering their mantissas according to size, so that the largest occurs first, the smallest last, and by then writing them down in that order from left to right, separating any two successive mantissas by $\rho+1$ zeros. *All* numbers obtainable in this way from ordinal numbers with the characteristic ρ, and no others, are ordinal numbers with the characteristic $\rho+1$.

Examples of ordinal numbers:

0.111, 1.1101, 1.2, 2.111, 2.2010011010011, 3.2010020001.

It can be determined uniquely of a given number with the characteristic $\rho+1$ from what numbers with the characteristic ρ it has been generated by the above rule. For, a number with the characteristic σ can obviously have no more than σ consecutive zeros in any one place.

Further details about the ordering of the ordinal numbers follow at 15.4.

15.2. *The correlation of ordinal numbers with derivations.*

With every given derivation (in the sense of 14.1) we can correlate a unique appropriate ordinal number calculated according to the following recursive rule:

The following observation must here be kept in mind: the maximum number (v) of consecutive zeros in the mantissa is larger than 1, and all of its *segments* that are separated by successions of v zeros, except for the last one, begin with the numeral 2; the *last* segment consists only of 1's.

If the endsequent of the derivation is a *basic sequent*, the derivation receives an ordinal number of the form 2.2001 ... 1, where the number of 1's must be chosen to be larger by one than the total number of *logical connectives* occurring in the sequent.

Now suppose that the endsequent is the conclusion of the application of a *rule of inference* and that for the derivation of the premisses their associated ordinal numbers are already known. From these the ordinal number of the whole derivation is calculated as follows:

If the endsequent is the conclusion of a ∀- or ¬-introduction, then the numeral 1 is adjoined to the ordinal number for the derivation of the premiss. By virtue of the stated properties of ordinal numbers for derivations, we have obviously another correct ordinal number in accordance with 15.1.

If the endsequent is the conclusion of a '*chain rule*' inference, we focus our attention on the mantissas of the ordinal numbers of the derivations for the premisses; suppose that v is the maximum number of consecutive zeros in all of these mantissas. Should there be *equal* mantissas among them, we *distinguish* these by adjoining to *one* of them $v+1$ zeros and one, 1, to the next one $v+1$ zeros and two 1's, etc.; this principle is applied to *every* occurrence of equal mantissas. The mantissas thus obtained are mutually *distinct*; they are then written down from left to right according to size (the largest one first) and two successive mantissas are in each case to be separated by $v+2$ zeros; finally $v+2$ zeros and one 1 are adjoined at the end. The result is the mantissa of the ordinal number for the whole derivation. For its *characteristic* we take the smallest natural number which *is larger* than the maximum number of consecutive zeros in the mantissa, it any, and which, *first*, exceeds by at least 2 the maximum number of consecutive zeros in any one of the ordinal numbers for the derivations of the premisses and which, *second*, is no smaller than twice the total number of *logical connectives* in the succedent formula of any one of the premisses preceding the major premiss (14.25).

If the endsequent is the conclusion of a '*complete induction*', then the ordinal number of the whole derivation receives a mantissa of the form 201 ... 10 ... 01; where the number of consecutive 1's is to be chosen greater by 1 than the total number of consecutive 1's in the corresponding place in the *larger* of the mantissas of the ordinal numbers for the derivations

§ 15, ORDINAL NUMBERS AND PROOF OF FINITENESS

of both *premisses* (*or either one of them*, if both are identical); i.e., if the latter mantissa begins with 200, one 1 is to be taken; otherwise the mantissa must begin with 201 . . . 10, in which case one more 1 is taken than in the number 201 . . . 10. The total number of consecutive *zeros* must be $v+2$, where v is the maximum number of consecutive zeros in the two mantissas mentioned. As *characteristic* we take the smallest natural number that *is larger* than the maximum number of consecutive zeros in the mantissa, if any, and which *first* exceeds by at least 2 the corresponding maximum number of zeros in any one of the two ordinal numbers used and which *second* is not smaller than twice the total number of logical connectives in the formula $\mathfrak{F}(1)$.

It is easily seen that this newly formed number is another correct ordinal number (15.1) and possesses, moreover, the special properties stated above.

15.3. *A reduction step diminishes the ordinal number of a derivation.*

We must now prove that with every reduction step on a derivation according to 14.2, the ordinal number of the newly resulting derivation becomes in general *smaller* than that of the old derivation. I shall show: the *characteristic* does not increase; the *mantissa* decreases in all cases in which the endsequent is not already in reduced *form* after the replacement of the free variables and terms (14.21, 14.22); the maximum number of consecutive zeros in the mantissa, furthermore, remains *unchanged except* in the case of a reduction according to 14.253, where it increases by exactly two.

I shall again proceed recursively, i.e., I shall prove the assertion by complete induction.

For derivations whose endsequent is a *basic sequent* the result follows from the method of correlating ordinal numbers with such derivations, together with the fact that in the reduction step the sequent undergoes a change according to 13.2 or 13.5, and here the total number of occurring logical connectives is *diminished.* (If the reduced form of the derivation is attained earlier, then the ordinal number remains unchanged.) What is important here is that in changes according to 13.5, the altered antecedent formula is always *omitted,* cf. 13.91–13.93.

Suppose now that the endsequent is the result of the application of a *rule of inference* and that the assertion has already been proved for the derivations of the *premisses.*

The *preliminary step* (14.22) has obviously no influence on the ordinal number of the derivation. If the reduced form of the endsequent results with this step, then the ordinal number remains unchanged. If this is not the case, the following holds:

If the endsequent is the conclusion of a $\forall$- or $\neg$-introduction, then the assertion follows at once from the method of correlating ordinal numbers with such a derivation.

Even if the endsequent is the conclusion of a '*complete induction*' the truth of the assertion follows easily. The 'complete induction' is, after all, transformed into a 'chain-rule' inference; this does not lead to an increase in the characteristic of the ordinal number; although the mantissa may become much *longer*, it nevertheless *diminishes*, since the mantissa of the ordinal number of one of the two original derivations of the premisses must always occur at the *beginning* of that mantissa. The maximum number of consecutive zeros $(v+2)$ remains unchanged.

Suppose, finally, that the endsequent is the conclusion of a '*chain-rule*' inference. The selection of the premiss of an earlier sequent as major premiss (14.25) does not alter the mantissa of the ordinal number; the *characteristic*, on the other hand, may diminish, because certain succedent formulae of the premisses no longer contribute to its calculation.

Suppose now that the reduction step takes the form of either 14.251 or 14.252. In this case one of the mantissas of the ordinal numbers for the derivation of the *premisses* is diminished without a change in the maximum *number of consecutive zeros* occurring in it. This has obviously a simultaneous diminishing effect on the mantissa for the ordinal number of the total derivation. The number of zeros is still $v+2$; the diminished mantissa may conceivably occur in a *later* place of the sequence, which is ordered by size; if the mantissa was one of several *identical* mantissas, then one fewer 1 is adjoined to the remaining mantissas; *in any case*, the first mantissa in the sequence of mantissas, each separated by $v+2$ zeros, which has not remained the same must be *smaller* than before; consequently the total mantissa has certainly also been diminished. The *characteristic* does not increase.

In a reduction step according to 14.253, the ordinal number of the derivation is altered as follows: let us first consider the ordinal numbers for the two derivations which conclude with the newly formed first or second 'chain-rule' inference. For these two derivations the situation is the *same* as that in the *previous* case, i.e.: the two mantissas are *smaller* than the mantissa of the ordinal number of the original derivation; the maximum number of consecutive zeros $(v+2)$ has remained the *same*; the characteristics have not increased. We now introduce the *third* 'chain-rule' inference and form the ordinal number of the new total derivation: Its *mantissa* begins with one of the two earlier mantissas followed by $v+3$ zeros (usually $v+4$); it is consequently smaller than that of the original ordinal number; the maximum number of consecutive zeros is $v+4$, hence larger by two than before; the *characteristic* of the total derivation, finally, *cannot have increased*, for the total number of logical connectives in the succedent formula $\mathfrak{F}(\mathfrak{n})$, $\mathfrak{A}$ or $\mathfrak{B}$, or $\mathfrak{A}$, resp., is *smaller* than that in the formula $\mathfrak{B}$, i.e., in $\forall \mathfrak{x}\, \mathfrak{F}(\mathfrak{x})$, $\mathfrak{A}\,\&\,\mathfrak{B}$, or $\neg\,\mathfrak{A}$, resp.; hence the sum of twice the number of logical connectives in the *former formulae* with $v+4$ zeros, which determines the new characteristic, is not larger than the sum of twice the number of logical connectives in the *latter formulae* with $v+2$ zeros; nor could the characteristic of the *original* derivation be smaller than the latter sum, since $\mathfrak{B}$ was one of the succedent formulae contributing to its calculation.

In a reduction step according to 14.254, the situation is the same as for 14.251 and 14.252, unless we are dealing with an *exception*. Such an exception can be handled without difficulty on the basis of our previous considerations; here one of the mantissas of the ordinal numbers for the derivations of the premises is no longer replaced by *one* smaller mantissa, as was done above, but by *two*; but the effect is the same in every desired respect. The characteristic is not increased; its maximum number of consecutive zeros before the reduction was not smaller by two than twice the total number of logical connectives in $\mathfrak{B}$ so that, after the reduction, the contributions of $\mathfrak{F}(\mathfrak{n})$, $\mathfrak{A}$ or $\mathfrak{B}$, or $\mathfrak{A}$, resp., cannot lead to an increase.

It has thus been *proved* that in a reduction step the ordinal number (usually) diminishes. The *most important point* was our consideration concerning the *characteristic* of the ordinal number in discussing the reduction steps 14.253 and 14.254; this is the idea which enables us to recognize a simplification of the derivation in such a reduction step in spite of the apparent increase in complexity. The *simplification* consists precisely in the fact that the permisses of the third 'chain-rule' inference are '*interwoven*' *to a lesser degree* (viz., to a degree corresponding to the total number of logical connectives in the succedent formula of the first premiss, which is also the antecedent formula of the second premiss) than the premisses of the first and second and the premisses of the original 'chain-rule' inference. The method of correlating an ordinal number with a 'chain-rule' inference (15.2) is formulated from the above point of view; all other details follow more or less automatically.

15.4. *Demonstration of the finiteness of the reduction procedure.*

Some facts – needed below – about the *ordering according to size* of the ordinal numbers:

With every number α having the characteristic $\rho(\rho \geqq 0)$ I correlate the system $\mathfrak{S}(\alpha)$ of those ordinal numbers with the characteristic $\rho+1$ in whose formation according to 15.1 the number α was the largest of the ordinal numbers with the characteristic ρ that were used. Every ordinal number with the characteristic $\rho+1$ belongs uniquely to one such system $\mathfrak{S}(\alpha)$. If α_1 is smaller than α_2, then every number of $\mathfrak{S}(\alpha_1)$ is also smaller than every number of $\mathfrak{S}(\alpha_2)$. The ordering of the systems $\mathfrak{S}(\alpha)$ corresponds therefore to the ordering of the numbers α. The following holds for the ordering of the numbers (with the characteristic $\rho+1$) within a system $\mathfrak{S}(\alpha)$: the *smallest* number within $\mathfrak{S}(\alpha)$ is the number $\alpha+1$. The *remaining* numbers of $\mathfrak{S}(\alpha)$ correspond order-isomorphically to the totality of those numbers with the characteristic $\rho+1$ which are *smaller* than $\alpha+1$ in the following way: Every number of $\mathfrak{S}(\alpha)$, except for $\alpha+1$, results from $\alpha+1$ through the adjunction of $\rho+1$ zeros, followed by the mantissa of one of the numbers with the characteristic $\rho+1$ which is smaller than $\alpha+1$. The *ordering* of these numbers is here preserved.

The correctness of all these assertions follows easily from the definition of the ordinal numbers. The reader may find it beneficial to examine the ordering of the ordinal numbers with the characteristics 1, as well as 2 and 3, for example, using this definition[50].

I now assert (*theorem of 'transfinite induction'*):

All ordinal numbers (15.1) are '*accessible*' in the following sense, by our running through them in the order of their increasing magnitude: the first number, 0.1, is considered 'accessible'; if all numbers smaller than a number β have furthermore been recognized as 'accessible', then β is also considered 'accessible'.

PROOF. 0.1 is accessible by hypothesis, hence also 0.11, hence also 0.111, etc., and it follows in general by complete induction that every number smaller than 0.2 is accessible. Hence 0.2 is also accessible, and thus are all numbers with the characteristic 0. I now apply a *complete induction*, i.e., I assume that the accessibility of all numbers up to and including those with the characteristic ρ ($\rho \geq 0$) has already been proved and that it is now to be proved for numbers with the characteristic $\rho+1$. The *first* of these numbers, i.e., the number with the mantissa 1, is accessible. It should be noted that we have already run through the numbers with the characteristic ρ. To every such number α corresponds a system $\mathfrak{S}(\alpha)$ of numbers with the characteristic $\rho+1$; this system consists of the number $\alpha+1$ and a system order-isomorphic with those numbers with the characteristic $\rho+1$ that are smaller than $\alpha+1$. To run through the numbers with the characteristic $\rho+1$ now amounts merely to a running through of the *systems* $\mathfrak{S}(\alpha)$ in the same way in which we ran through the numbers α with the characteristic ρ; for if a number $\alpha+1$ has been recognized as accessible, then all remaining numbers of the system $\mathfrak{S}(\alpha)$ obviously become accessible at the same time; we need merely run through this system *in exactly the same way* in which we have already run through the *isomorphic* system of the numbers (with the characteristic $\rho+1$) *smaller* than $\alpha+1$. In this way we can run through all numbers with the characteristic $\rho+1$ by virtue of having run through the numbers with the characteristic ρ. To the totality of numbers α (with the characteristic ρ) smaller than a number α_0 there corresponds, in the case of the number α_0+1 (with the characteristic $\rho+1$), the totality of numbers *belonging to the systems* $\mathfrak{S}(\alpha)$ (where $\alpha < \alpha_0$).

CONCLUSION. By means of the 'theorem of transfinite induction' the *finiteness of the reduction procedure* for arbitrary derivations now follows at once. If the finiteness of the reduction procedure has already been proved for all derivations whose ordinal number is smaller than a number β, then this also holds for every derivation with the ordinal number β; for by a *single* reduction step the latter derivation is transformed into a derivation with a *smaller* ordinal number or a derivation in reduced form. (If the derivation was already in reduced form, then there was nothing more to prove.) Thus the property of the finiteness of the reduction procedure carries over from the totality of the derivations with the ordinal numbers *smaller* than β to the derivations with the ordinal number β; by the theorem of transfinite induction this property therefore holds for *all* derivations with *arbitrary* ordinal numbers. This *concludes* the consistency proof.

SECTION V. REFLECTIONS ON THE CONSISTENCY PROOF

§ 16. The forms of inference used in the consistency proof

I shall review in the following the inferences and derived concepts used in the consistency proof *from two aspects*: First I shall examine to what extent they can be considered as *indisputable* (16.1); second, in connection with the *theorem of Gödel* (2.32), to what extent they correspond to the techniques of proof contained in formalized elementary number theory and in what way they *go beyond* these techniques (16.2).

16.1. In terms of the *indisputability* of the methods of proof used, the critical point is the *proof of the finiteness* of the reduction procedure (15.4). We shall come back to this point later. All other techniques of proof used in the consistency proof can certainly be considered as '*finitist*' in the sense outlined in detail in section III. This cannot be 'proved', if for no other reason than the fact that the notion of 'finitist' is not unequivocally formally defined and cannot in fact be delimited in this way. All we can do is to examine every individual inference from this point of view and try to assess whether that inference is in harmony with the finitist sense of the concepts that occur, and make sure that it does not rest on an inadmissible 'actualist' interpretation of these concepts. I shall discuss briefly the most relevant passages of the consistency proof:

The *objects* of the consistency proof, as of proof theory in general, are certain *symbols* and *expressions*, such as terms, formulae, sequents, derivations, ordinal numbers, not to forget the natural numbers. All these objects are defined (3.2, 5.2, 14.1, 15.1) by *construction rules* corresponding to the definition of the natural numbers (8.11); in each case such a rule indicates how more and more such objects can be constructed step by step. – Here it must be presupposed that in formalized elementary number theory certain definite 'functions', 'predicates' and 'axioms' have been stipulated which satisfy the conditions laid down for these objects (13.12, 13.3, 13.91). Strictly speaking, this presupposition introduces a transfinitely used 'if – then' into the consistency proof; but this 'if – then' is obviously harmless, since the proof need not be regarded as meaningful at all until that presupposition has actually been justified and the above conditions have been shown to hold.

A number of *functions* and *predicates* were furthermore applied to these objects and they were *decidably defined* in the sense of 8.12. E.g., the *function* 'the endformula of a derivation', the *predicate* 'containing at least one $\forall$- or $\exists$-symbol' and many others. The following functions, in particular, were also decidably defined, as is easily verified: 'the derivation resulting from a derivation by a *transformation* according to § 12', 'the derivation resulting from a derivation by a *reduction step* in which the conditions of a possible choice were unequivocally specified' (14.2), 'the ordinal number of a derivation' (15.2).

Furthermore, propositions of the following kind were proved by *complete induction*: 'for all sequents', 'for all derivations' etc., whose validity for each *individual* sequent or derivation was *decidable*. E.g.: 'The figure resulting from a derivation by a reduction step is another derivation and the transformation of the endsequent fulfils certain conditions' (14.2); 'in carrying out a reduction step we diminish the ordinal number' (15.3).

In applying the concept 'all' in the consistency proof, I have not used the unwieldy finitist expression given for it in 10.11; here the distinction between the actualist and finitist interpretations has, in any case, no bearing on our reasoning.

The *negation* of a transfinite proposition occurs only once in the entire proof (at 13.90) and only in a harmless form in which the proposition concerned leads to a quite elementary contradiction. The negation can actually be *avoided* altogether if for 'consistency' the following positive expression is used: 'every derivation has an endformula which does not have the form $\mathfrak{A} \ \& \ \neg \ \mathfrak{A}$'. Here the 'not' is no longer transfinite.

16.11. *What can be said, finally, about* the *proof of the finiteness* of the reduction procedure (15.4)?

The concept of *'accessibility'* in the 'theorem of transfinite induction' is of a very special kind. It is certainly not *decidable* in advance whether it is going to apply to an arbitrary given number; from the point of view explained in § 9, this concept therefore has *no immediate sense*, since an 'actualist sense' has after all been rejected. It acquires a *sense* merely by being predicated of a definite number for which *its validity* is simultaneously *proved*. It is quite permissible to introduce concepts in this way; the same situation arises, after all, in the case of *all transfinite propositions* if a finitist sense is to be ascribed to them, cf. § 10. With the statement that 'if all numbers smaller than β have already been *recognized* as accessible, then β is also accessible', the definition of the 'accessibility' is already formulated in conformity with this interpretation. No *circularity* is involved in this formulation; the definition is, on the contrary, entirely *constructive*; for β is counted as accessible only when all numbers smaller than β have *previously* been recognized as accessible. The 'all' occurring here is of course to be interpreted *finitistically* (10.11); in each case we are dealing with a totality for which a *constructive* rule for generating all elements is given.

About the *proof* of the theorem of transfinite induction the following must be said: From the way the concept of 'accessibility' was defined it follows that in proving this theorem, a 'running through' of all ordinal numbers in ascending magnitude must take place. In dealing with the numbers with the characteristic 0, the following is to be observed: the *infinite* totality of the numbers smaller than 0.2 is transcended by one single idea: the proof *can* be extended *arbitrarily* far into this totality; hence it may be considered as completed for the *entire* totality. This 'potential' interpretation of the 'running through' of an infinite totality must be applied throughout the *entire proof*.

The occurrence of a *transfinite induction hypothesis* in the complete induction on ρ is to be interpreted in the sense of 10.5 and is therefore indisputable. In the inference 'if the number $\alpha+1$ has been recognized as accessible, then all remaining numbers of the system $\mathfrak{S}(\alpha)$ are accessible', a *transfinite* 'if – then' occurs. Objections were raised against this concept at 11.1; but these do not apply to the present case for the very reason that the hypothesis is here not to be interpreted *hypothetically*, but rather as follows: *only after* having reached $\alpha+1$ can we successfully run through the numbers of $\mathfrak{S}(\alpha)$ (viz., in exact correspondence with the way in which we ran through the numbers up to $\alpha+1$).

Now let us consider the induction step *as a whole*, i.e., the reinterpretation of the running through of the $(\rho+1)$-systems in terms of the running through of the ρ-systems. This is undoubtedly the most critical point of the argument. Yet I believe that if we think about it deeply enough we cannot dispute that the argument here used has considerable plausibility. We might, for example, visualize the initial cases with the characteristics 1, 2, 3 in detail. As the characteristic grows, nothing new is *basically* added; the method of progression always remains the same. It must of course be admitted that the complexity of the multiply-nested infinities which must be 'run through' grows considerably; this running through must always be regarded as 'potential', as was done in the case of the characteristic 0. The difficulty lies in the fact that although the *precise finitist sense* of the 'running through' of the ρ-numbers is reasonably perspicuous in the initial cases, it becomes of such great *complexity* in the general case that it is only remotely visualizable; yet this surely constitutes an adequate foundation justifying the possibility of running through the $(\rho+1)$-numbers.

The 'conclusion', finally, adds nothing essentially new. The proposition that the reduction procedure for a derivation is *finite* regardless of how possible *choices* may be made, contains a transfinite 'there is', viz., with respect to the *total number* of reduction steps. This proposition is of the same kind as the proposition concerning the 'accessibility'; in each special case it receives its definite *sense* only through the *proof* of its validity for this case; this corresponds to the finitist interpretation (10.3). For the purpose of the *consistency proof* alone, incidentally, the notion of a 'choice' is dispensable, since we are here dealing only with the reduction of a derivation with the endsequent $\to 1 = 2$, and since all reduction steps are *unequivocal* and do not depend on choices. The total number of steps is not specified in advance; we can merely make certain statements about it and these become *more and more indefinite* as the ordinal number of the derivation increases. (The place of a direct statement of such a number is taken by its 'statability'. This can undoubtedly still be regarded as being in harmony with the finitist view.

§ 16, THE FORMS OF INFERENCE USED IN THE CONSISTENCY PROOF

Altogether, I am inclined to believe that in terms of the fundamental distinction between disputable and indisputable methods of proof (§ 9), the *proof of the finiteness* of the reduction procedure (15.4) can still be considered as indisputable, so that the consistency proof represents a real *vindication* of the disputable parts of elementary number theory.

16.2. In order to examine the extent to which the consistency proof *coincides with the theorem of Gödel* (2.32) we would first have *to correlate natural numbers* with the objects of proof theory (formulae, derivations, etc.) corresponding to the way in which this was done in Gödel's paper cited in footnote 3, and would also have to introduce the required functions and predicates for these objects as functions and predicates for the corresponding natural numbers. Then the consistency proof becomes a proof with the natural numbers as objects. In order to obtain a formally delimited formalism we would have to limit the possibilities of definition provided for above to definite schemata; these can easily be chosen general enough to allow for the definition of all functions and predicates required in proof theory; cf., for example, Gödel's version.

The forms of inference in the consistency proof are then *none other* than those presented in our formalization of number theory; *only* the proof of finiteness (15.4) occupies again a special position. It is impossible to see how the latter proof could be carried out with the techniques of elementary number theory. For this reason the consistency proof is in harmony with Gödel's theorem.

In this connection the following two facts, which will not be proved since their proof would lead us too far afield, are of interest:

1. If the inference of *complete induction* is omitted from formalized elementary number theory, then the consistency proof can be formulated without essential change in such a way that – after having carried out the mentioned translation into a proof about natural numbers – the techniques of elementary number theory (*including* complete induction) suffice.

2. The consistency proof for *the whole of* elementary number theory, translated into a theory with the natural numbers as objects, can be carried out with techniques from *analysis*[51].

The special position of the inference of *complete induction* is due to the following fact: if this inference *is omitted*, then a *definite upper bound* can be given for the *total number of reduction steps* required for the reduction of a given sequent. Yet if the inference of complete induction is included, then this number, in its dependence on *choices*, can become *arbitrarily large*. This is so since in the reinterpretation of this rule of inference (14.24) the total number of required reduction steps for the sequent $\Gamma, \Delta \to \mathfrak{F}(\mathfrak{t})$ obviously depended on the number $\mathfrak{n}$ (the value of $\mathfrak{t}$) and this number may *depend on a choice*, as is the case if $\mathfrak{t}$ is a *free variable*, and must therefore first be replaced by an arbitrarily *chosen* numeral $\mathfrak{n}$. In this case it may happen that there exists *no* general bound for the total number of reduction steps required in the reduction of the sequent $\Gamma, \Delta \to \mathfrak{F}(\mathfrak{t})$.

This fact seems to be the reason why the rule of complete induction could not be included in the *earlier* consistency proofs (2.4).

§ 17. Consequences of the consistency proof

First I shall discuss the question to what extent the consistency proof remains applicable if the 'elementary number theory' formulated in section II is *extended* by the addition of new concepts and methods (17.1; then I shall point out its *transferability* to other branches of mathematics (17.2), and shall finally examine certain *objections* by the 'intuitionists' against the significance of consistency proofs as such (17.3).

17.1. For the value of a consistency proof it is essential to know whether the stipulated formalism for the particular mathematical theory involved, in our case elementary number theory, really *fully* encompasses that theory (cf. 3.3, 5.3). Now in practice elementary number theory is not subject to any formal restrictions; it can always be extended further by new kinds of specific concepts, possibly also by the application of new kinds of forms of inference. How does this affect the consistency proof? Whenever the *present framework is transcended, an extension of the consistency proof to the newly incorporated techniques is required.* The consistency proof is already designed in such a way that this is to a very large degree possible without difficulties.

If new *functions* or *predicates* for natural numbers are introduced, for example, then a *decision rule* in accordance with 8.12 must be given for them; if additional mathematical *axioms* are introduced, then a *reduction rule* must be given for them in accordance with 13.91 (cf. § 6 and 10.14). *Not effectively decidable* derived concepts, in the sense of 6.3, present no difficulties either, since they *can be eliminated* by the method described at that point. All these requirements are easily fulfilled, as long as the introductions are in the generally accepted sense 'correct' and the axioms 'true'.

Even new kinds of *inferences* may be carried out which are not representable in the present formalism. In fact, *every formally defined* system containing elementary number theory is *necessarily incomplete* in the sense that there are number-theoretical theorems of an elementary character whose truth can be proved by plausible finitist inferences, but not by means of methods of proof of the system proper[52]. This fact was advanced as an argument against the value of consistency proofs[53]. Yet my consistency proof remains unaffected by it; quite generally, it can here be said: if an elementary number-theoretical theorem can be proved by means of inferences not belonging to my formalism, then the statement of a *reduction rule* for this theorem according to 13.91 will include the theorem in the consistency proof. The theorem given as an example by Gödel has the quite elementary form $\forall \mathfrak{x}\, \mathfrak{P}(\mathfrak{x})$, where $\mathfrak{P}$ represents a *decidable predicate* about the natural numbers; the fact that the finitist truth of this theorem has been recognized means that $\mathfrak{P}(\mathfrak{n})$ is true for *each definite* $\mathfrak{n}$, and from this the *reducibility* of the sequent $\to \forall \mathfrak{x}\, \mathfrak{P}(\mathfrak{x})$ according to 13.21, 13.4 follows at once.

The concept of the *reduction rule* has in fact been kept *general* enough so that it is not tied to any definite logical formalism, but corresponds rather to the general concept of 'truth', certainly to the extent to which that concept has any clear sense at all (cf. 13.8).

If a new *form of inference* is to be included as such in elementary number theory formalized up to now, it must be suitably incorporated in the reduction procedure. (An example might be a 'transfinite induction' up to a fixed 'number of the second number class'.)

To be sure, if derived concepts and forms of inference *from analysis* (which are also used in proofs of number-theoretical theorems) are to be included in elementary number theory, then the consistency proof can in general not be extended to these additions in a straightforward way; here difficulties arise which still await solution.

17.2. The *consistency proof for elementary number theory* carries over without difficulty to *a number of other branches of mathematics*. This holds quite generally for such mathematical theories whose *objects* are given by a *construction rule* corresponding to that for the natural numbers (8.11). A particularly simple and universally applicable example is the following: first a definite number of primitive symbols is given and it is then stated that each one of these symbols designates an object; if a primitive symbol is adjoined to the designation of an object, then this again results in the designation of an object. (In short: "Every finite sequence of primitive symbols designates an object of the theory".)

In such theories *functions* and *predicates* are then introduced by decidable definitions (8.12) and the same logical *forms of inference* are used as those in elementary number theory. The *consistency proof* carries over at once, the only difference being that the place of the 'numerals' is taken by the 'object symbols' of the theory and this changes nothing *essential*.

Such branches of mathematics are, for example, extensive parts of *algebra* (*polynomials* as objects are indeed finite combinations of symbols); from the realm of geometry, e.g., *combinatorial topology*; even large *parts of analysis* may be represented in this way, if the concept of a real number is not used in its *most general* form. Finally important parts of *proof theory* also belong here (cf. 16.1).

The addition of negative numbers, fractional numbers, diophantine equations, etc., to the natural numbers as objects in elementary number theory proper (3.31) can be incorporated in the consistency proof *in the same way*. All propositions about these objects can of course also be *translated* into propositions about the natural numbers, as mentioned at 3.31, by correlating these new objects in an appropriate way with the natural numbers. The same is also true of all other theories of the kind mentioned; a one-to-one correspondence can always be established between 'finite combinations of symbols' and the natural numbers ('denumerability'). This, however, is unnecessarily cumbersome and unnatural for the purpose of consistency proofs.

17.3. (Cf. § 9.) On the part of the *intuitionists*, the following *objection* is raised against the significance of consistency proofs[54]: even if it had been demonstrated that the disputable forms of inference cannot lead to mutually contradictory results, these results would nevertheless be propositions *without sense* and their investigation therefore an idle pastime; real *knowledge* could be gained only by means of indisputable intuitionist (or finitist, as the case may be) forms of inference.

Let us, for example, consider the existential proposition cited at 10.6, for which the *statement* of a number whose existence is asserted is not possible. According to the intuitionist view, this proposition is therefore *without sense*; an existential proposition can after all be significantly asserted only if a numerical example is available.

What can we say to this?

Does such a proposition have any *cognitive value*? To be sure, a certain *practical value* of propositions of this kind lies first of all in the following possibility of application, advanced by opponents of the untuitionist interpretation:

They might possibly serve as a source for the derivation of simple propositions, possibly representable by minimal formulae (3.24), which are themselves finitist and intuitionistically significant and which must be *true* by virtue of the consistency proof.

Furthermore, an existential proposion $\exists \mathfrak{x}\, \mathfrak{F}(\mathfrak{x})$, e.g., for which no example is given, nevertheless serves the purpose of making a search for a proof for the proposition $\forall \mathfrak{x} \neg \mathfrak{F}(\mathfrak{x})$ unnecessary; for there can be no such proof, since a contradiction would otherwise result.

These are certainly reasons which make proofs of theorems by means of 'actualist' forms of inference seem *not entirely useless*, apart from the 'aesthetic value' of mathematical research as such.

Thus propositions of actualist mathematics seem to have a certain utility, but no *sense*. The major part of my consistency proof, however, consists precisely in *ascribing a finitist sense* to actualist propositions, viz.: for every arbitrary proposition, as long as it is provable, a *reduction rule* according to 13.6 *can be stated*, and this fact represents the finitist sense of the proposition concerned and this sense is gained precisely through the consistency proof.

This 'finitist sense' can admittedly be rather *complicated* for even simply formed propositions and has in general a looser connection with the (actualistically determined) *form* of the proposition than is the case in the realm of finitest reasoning.

In this way the above mentioned existential proposition, e.g., also receives a finitist sense, but this sense is *weaker* than that of a finitistically proved existential proposition, since it does *not* assert that an example can be given.

A quite different question is what *significance* can still be attached to the *actualist sense* of the propositions. The proof certainly reveals that it is possible to reason consistently 'as if' everything in the infinite domain of objects were as actualistically determined as in finite domains (cf. § 9). Whether and to what extent, however, anything 'real' corresponds to the actualist sense of a transfinite proposition – apart from what its restricted *finitist* sense expresses – is a question which the *consistency proof does not* answer.

APPENDIX TO # 4

In footnote 49 Gentzen states that articles 14.1–16.11 were inserted in February 1936 in place of an earlier text. Prof. Paul Bernays kindly communicated to the editor a galley proof containing the earlier passages, and they have been collected together in this appendix. The following table should enable the reader to locate the relevant places in # 4:

#4		Galley proof
13.83	coincides with	13.83
13.90	coincides essentially with	14.0
13.91	coincides essentially with	14.1
13.92	coincides essentially with	14.2
13.93–15.4	replaces	14.3–14.63
16.1	replaces	15.1
16.2	coincides with	15.2 (except for the passage appearing in this appendix)
17.1	coincides with	16.1 (except for the footnote stating that the ω-consistency of elementary number theory follows from the consistency proof; this note was omitted from the final version of the proof)
17.2	coincides with	16.2
17.3	coincides with	16.3

14.3. If a reduction rule is known for a sequent, then a reduction rule can also be stated for every sequent which has resulted from the former by structural transformation. Viz.: An interchange of antecedent formulae (5.241) does not affect the reduction procedure. If an antecedent formula was omitted which was identical with another antecedent formula (5.242), we reduce the new sequent in the same way as the old one; but if the omitted formula would have been subject to a reduction step according to 13.5, we apply this reduction step to the formula identical with the omitted formula and then retain the latter formula – this is permissible.

If a formula was adjoined to the antecedent formulae (5.243), we first carry out the required reductions on it according to 13.11 and 13.12 and continue the rest of the reduction up to the reduced form as if this formula were not even present.

A redesignating of a bound variable (5.244) does not necessitate a change in the reduction rule.

In the following I shall repeatedly make tacit use of the fact that a reduction rule for a sequent which results from another sequent by a structural transformation can be obtained from the reduction rule of the former sequent.

14.4. Now it still remains to be shown that a reduction rule can always be given for a sequent which results from those sequents by the application of a rule of inference for which reduction rules are already known. After the transformation according to § 12, the following rules of inference can still be applied in the derivation: ∀-introduction, ∀-elimination, &-introduction, &-elimination, 'reductio', 'elimination of the double negation' and 'complete induction'. I shall deal with them in that order.

14.41. Suppose that we are given a ∀-*introduction*: 'From $\Gamma \to \mathfrak{F}(\mathfrak{a})$ follows $\Gamma \to \forall \mathfrak{x}\, \mathfrak{F}(\mathfrak{x})$. Assume a reduction rule to be known for the sequent $\Gamma \to \mathfrak{F}(\mathfrak{a})$. The reduction of $\Gamma \to \forall \mathfrak{x}\, \mathfrak{F}(\mathfrak{x})$ must begin with the replacement of any free variables that may occur by arbitrarily chosen numerals (13.11). Suppose that $\Gamma^* \to \forall \mathfrak{x}\, \mathfrak{F}^*(\mathfrak{x})$ results. If no free variables occurred, then $\Gamma^* \to \forall \mathfrak{x}\, \mathfrak{F}^*(\mathfrak{x})$ stands again for $\Gamma \to \forall \mathfrak{x}\, \mathfrak{F}(\mathfrak{x})$. (Correspondingly in what follows.) Then all minimal terms that occur must be replaced by their numerical values (13.12), and this results in $\Gamma^{**} \to \forall \mathfrak{x}\, \mathfrak{F}^{**}(\mathfrak{x})$. This sequent is reduced according to 13.21 to $\Gamma^{**} \to \mathfrak{F}^{**}(\mathfrak{n})$, where $\mathfrak{n}$ is to be chosen arbitrarily. Any new minimal terms must again be replaced by their numerical values in accordance with 13.12, and this results in $\Gamma^{**} \to \mathfrak{F}^{***}(\mathfrak{n})$.

The reduction of the sequent $\Gamma \to \mathfrak{F}(\mathfrak{a})$ must also begin with the replacement of the free variables. For this replacement we may, in particular, use the same numerals that were chosen in the reduction of $\Gamma \to \forall \mathfrak{x}\, \mathfrak{F}(\mathfrak{x})$, as well as the symbol $\mathfrak{n}$ for the replacement of $\mathfrak{a}$, so that the sequent $\Gamma^* \to \mathfrak{F}^*(\mathfrak{n})$ results. Then follows the replacement of possible minimal terms and from this $\Gamma^{**} \to \mathfrak{F}^{***}(\mathfrak{n})$ obviously results, i.e., the same sequent as above. By virtue of the reduction rule for $\Gamma \to \mathfrak{F}(\mathfrak{a})$, consequently, a reduction rule is now statable for this sequent; hence a reduction rule has also been obtained for $\Gamma \to \forall \mathfrak{x}\, \mathfrak{F}(\mathfrak{x})$.

14.42. Suppose we are given a ∀-*elimination*: 'From $\Gamma \to \forall \mathfrak{x}\, \mathfrak{F}(\mathfrak{x})$ results $\Gamma \to \mathfrak{F}(\mathfrak{t})$'. The sequent $\Gamma \to \mathfrak{F}(\mathfrak{t})$ is again subjected to reduction steps according to 13.11 and 13.12, if necessary; suppose that $\Gamma^* \to \mathfrak{F}^*(\mathfrak{n})$ results. In the reduction of $\Gamma \to \forall \mathfrak{x}\, \mathfrak{F}(\mathfrak{x})$, which must begin with reduction steps according to 13.11 (if necessary), 13.12 (if necessary), and then a step according to 13.21, possibly followed by further steps according to 13.12, the numerals to be substituted may obviously be chosen so that these steps also yield the sequent $\Gamma^* \to \mathfrak{F}^*(\mathfrak{n})$. We therefore have a reduction rule for that sequent and hence also for $\Gamma \to \mathfrak{F}(\mathfrak{t})$.

14.43. The &-*introduction* and the &-*elimination* are dealt with analogously to the ∀-introduction and ∀-elimination. Here the reduction step according to 13.21 is replaced by a step according to 13.22.

14.44. In dealing with the three rules of inference still remaining, I make use of the following *lemma*: 'If reduction rules are known for two sequents of the form $\Gamma \to \mathfrak{D}$ and $\mathfrak{D}, \Delta \to \mathfrak{C}$ in which no free variables and no minimal terms occur, then a reduction rule can also be given for the sequent $\Gamma, \Delta \to \mathfrak{C}$.' (The meaning of the symbols $\Gamma, \Delta, \mathfrak{C}$ and $\mathfrak{D}$ is the same as that defined at 5.250, $\mathfrak{D}$ also stands for an arbitrary formula.) The *proof* of this lemma, which represents the major part of the consistency proof, *follows* at 14.6. Here I shall *first* show how the lemma is applied to the 'reductio', the 'elimination of the double negation' and to 'complete induction'.

14.441. Suppose that a *'reductio'* is given: 'From $\mathfrak{A}, \Gamma \to \mathfrak{B}$ and $\mathfrak{A}, \Delta \to \neg \mathfrak{B}$ follows $\Gamma, \Delta \to \neg \mathfrak{A}$.' We first reduce $\Gamma, \Delta \to \neg \mathfrak{A}$ (if required) according to 13.11 and 13.12; suppose that the result is $\Gamma^*, \Delta^* \to \neg \mathfrak{A}^*$. This we reduce according to 13.23 to $\Gamma^*, \Delta^*, \mathfrak{A}^* \to 1 = 2$. In the reduction of $\mathfrak{A}, \Gamma \to \mathfrak{B}$, on the other hand, we can choose the numerals to be substituted in the reduction steps according to 13.11 and 13.12, which are carried out first, so that from these steps a sequent of the form $\mathfrak{A}^*, \Gamma^* \to \mathfrak{B}^*$ results. In the same way it can be achieved that $\mathfrak{A}, \Delta \to \neg \mathfrak{B}$ assumes the form $\mathfrak{A}^*, \Delta^* \to \neg \mathfrak{B}^*$, after the appropriate reduction steps. This then yields the sequent $\mathfrak{A}^*, \Delta^*, \mathfrak{B}^* \to 1 = 2$ by 13.23. Reduction rules are therefore known for the sequents $\mathfrak{A}^*, \Gamma^* \to \mathfrak{B}^*$ and $\mathfrak{A}^*, \Delta^*, \mathfrak{B}^* \to 1 = 2$ and, *by the lemma*, therefore also for $\mathfrak{A}^*, \Gamma^*, \mathfrak{A}^*, \Delta^* \to 1 = 2$, i.e., (14.3) also for $\Gamma^*, \Delta^*, \mathfrak{A}^* \to 1 = 2$. We have thus a reduction rule for $\Gamma, \Delta \to \neg \mathfrak{A}$.

14.442. Suppose that we are given an *'elimination of the double negation'*: 'From $\Gamma \to \neg \neg \mathfrak{A}$ follows $\Gamma \to \mathfrak{A}$.' The reductions of $\Gamma \to \mathfrak{A}$ according to 13.11 and 13.12 which may first be necessary, can be carried out analogously on $\Gamma \to \neg \neg \mathfrak{A}$. We must therefore still reduce a sequent $\Gamma^* \to \mathfrak{A}^*$ in which free variables and minimal terms no longer occur, and this will simultaneously yield a reduction rule for $\Gamma^* \to \neg \neg \mathfrak{A}^*$.

It is sufficient to state a reduction rule for the sequent $\neg \neg \mathfrak{A}^* \to \mathfrak{A}^*$. For we can then apply the *lemma* and from the availability of reduction rules for $\Gamma^* \to \neg \neg \mathfrak{A}^*$ and $\neg \neg \mathfrak{A}^* \to \mathfrak{A}^*$ conclude the statability of a reduction rule for $\Gamma^* \to \mathfrak{A}^*$.

The sequent $\neg \neg \mathfrak{A}^{**} \to \mathfrak{A}^*$ can be reduced easily according to the following rule (cf. 14.1): We reduce the succedent formula according to 13.21, 13.22, and 13.12 until it has the form $\neg \mathfrak{C}$ or is a minimal formula. If it has become a correct minimal formula the reduction is finished. If it has assumed the form $\neg \mathfrak{C}$, we continue the reduction according to 13.23 and obtain $\neg \neg \mathfrak{A}^*, \mathfrak{C} \to 1 = 2$, further (by 13.53) we obtain $\mathfrak{C} \to \neg \mathfrak{A}^*$, then (by 13.23) $\mathfrak{C}, \mathfrak{A}^* \to 1 = 2$. In the case where the succedent formula has become a false minimal formula we proceed in the same way; in the latter case we first obtain $\to \neg \mathfrak{A}^*$ and then $\mathfrak{A}^* \to 1 = 2$.

In both cases we have thus obtained a sequent which *also* occurs in the reduction of the *basic logical sequent* $\mathfrak{A}^* \to \mathfrak{A}^*$ according to the procedure stated at 14.1. We need therefore merely follow the procedure stated at 14.1 in order to complete the reduction of the sequent.

14.443. Suppose that a *'complete induction'* is given: 'From $\Gamma \to \mathfrak{F}(1)$ and $\mathfrak{F}(\mathfrak{a}), \Delta \to \mathfrak{F}(\mathfrak{a}+1)$ follows $\Gamma, \Delta \to \mathfrak{F}(\mathfrak{t})$.' In $\Gamma, \Delta \to \mathfrak{F}(\mathfrak{t})$ we first replace all free variables that may occur by arbitrarily chosen numerals (13.11) and obtain $\Gamma^*, \Delta^* \to \mathfrak{F}^*(\mathfrak{t}^*)$. Then we carry out reduction steps according to 13.12 (if necessary) and achieve in this way that finally every occurrence of $\mathfrak{t}^*$ has been replaced by the *numeral* $\mathfrak{n}$, which represents the value of the term (the term now no longer contains a variable). The sequent has thus become $\Gamma^*, \Delta^* \to \mathfrak{F}^*(\mathfrak{n})$. Now we carry out further reduction steps according to 13.12 (if necessary), until all minimal terms have been eliminated. The sequent then has the form $\Gamma^{**}, \Delta^{**} \to (\mathfrak{F}^*(\mathfrak{n}))^*$.

In the reduction of $\Gamma \to \mathfrak{F}(1)$ and $\mathfrak{F}(\mathfrak{a}), \Delta \to \mathfrak{F}(\mathfrak{a}+1)$, which begins with the replacement of free variables, we can actually choose the numerals to be substituted so that they agree with the numerals chosen previously and can replace the variable $\mathfrak{a}$, which did not occur in $\Gamma, \Delta \to \mathfrak{F}(\mathfrak{t})$, by any one of the numerals from 1 to $\mathfrak{m}$, where $\mathfrak{m}$ denotes the number 1 smaller than $\mathfrak{n}$. It then follows that for each one of the sequents

$$\Gamma^* \to \mathfrak{F}^*(1)$$

$$\mathfrak{F}^*(1), \Delta^* \to \mathfrak{F}^*(1+1)$$

$$\mathfrak{F}^*(2), \Delta^* \to \mathfrak{F}^*(2+1)$$

$$\cdots$$

$$\mathfrak{F}^*(\mathfrak{m}), \Delta^* \to \mathfrak{F}^*(\mathfrak{m}+1)$$

reduction rules are statable. If these sequents are then reduced by the reduction steps prescribed in 13.12, there obviously result sequents of the following form, for which reduction rules are therefore also statable:

$$\Gamma^{**} \to (\mathfrak{F}^*(1))^*$$

$$(\mathfrak{F}^*(1))^*, \Delta^{**} \to (\mathfrak{F}^*(2))^*$$

$$(\mathfrak{F}^*(2))^*, \Delta^{**} \to (\mathfrak{F}^*(3))^*$$

$$\cdots$$

$$(\mathfrak{F}^*(\mathfrak{m}))^*, \Delta^{**} \to (\mathfrak{F}^*(\mathfrak{n}))^*.$$

We now apply the *lemma*: From the reduction rules for $\Gamma^{**} \to (\mathfrak{F}^*(1))^*$ and $(\mathfrak{F}^*(1))^*, \Delta^{**} \to (\mathfrak{F}^*(2))^*$, we obtain a reduction rule for the sequent $\Gamma^{**}, \Delta^{**} \to (\mathfrak{F}^*(2))^*$ from it, and from the reduction rule for $(\mathfrak{F}^*(2))^*$, $\Delta^{**} \to (\mathfrak{F}^*(3))^*$, we obtain a reduction rule for $\Gamma^{**}, \Delta^{**} \to (\mathfrak{F}^*(3))^*$, etc.; finally, it follows that a *reduction rule is statable* for the sequent Γ^{**}, $\Delta^{**} \to (\mathfrak{F}(\mathfrak{n}))^*$, hence also for $\Gamma, \Delta \to \mathfrak{F}(\mathfrak{t})$, since this sequent had actually already been reduced above to the form of the former sequent.

14.5. Now the proof of the 'lemma' is still outstanding. At this point I should like to add a few remarks which may contribute to an easier understanding of the proof.

What is the reason for the special position of the '*lemma*'? Let us examine the kind of *finitist interpretation* afforded by the reduction concept in place of the 'actualist truth': the concepts ∀ and & are interpreted in a quite *natural* way ('reduced', 13.21 and 13.22), and the associated rules of inference (14.41–14.43) are dealt with in a correspondingly effortless manner. Not so for ¬; the formula ¬ $\mathfrak{A}$ is interpreted as $\mathfrak{A} \to 1 = 2$ (13.23) and in order to reduce *this* form further the reduction steps on *antecedent* formulae (13.5) are necessary. To the informal sense of the → there therefore corresponds a comparatively *artificial* and less immediately comparable reduction procedure. The difficulties which the ⊃ and ¬ present to a finitist interpretation (§ 11) make it indeed impossible to state a more '*natural*' procedure.

A basic *form of inference* which genuinely reflects the informal meaning of the → is the following: 'From the assumptions Γ follows $\mathfrak{D}$. From the assumption $\mathfrak{D}$ and further assumptions Δ follows $\mathfrak{E}$. Then $\mathfrak{E}$ also follows from the assumptions Γ, Δ.' This form of inference is implicit both in the '*reductio*' and in '*complete induction*'. Hence the reliance on the *lemma* (14.44) in dealing with these two rules of inference.

In the proof of the lemma the difficulty now consists in bridging the gap between the *actualist* meaning of the →, according to which the mentioned form of inference is trivially 'true', and the *dissimilar finitist* interpretation given by the reduction concept. The fundamental idea of the proof is this: in reducing $\Gamma \to \mathfrak{D}$ the $\mathfrak{D}$ is referred back to 'something simpler' (13.21–13.23). The same is done with the *antecedent* formula $\mathfrak{D}$ in the reduction of $\mathfrak{D}, \Delta \to \mathfrak{E}$ (13.51–13.53). From this we generally obtain two new sequents $\Gamma \to \mathfrak{D}^*$ and $\mathfrak{D}^*, \Delta \to \mathfrak{E}$; this method can be continued (complete induction on the number of logical connectives in $\mathfrak{D}$) until a minimal formula takes the place of $\mathfrak{D}$, and we have thus a trivial case. Yet this method does not suffice if in the reduction of the antecedent formula $\mathfrak{D}$ that formula *is retained.* The consideration of this possibility requires a *further* reduction argument of a *special* kind (14.63).

14.6. Proof of the lemma.

The lemma runs: 'If reduction rules are known for two sequents of the form $\Gamma \to \mathfrak{D}$ and $\mathfrak{D}, \Delta \to \mathfrak{C}$ in which no free variables and no minimal terms occur, then a reduction rule can also be stated for the sequent $\Gamma, \Delta \to \mathfrak{C}$.'

The latter sequent will be called the *mix sequent* of the two other sequents; the formula $\mathfrak{D}$ its *mix formula.*

In order to prove the lemma, I apply a *complete induction* on the number of logical connectives occurring in the mix formula. I therefore assume that the total number of these connectives is equal to a fixed number ρ and that the lemma has already been proved for smaller ρ or that ρ is equal to 0.

14.60. Suppose therefore that two definite sequents $\Gamma \to \mathfrak{D}$ and $\mathfrak{D}, \Delta \to \mathfrak{C}$ without free variables and minimal terms, with ρ logical connectives in the formula $\mathfrak{D}$, are given and that for each sequent a reduction rule is known. It must then be shown that a reduction rule can also be given for the mix sequent $\Gamma, \Delta \to \mathfrak{C}$.

14.61. I shall first deal with the case where the sequent $\mathfrak{D}, \Delta \to \mathfrak{C}$ is already in *reduced form.* If $\mathfrak{C}$ is a true minimal formula, then $\Gamma, \Delta \to \mathfrak{C}$ is also in reduced form. The same holds if $\mathfrak{C}$ is a false minimal formula and if in Δ a false minimal formula occurs. The case remains where $\mathfrak{C}$ and $\mathfrak{D}$ are false minimal formulae. In that case $\Gamma, \Delta \to \mathfrak{C}$ is reduced according to precisely the same rule as that prescribed for $\Gamma \to \mathfrak{D}$. Since $\mathfrak{C}$ and $\mathfrak{D}$ are both false minimal formulae, their difference is here immaterial; and the formulae designated by Δ may be ignored altogether in the reduction (cf. 14.3).

14.62. Suppose that the sequent $\mathfrak{D}, \Delta \to \mathfrak{C}$ is *not* yet in *reduced form.* Relative to the *first reduction step* to be carried out on the sequent, I then distinguish *three cases*:

1. Suppose that $\mathfrak{C}$ is not a minimal formula.

2. Suppose that $\mathfrak{C}$ is a false minimal formula and that the first prescribed reduction step for the sequent $\mathfrak{D}, \Delta \to \mathfrak{C}$ (according to 13.5) does *not* affect the antecedent formula $\mathfrak{D}$.

3. Suppose that $\mathfrak{C}$ is a false minimal formula and that the mentioned reduction step (according to 13.5) affects the antecedent formula $\mathfrak{D}$.

I shall deal with each of the three cases separately.

14.621. Suppose that the *first case* arises. The first reduction step to be carried out on the mix sequent $\Gamma, \Delta \to \mathfrak{C}$ is the appropriate step from 13.21, 13.22, 13.23, where the *choise* of $\mathfrak{n}$, or of $\mathfrak{A}$ or $\mathfrak{B}$, resp., *is free* if $\mathfrak{C}$ has the form $\forall \mathfrak{x} \, \mathfrak{F}(\mathfrak{x})$ or $\mathfrak{A} \,\&\, \mathfrak{B}$. Suppose that after this reduction step (and, if necessary, successive steps according to 13.12, until no further *minimal terms* occur) the sequent runs $\Gamma, \Delta^* \to \mathfrak{C}^*$. The first reduction step on the sequent $\mathfrak{D}, \Delta \to \mathfrak{C}$ must necessarily be of the *same* kind, and in the case of a choice, the same choice may be made as above, so that after the first reduction step (and possibly further necessary steps according to 13.12) this sequent assumes the form $\mathfrak{D}, \Delta^* \to \mathfrak{C}^*$. Now the following *assertion* still remains *to be proved*, which is again a special case of the lemma: 'On the basis of the known reduction rules for the sequents $\Gamma \to \mathfrak{D}$ and $\mathfrak{D}, \Delta^* \to \mathfrak{C}^*$ a reduction rule for their *mix sequent* $\Gamma, \Delta^* \to \mathfrak{C}^*$ is also statable.' I shall postpone the proof of this assertion for the time being.

14.622. Suppose that the *second case* arises. After its first reduction step (according to 13.5) (and possibly successive steps according to 13.12, until no further *minimal terms* occur) the sequent $\mathfrak{D}, \Delta \to \mathfrak{C}$ runs $\mathfrak{D}, \Delta^* \to \mathfrak{C}^*$. The reduction of $\Gamma, \Delta \to \mathfrak{C}$ must then begin with *corresponding* steps so that it yields $\Gamma, \Delta^* \to \mathfrak{C}^*$ from the former sequent. In that case the following *assertion* still remains *to be proved*, which once again is a special case of the lemma: 'On the basis of the known reduction rules for the sequents $\Gamma \to \mathfrak{D}$ and $\mathfrak{D}, \Delta^* \to \mathfrak{C}^*$, a reduction rule for their mix sequent $\Gamma, \Delta^* \to \mathfrak{C}^*$ is also statable.'

14.623. Suppose that the *third case* arises. I distinguish *three subcases*, depending on whether $\mathfrak{D}$ has the form $\forall \mathfrak{x} \mathfrak{F}(\mathfrak{x})$, $\mathfrak{A} \& \mathfrak{B}$ or $\neg \mathfrak{A}$, i.e., depending on whether the first prescribed reduction step on $\mathfrak{D}, \Delta \to \mathfrak{C}$ takes the form of 13.51, 13.52 or 13.53. The treatment of these three cases is not essentially different.

14.623.1. Suppose that $\mathfrak{D}$ has the form $\forall \mathfrak{x} \, \mathfrak{F}(\mathfrak{x})$. In that case the first reduction step turns the sequent $\mathfrak{D}, \Delta \to \mathfrak{C}$, i.e., $\forall \mathfrak{x} \, \mathfrak{F}(\mathfrak{x}), \Delta \to \mathfrak{C}$, into $\mathfrak{F}(\mathfrak{n})$, $\forall \mathfrak{x} \, \mathfrak{F}(\mathfrak{x}), \Delta \to \mathfrak{C}$ or $\mathfrak{F}(\mathfrak{n}), \Delta \to \mathfrak{C}$. The sequent $\Gamma \to \mathfrak{D}$ is identical with $\Gamma \to \forall \mathfrak{x} \, \mathfrak{F}(\mathfrak{x})$ and *its* first reduction step must therefore yield $\Gamma \to \mathfrak{F}(\mathfrak{m})$ (according to 13.21), with arbitrarily chosen $\mathfrak{m}$. In particular, we can choose the numeral $\mathfrak{n}$ for $\mathfrak{m}$ and obtain $\Gamma \to \mathfrak{F}(\mathfrak{n})$.

If $\mathfrak{F}(\mathfrak{n})$ contains minimal terms we subject it, and the sequent dealt with before, to further reductions according to 13.12, as prescribed, until no further minimal terms occur. The two sequents then run $\Gamma \to (\mathfrak{F}(\mathfrak{n}))^*$ and $(\mathfrak{F}(\mathfrak{n}))^*, \forall \mathfrak{x} \, \mathfrak{F}(\mathfrak{x}), \Delta \to \mathfrak{C}$ or $(\mathfrak{F}(\mathfrak{n}))^*, \Delta \to \mathfrak{C}$. If no minimal terms had occurred, $(\mathfrak{F}(\mathfrak{n}))^*$ shall stand for the formula $\mathfrak{F}(\mathfrak{n})$. First, I consider the case where $\mathfrak{D}, \Delta \to \mathfrak{C}$ has assumed the *second* form, viz., $(\mathfrak{F}(\mathfrak{n}))^*, \Delta \to \mathfrak{C}$. Here reduction rules for the sequents $\Gamma \to (\mathfrak{F}(\mathfrak{n}))^*$ and $(\mathfrak{F}(\mathfrak{n}))^*, \Delta \to \mathfrak{C}$ are known; I now apply the *induction hypothesis*, according to which the lemma is assumed to be proved for mix formulae with fewer logical connectives than those contained in $\mathfrak{D}$; from this it follows that a reduction rule is also statable for the mix sequent of the two given sequents, i.e., for the sequent $\Gamma, \varDelta \to \mathfrak{C}$. For the mix formule $(\mathfrak{F}(\mathfrak{n}))^*$ obviously contains one fewer logical connective, viz., the $\forall$, than the formula $\mathfrak{D}$ which, as we know, is identical with $\forall \mathfrak{x}\, \mathfrak{F}(\mathfrak{x})$. This completes the present case.

If $\mathfrak{D}, \varDelta \to \mathfrak{C}$ has assumed the more complicated form $(\mathfrak{F}(\mathfrak{n}))^*, \forall \mathfrak{x}\mathfrak{F}(\mathfrak{x}), \varDelta \to \mathfrak{C}$, however, the following assertion still remains to be proved: 'On the basis of the known reduction rules for the sequents $\Gamma \to \forall \mathfrak{x}\mathfrak{F}(\mathfrak{x})$ and $\forall \mathfrak{x}\, \mathfrak{F}(\mathfrak{x}), (\mathfrak{F}(\mathfrak{n}))^*, \varDelta \to \mathfrak{C}$ a reduction rule is also statable for their mix sequent $\Gamma, (\mathfrak{F}(\mathfrak{n}))^*, \varDelta \to \mathfrak{C}$.' The proof for this will be postponed for the time being; once it has been carried out, the induction hypothesis can be applied as before, and from the fact that reduction rules are known for the sequents $\Gamma \to (\mathfrak{F}(\mathfrak{n}))^*$ and $(\mathfrak{F}(\mathfrak{n}))^*, \Gamma, \varDelta \to \mathfrak{C}$ it can be inferred that a reduction rule is also statable for their mix sequent $\Gamma, \Gamma, \varDelta \to \mathfrak{C}$, and hence for $\Gamma, \varDelta \to \mathfrak{C}$.

14.623.2. Suppose that $\mathfrak{D}$ has the form $\mathfrak{A} \& \mathfrak{B}$. Then the first reduction step on the sequent $\mathfrak{D}, \varDelta \to \mathfrak{C}$ yields $\mathfrak{A}, \mathfrak{A} \& \mathfrak{B}, \varDelta \to \mathfrak{C}$ (or $\mathfrak{B}, \mathfrak{A} \& \mathfrak{B}, \varDelta \to \mathfrak{C}$), or $\mathfrak{A}, \varDelta \to \mathfrak{C}$ (or $\mathfrak{B}, \varDelta \to \mathfrak{C}$). In the first reduction step on the sequent $\Gamma \to \mathfrak{A} \& \mathfrak{B}$, a choice can be made in such a way that $\Gamma \to \mathfrak{A}$ (or $\Gamma \to \mathfrak{B}$) results (according to 13.22).

If $\mathfrak{D}, \varDelta \to \mathfrak{C}$ has assumed the form without $\mathfrak{A} \& \mathfrak{B}$, we apply the *induction hypothesis* at once: Since reduction rules are known for the sequents $\Gamma \to \mathfrak{A}$ (or $\Gamma \to \mathfrak{B}$) and $\mathfrak{A}, \varDelta \to \mathfrak{C}$ (or $\mathfrak{B}, \varDelta \to \mathfrak{C}$) and since the mix formula $\mathfrak{A}$ (or $\mathfrak{B}$) contains fewer logical connectives than $\mathfrak{A} \& \mathfrak{B}$, a reduction rule is also statable for the mix sequent $\Gamma, \varDelta \to \mathfrak{C}$.

In the other case the following *assertion* is still *to be proved*: 'On the basis of the known reduction rules for the sequents $\Gamma \to \mathfrak{A} \& \mathfrak{B}$ and $\mathfrak{A} \& \mathfrak{B}, \mathfrak{A}, \varDelta \to \mathfrak{C}$ (or $\mathfrak{A} \& \mathfrak{B}, \mathfrak{B}, \varDelta \to \mathfrak{C}$) a reduction rule is also statable for their mix sequent $\Gamma, \mathfrak{A}, \varDelta \to \mathfrak{C}$ (or $\Gamma, \mathfrak{B}, \varDelta \to \mathfrak{C}$).' For, if this has been proved, it follows once again by an application of the induction hypothesis that, given reduction rules for $\Gamma \to \mathfrak{A}$ (or $\Gamma \to \mathfrak{B}$) and $\mathfrak{A}, \Gamma, \varDelta \to \mathfrak{C}$ (or $\mathfrak{B}, \Gamma, \varDelta \to \mathfrak{C}$), a reduction rule is also statable for the mix sequent $\Gamma, \Gamma, \varDelta \to \mathfrak{C}$, and hence for $\Gamma, \varDelta \to \mathfrak{C}$.

14.623.3. Suppose that $\mathfrak{D}$ has the form $\neg \mathfrak{A}$. The first reduction step then turns the sequent $\mathfrak{D}, \varDelta \to \mathfrak{C}$ into $\neg \mathfrak{A}, \varDelta \to \mathfrak{A}$ or $\varDelta \to \mathfrak{A}$. In its first reduction step (according to 13.23) the sequent $\Gamma \to \neg \mathfrak{A}$ then becomes $\mathfrak{A}, \Gamma \to 1 = 2$.

If $\mathfrak{D}, \varDelta \to \mathfrak{C}$ has assumed the form $\varDelta \to \mathfrak{A}$, we apply the *induction hypothesis* at once: Since reduction rules are known for the sequents $\varDelta \to \mathfrak{A}$ and $\mathfrak{A}, \Gamma \to 1 = 2$, and since the mix formula $\mathfrak{A}$ contains fewer logical connectives than $\neg \mathfrak{A}$, a reduction rule is also statable for the mix sequent

$\varDelta, \varGamma \to 1 = 2$. The same therefore also holds for $\varGamma, \varDelta \to \mathfrak{C}$; for $\mathfrak{C}$, like $1 = 2$, is a false minimal formula.

In the other case the following *assertion* is still *to be proved*: 'On the basis of the reduction rules known for the sequents $\varGamma \to \neg \mathfrak{A}$ and $\mathfrak{A}, \varDelta \to \mathfrak{A}$, a reduction rule is also statable for their mix sequent $\varGamma, \varDelta \to \mathfrak{A}$.' If this has been proved, it follows again by the use of the *induction hypothesis* that, given the reduction rules for $\varGamma, \varDelta \to \mathfrak{A}$ and $\mathfrak{A}, \varGamma \to 1 = 2$, a reduction rule is also statable for the mix sequent $\varGamma, \varDelta, \varGamma \to 1 = 2$, and hence also for $\varGamma, \varDelta \to \mathfrak{C}$.

14.63. CONCLUSION OF THE PROOF. In several of the cases discussed an *assertion* was made whose *proof* had been *postponed*. In each case this assertion was of the following form: 'On the basis of the known reduction rules for the sequent $\varGamma \to \mathfrak{D}$ and a sequent of the form $\mathfrak{D}, \varDelta^* \to \mathfrak{C}^*$ which has resulted from $\mathfrak{D}, \varDelta \to \mathfrak{C}$ by one or several *reduction steps* carried out according to the appropriate reduction rule, a reduction rule is also statable for their *mix sequent* $\varGamma, \varDelta^* \to \mathfrak{C}^*$.' Here the sequents $\varGamma \to \mathfrak{D}$ and $\mathfrak{D}, \varDelta^* \to \mathfrak{C}^*$ contained no free variables and no minimal terms.

This assertion is quite obviously of the same kind as that made at 14.60 and it is precisely for this that the entire proof was intended. The mix formula $\mathfrak{D}$ is *the same* as that in the earlier assertion; the sequent $\varGamma \to \mathfrak{D}$ plays *the same role*; in place of $\mathfrak{D}, \varDelta \to \mathfrak{C}$, however, there now occurs a sequent *obtained* from the latter *by one or several reduction steps*.

In order to prove the new assertion we now apply exactly the *same* inferences as before (14.61 to 14.623.3); hence there (possibly) remains to be proved *another* assertion of the *same* kind, where the second sequent *once again results* from $\mathfrak{D}, \varDelta^* \to \mathfrak{C}^*$ by *at least one reduction step*.

Continuing in this way, we must *reach the end in finitely many steps*, i.e., the completion of the proof. This is so since the continual reduction of the sequent $\mathfrak{D}, \varDelta \to \mathfrak{C}$, which proceeds according to the *reduction rule* stated for that sequent, must (13.6) lead to the reduced form in finitely many steps, so that no further reduction is required (14.61) (assuming that a further reduction did not become unnecessary in one of the earlier steps).

15.1. The techniques of proof used in the consistency proof can certainly be considered '*finitist*' in the sense outlined in detail in section III. This cannot be 'proved' if for no other reason than the fact that the notion of 'finitist' is not unequivocably formally defined and cannot in fact be delimited in this way. All we can do is to examine every individual inference from this point of view and try to assess whether that inference is in harmony with the finitist sense of the concepts that occur and make sure that it does not rest on an inadmissible 'actualist' interpretation of these concepts. I shall discuss briefly the most relevant passages of the consistency proof:

The *objects* of the consistency proof, as of proof theory in general, are certain *symbols* and *expressions*, such as terms, formulae, sequents, derivations, not to forget the natural numbers. All these objects are defined (3.2, 5.2) by *construction rules* corresponding to the definition of the natural numbers (8.11); in each case such a rule indicates how more and more such objects can be constructed step by step.

Several *functions* and *predicates* were furthermore applied to these objects and these were *decidably defined* in the sense of 8.12. E.g., the *function* 'the endformula of a derivation', the *predicate* 'containing at least one ∀ or ∃-symbol' and many others.

In the transformation of the derivation in § 12 only quite harmless, entirely finitist concepts and inferences were required.

The concept of the *'reduction rule'* which is central to the consistency proof is of a *special kind*. The proposition 'for a certain sequent a reduction rule is known' contains the concepts 'all' and 'there exists' in that it asserts that the reduction rule *exists*, and that the reduction procedure to be carried out according to the rule is defined for *all* possible choices of numerals to be substituted in the case where a choice arises in the reduction (13.6), and that the procedure terminates in finitely many steps, i.e., that once again *there exists* a natural number in each case which indicates the total number of steps. (This number generally depends on the choices made.)

The two instances of 'there exists' in the reducibility proof were actually always used *finitistically* in the sense of 10.3. Hence the expressions: 'a rule is *known, given, statable*'. At 14.2, e.g., the reduction rule for basic logical sequents was *stated* precisely and the *total number* of required reduction steps can be inferred at once. In 14.3–14.44 it was stated in each case how an already existing reduction rule must be modified *in order to obtain* from it a reduction rule for a further sequent. In the remaining proof the transfinite 'there exists', in connection with 'there exists a reduction rule', was always used in the *finitist* sense that such a rule was given or (in the case of 'introduction' inferences) a new rule could be *stated*.

Corresponding remarks hold for the 'there exists' in relation to the *total number of reduction steps*; with the formulation of a reduction rule on the basis of known reduction rules, there is always connected the possibility of determining the total number of newly arising (or disappearing) reduction steps.

In the *lemma* an essentially novel element arises with the transfinite use of the concept 'follows' in expressions of the form 'if a certain proposition holds then a certain other proposition also holds'. Here we must recall the objections which were raised in 11.1 against the unrestricted use of his concept. It turns out however that in the consistency proof the 'follows' occurs only in *one* context: 'If reduction rules are known for two particular sequents, then a reduction rule is also statable for a certain third sequent formed from the former sequent.' From the finitist standpoint this use of 'follows' is unobjectionable; after all, no *nesting* whatever of 'follows' concepts occurs; here the 'follows' is to be understood simply as an expression for the fact that by means of finitistically correct inferences the validity of a proposition (not involving 'follows' concepts) is derivable from the validity of another proposition (not involving 'follows' concepts). (The 'follows' is interpreted 'metatheoretically', as it were.) The forms of inference of the 'follows'-introduction and 'follows'-elimination are in harmony with this interpretation (cf. 11.1), and these are precisely the inferences occurring in the proof of the lemma (14.6) and in its applications (at 14.441, 14.442 and 14.443).

In applying the concept 'all' in the consistency proof, I have not used the unwieldy finitist expression given for it in 10.11; here the distinction between the actualist finitist interpretations has no bearing on our reasoning in any case.

Complete inductions occurred repeatedly in the consistency proof (at 14.6, 14.63, 14.443, and elsewhere). These are to be interpreted according to 10.5 and in this sense they are quite unobjectionable even *in the case* where the induction hypothesis is a *transfinite* proposition.

The *negation* of a transfinite proposition occurs only once in the entire proof (at 13.90) and only in a harmless form in which the proposition concerned leads to a quite elementary contradiction. The negation can actually be *avoided* altogether if for 'consistency' the following positive expression is used: 'Every derivation has an endformula which does not have the form $\mathfrak{A} \ \& \ \neg \ \mathfrak{A}$.' Here the 'not' is no longer transfinite.

I hope that these reflections have helped to make the *finitist character* of the techniques of proof used in the consistency proof sufficiently credible.

15.11. I consider it not impossible that the inferences used in the consistency proof can be reduced to *still more elementary ones*, so that the techniques of proof that have to be presupposed as correct without further justification *can be restricted* further.

15.2. In order to examine the extent to which the consistency proof ... example, Gödel's version.

The forms of inference in the consistency proof are then *none other* than those presented in our formalization of number theory; only the *concept of the reduction rule* occupies a special position. In its general form it can *not* be formulated ((dargestellt)) by the techniques of elementary number theory. For this reason the consistency proof is in harmony with Gödel's theorem.

In this connection ... could not be included in the *earlier* consistency proofs (2.4).

5. THE CONSISTENCY OF THE SIMPLE THEORY OF TYPES

For the following I presuppose a knowledge of the basic facts of mathematical logic, in particular, a knowledge of propositional and predicate logic and of type theory[55].

The consistency of *propositional logic* is usually proved by the calculation of truth values[56]. The consistency of *predicate logic*[57] is provable by a simple extension of this procedure[58]. Here the basic idea is to specialize the domain of objects so that it contains only a single element.

Starting with the same basic idea, we shall prove the consistency of the *simply theory of types* in an elementary way. By the 'simple theory of types' I mean the 'extended predicate calculus' in the sense of H.-A.[59], together with Russell's hierarchy of types of predicates, but without the finer subdivision of predicates in the so-called 'ramified theory of types'[60]. More about this in § 1. The simple theory of types comprises essentially the system of the 'Principia Mathematica' together with the 'axiom of choice', but without the 'axiom of infinity'[61]. It is only by the inclusion of this axiom, i.e., by the stipulation of an *infinite* domain of objects, that the essential difficulties, whose resolution constitutes the main task of Hilbert's proof theory, arise. It is noteworthy that an omission of the hierarchy of types leads to *contradictions* even *without* the axiom of infinity, e.g., to 'Russell's antinomy'[62]; the following proof will show that this is no longer possible if the distinction of types is made.

§ 1. The formal structure of the simple theory of types[63]

German and Greek letters will be used as syntactic variables. Subst $\mathfrak{A} \begin{pmatrix} \mathfrak{b} \\ \mathfrak{c} \end{pmatrix}$ stands for the expression which results from an expression $\mathfrak{A}$ if every occurrence of the symbol $\mathfrak{b}$ in $\mathfrak{A}$ is replaced by the symbol $\mathfrak{c}$.

1.1. Definition of a '*formula*' (formal counterpart of a proposition in the simple theory of types).

1.11. We shall need *free* and *bound* variables of different *types*. As variables we shall admit arbitrary symbols that have not yet been assigned a different use, but it must be stated in each case whether the symbol concerned is to represent a free or a bound variable; the type of the variables will be indicated by a numerical subscript. Types are: 0, 1, 2, 3, ... (Informally, variables of type 0 stand for arbitrary elements of the domain of objects, variables of type 1 for arbitrary properties (= one-place predicates) of objects and, in general (for $v \geq 1$), variables of type $v+1$ for arbitrary properties of properties of type v. The properties represent at the same time the 'classes' (= sets) of elements for which the property concerned holds.)

Usually, *multi-place* predicates are used, and so are predicates of predicates, etc., necessitating a further hierarchy of predicates within the various types[64].) As observed by Gödel[65], however, this second hierarchy is dispensable.

1.12. An expression of the form $\mathfrak{a}_{v+1}\mathfrak{b}_v$, where $\mathfrak{a}_{v+1}$ and $\mathfrak{b}_v$ are arbitrary free variables (with the subscripts designating their types, $v \geq 0$), is a formula. (Informal meaning: 'The property $\mathfrak{a}_{v+1}$ holds for $\mathfrak{b}_v$'; or: '$\mathfrak{b}_v$ is an element of the set $\mathfrak{a}_{v+1}$'.)

If $\mathfrak{A}$ is a formula, then so is $\neg \mathfrak{A}$. (Informal meaning: '$\mathfrak{A}$ does not hold'.) If $\mathfrak{A}$ and $\mathfrak{B}$ are formulae, then so is $\mathfrak{A} \& \mathfrak{B}$ ('$\mathfrak{A}$ holds and $\mathfrak{B}$ holds'), $\mathfrak{A} \vee \mathfrak{B}$ ('$\mathfrak{A}$ holds or $\mathfrak{B}$ holds'), $\mathfrak{A} \supset \mathfrak{B}$ ('if $\mathfrak{A}$ holds, then $\mathfrak{B}$ holds') and $\mathfrak{A} \supset \subset \mathfrak{B}$ ('$\mathfrak{A}$ holds if and only if $\mathfrak{B}$ holds').

If $\mathfrak{A}$ is a formula in which the free variable $\mathfrak{a}_v$ occurs and in which the bound variable $\mathfrak{x}_v$ does not occur, then $\forall \mathfrak{x}_v \, \text{Subst} \, \mathfrak{A} \begin{pmatrix} \mathfrak{a}_v \\ \mathfrak{x}_v \end{pmatrix}$ and $\exists \mathfrak{x}_v \, \text{Subst} \, \mathfrak{A} \begin{pmatrix} \mathfrak{a}_v \\ \mathfrak{x}_v \end{pmatrix}$ are also formulae. (Informal meaning: 'Subst $\mathfrak{A} \begin{pmatrix} \mathfrak{a}_v \\ \mathfrak{x}_v \end{pmatrix}$ holds for all $\mathfrak{x}_v$'; or: 'there is an $\mathfrak{x}_v$, so that Subst $\mathfrak{A} \begin{pmatrix} \mathfrak{a}_v \\ \mathfrak{x}_v \end{pmatrix}$ holds'.)

1.13. No expressions other than those formed in accordance with 1.12 are formulae. Brackets will serve to display the structure of a formula unambiguously. *Example* of a *formula*:

$$\neg \exists y_2 \, [a_1 b_0 \, \& \, \forall z_4 \, (c_3 y_2 \supset \subset z_4 a_3)].$$

a_1, b_0, c_3 and a_3 are free variables, y_2 and z_4 bound variables.

A formula, or part of a formula, of the form

$$\forall \mathfrak{x}_{v+1} \, (\mathfrak{x}_{v+1} \mathfrak{r}_v \supset \mathfrak{x}_{v+1} \mathfrak{s}_v),$$

where $\mathfrak{r}_v, \mathfrak{s}_v$ and $\mathfrak{x}_{v+1}$ are variables, may be abbreviated by: $\mathfrak{r}_v = \mathfrak{s}_v$.

1.2. Definition of a *'derivation'* (formal counterpart of a *proof* in the simple theory of types).

A derivation is a sequence of formulae of which each is either a 'basic formula' or has resulted from one of the preceding formulae in the sequence by the application of a 'rule of inference'. The admissible basic formulae and rules of inference will now be stated.

1.21. (Propositional logic.) Among the basic formulae are all formulae that result from 'logically true combinations of propositions', in the sense of H.-A.[66], by the substitution of arbitrary formulae, in our sense (1.1), for the 'basic propositions'. In addition, we use the following rule of inference: From two formulae of the form $\mathfrak{A}$ and $\mathfrak{A} \supset \mathfrak{B}$ the formula $\mathfrak{B}$ may be derived.

1.22. (Predicate logic.) Among the basic formulae are all formulae of the form $(\forall \mathfrak{x}_\nu \mathfrak{A}) \supset \text{Subst } \mathfrak{A} \begin{pmatrix} \mathfrak{x}_\nu \\ \mathfrak{a}_\nu \end{pmatrix}$ and $\text{Subst } \mathfrak{A} \begin{pmatrix} \mathfrak{x}_\nu \\ \mathfrak{a}_\nu \end{pmatrix} \supset \exists \mathfrak{x}_\nu \mathfrak{A}$, where $\forall \mathfrak{x}_\nu \mathfrak{A}$ and $\exists \mathfrak{x}_\nu \mathfrak{A}$ stand for arbitrary formulae of that form and $\mathfrak{a}_\nu$ for an arbitrary free variable; ν continues to stand for an integer ≥ 0.

Rules for $\forall$ and $\exists$: From a formula of the form $\mathfrak{A} \supset \mathfrak{B}$ the formula $\mathfrak{A} \supset \forall \mathfrak{x}_\nu \text{ Subst } \mathfrak{B} \begin{pmatrix} \mathfrak{a}_\nu \\ \mathfrak{x}_\nu \end{pmatrix}$ may be derived, where $\mathfrak{a}_\nu$ stands for a free variable occurring in $\mathfrak{B}$ but not in $\mathfrak{A}$, and $\mathfrak{x}_\nu$ stands for a bound variable not occurring in $\mathfrak{B}$. From a formula of the form $\mathfrak{B} \supset \mathfrak{A}$ the formula $\left(\exists \mathfrak{x}_\nu \text{ Subst } \mathfrak{B} \begin{pmatrix} \mathfrak{a}_\nu \\ \mathfrak{x}_\nu \end{pmatrix}\right) \supset \mathfrak{A}$ may be derived, where $\mathfrak{a}_\nu$ stands for a free variable occurring in $\mathfrak{B}$, but not in $\mathfrak{A}$, and $\mathfrak{x}_\nu$ for a bound variable not occurring in $\mathfrak{B}$.

1.23. (*Equality*.) Every formula of the form

$$[\forall \mathfrak{x}_\nu (\mathfrak{a}_{\nu+1} \mathfrak{x}_\nu \supset \subset \mathfrak{b}_{\nu+1} \mathfrak{x}_\nu)] \supset \mathfrak{a}_{\nu+1} = \mathfrak{b}_{\nu+1}$$

is a basic formula, where $\mathfrak{a}_{\nu+1}$ and $\mathfrak{b}_{\nu+1}$ are to be replaced by arbitrary free variables and $\mathfrak{x}_\nu$ by an arbitrary bound variable.

1.24. (Axioms of *set formation*.) Every formula of the form

$$\exists \mathfrak{x}_{\nu+1} \forall \mathfrak{y}_\nu \left[\mathfrak{x}_{\nu+1} \mathfrak{y}_\nu \supset \subset \text{Subst } \mathfrak{A} \begin{pmatrix} \mathfrak{a}_\nu \\ \mathfrak{y}_\nu \end{pmatrix} \right]$$

is a basic formula; here $\mathfrak{A}$ is to be replaced by an arbitrary formula in which the (arbitrary) free variable $\mathfrak{a}_\nu$ occurs; $\mathfrak{x}_{\nu+1}$ and $\mathfrak{y}_\nu$ are to be replaced by arbitrary bound variables not occurring in $\mathfrak{A}$. (Informal meaning ('axiom of comprehension'): Any arbitrarily formed proposition $\mathfrak{A}$ in which the free variable $\mathfrak{a}_\nu$ occurs, determines a *set* $(\mathfrak{x}_{\nu+1})$, viz., the set of those $\mathfrak{a}_\nu$, for which $\mathfrak{x}_{\nu+1}$ holds).

Furthermore, every formula of the form

$\{[\forall \mathfrak{x}_{\nu+1}(\mathfrak{a}_{\nu+2}\mathfrak{x}_{\nu+1} \supset \exists \mathfrak{y}_\nu \mathfrak{x}_{\nu+1}\mathfrak{y}_\nu)] \,\&\, \forall \mathfrak{z}_{\nu+1} \forall \mathfrak{u}_{z+1}\,[(\mathfrak{a}_{\nu+2}\,\mathfrak{z}_{\nu+1}$
$\&\,\mathfrak{a}_{\nu+2}\mathfrak{u}_{\nu+1}\,\&\,\neg\,\mathfrak{z}_{\nu+1} = \mathfrak{u}_{\nu+1}) \supset \neg\,\exists \mathfrak{v}_\nu\,(\mathfrak{z}_{\nu+1}\mathfrak{v}_\nu\,\&\,\mathfrak{u}_{\nu+1}\mathfrak{v}_\nu)]\}$
$\supset \exists \mathfrak{w}_{\nu+1} \forall \mathfrak{r}_{\nu+1}\,\{\mathfrak{a}_{\nu+2}\mathfrak{r}_{\nu+1} \supset \exists \mathfrak{s}_\nu\,[\mathfrak{w}_{\nu+1}\mathfrak{s}_\nu\,\&\,\mathfrak{r}_{\nu+1}\mathfrak{s}_\nu$
$\&\,\neg\,\exists \mathfrak{t}_\nu\,(\mathfrak{w}_{\nu+1}\mathfrak{t}_\nu\,\&\,\mathfrak{r}_{\nu+1}\mathfrak{t}_\nu\,\&\,\neg\,\mathfrak{t}_\nu = \mathfrak{s}_\nu)]\}$

is a basic formula; here $\mathfrak{a}_{\nu+2}$ is to be replaced by an arbitrary free variable, and $\mathfrak{x}_{\nu+1}, \mathfrak{y}_\nu, \mathfrak{z}_{\nu+1}, \mathfrak{u}_{\nu+1}, \mathfrak{v}_\nu, \mathfrak{w}_{\nu+1}, \mathfrak{r}_{\nu+1}, \mathfrak{s}_\nu$ and $\mathfrak{t}_\nu$ are to be replaced by arbitrary mutually distinct bound variables. (Informal meaning ('axiom of choice'): For every arbitrary totality $(\mathfrak{a}_{\nu+2})$ of non-empty pairwise disjoint sets, there exists a *choice set* $(\mathfrak{w}_{\nu+1})$ which has one and only one element $(\mathfrak{s}_\nu)$ in common with each set in this totality.)

§ 2. The consistency proof

2.1. *Underlying ideas.* We shall show that the formal system developed in § 1 is *consistent*, i.e., that in it no formula of the form $\mathfrak{A}\,\&\,\neg\,\mathfrak{A}$ can be derived.

Informally interpreted, the simple theory of types is intended to be valid for every arbitrarily stipulated (non-empty) domain of objects. If it were contradictory, then this would already have to become apparent in the special case where the domain of objects consists only of a *single* element. In this case, however, all propositions are decidable; this is due to the fact that every type consists of only *finitely many* objects, viz.: type 0 contains one object; type 1 contains the set containing that object and the set not containing that object, hence two different sets; correspondingly, type 2 contains four different sets, etc. In this special case it can easily be verified that all propositions are 'true' and that, *a fortiori*, no contradiction can be proved.

2.2. THE PROOF PROPER. Suppose that an arbitrary derivation in the simple theory of types (1.2) is given. We shall *transform* the derivation in four steps.

We first introduce the following *auxiliary symbols* (standing for 'the single object of type 0', the two 'sets of type 1', etc.):

$$\gamma_0^{(1)};\ \gamma_1^{(1)},\ \gamma_1^{(2)};\ \gamma_2^{(1)},\ \gamma_2^{(2)},\ \gamma_2^{(3)},\ \gamma_2^{(4)};\ \ldots;$$

in general, ρ_ν symbols with the subscript ν: $\gamma_\nu^{(1)}\ldots\gamma_\nu^{(\rho_\nu)}$, where $\rho_\nu = 2^{\rho_{\nu-1}}$ (for $\nu \geqq 1$).

2.21. *First transformation step* (replacement of the free variables by the γ's).

We take an arbitrary formula $\mathfrak{A}$ of the derivation containing the *free variable* $\mathfrak{a}_\nu$, and replace it by the following sequence of expressions:

Subst $\mathfrak{A} \begin{pmatrix} \mathfrak{a}_\nu \\ \gamma_\nu^{(1)} \end{pmatrix}, \ldots,$ Subst $\mathfrak{A} \begin{pmatrix} \mathfrak{a}_\nu \\ \gamma_\nu^{(\rho_\nu)} \end{pmatrix}$. This transformation must be repeated until all free variables have been eliminated; in other words, the newly resulting expressions must also be dealt with in the same way if they still contain free variables.

2.22. *Second transformation step* (replacement of the bound variables by the γ's).

Suppose that a certain part of one of the expressions that has resulted by step 1 (or of one of the initial formulae, if these were left unaffected by step 1) has the form $\forall \mathfrak{x}_\nu \mathfrak{A}$ or $\exists \mathfrak{x}_\nu \mathfrak{A}$. In that case we replace it by the conjunction (disjunction) of the expressions

$$\text{Subst } \mathfrak{A} \begin{pmatrix} \mathfrak{x}_\nu \\ \gamma_\nu^{(1)} \end{pmatrix}, \ldots, \text{Subst } \mathfrak{A} \begin{pmatrix} \mathfrak{x}_\nu \\ \gamma_\nu^{(\rho_\nu)} \end{pmatrix}.$$

This transformation is repeated (and is applied to newly resulting expressions, if necessary) until all bound variables have been eliminated. It is easily seen that *the order* in which these transformations are carried out has no bearing whatever on the final result.

2.23. *Third transformation step* (substitution of truth-values for the γ's).

We introduce two further auxiliary symbols: The 'truth-values' $\vee$ (for 'truth') and $\wedge$ (for 'falsity').

Suppose now that some part of one of the expressions that has resulted from the derivation formulae by the first and second transformation steps has the form $\gamma_{\nu+1}^{(\sigma)} \gamma_\nu^{(\tau)}$. It is replaced by $\vee$ or $\wedge$ according to the following instruction: $\gamma_1^{(1)} \gamma_0^{(1)}$ is replaced by $\vee$, $\gamma_1^{(2)} \gamma_0^{(1)}$ by $\wedge$; in general: The symbols $\vee$ and $\wedge$ must be correlated with the expressions $\gamma_{\nu+1}^{(\lambda)} \gamma_\nu^{(\mu)}$ ($\nu \geq 0; \lambda = 1, \ldots, \rho_{\nu+1}; \mu = 1, \ldots, \rho_\nu$) in such a way that to every possible distribution of the symbols $\vee$ and $\wedge$ over the symbols γ_ν there corresponds exactly one $\gamma_{\nu+1}$ whose combinations with the γ_ν's receive the corresponding values $\vee$ and $\wedge$. This can be done since the total number of the γ_ν's equals ρ_ν, and since the total number of possible distributions of $\vee$'s and $\wedge$'s over the γ_ν's is therefore $2^{\rho_\nu} = \rho_{\nu+1}$ and thus equal to the total number of the $\gamma_{\nu+1}$'s. (Informal sense: The $\gamma_{\nu+1}$'s are intended to represent all possible sets of γ_ν's.)

The same transformation is repeated until all γ's have been eliminated. After all, the γ's could occur nowhere except in one of the $\gamma_{\nu+1}^{(\sigma)} \gamma_\nu^{(\tau)}$'s (1.12).

2.24. *Fourth transformation step* (calculation of truth-values).

All formulae of the original derivation have now been replaced by expressions containing only the symbols $\vee$ and $\wedge$, linked by the propositional connectives $\neg$, &, $\vee$, $\supset$, $\supset \subset$. These expressions must now be 'evaluated' in the way familiar from propositional logic (i.e., $\vee$ & $\vee$ is replaced by $\vee$; $\vee$ & $\wedge$ is replaced by $\wedge$, etc.) so that every expression is finally reduced to only a single symbol $\vee$ or $\wedge$.

2.25. Conclusion of the proof. I now assert: The derivation which has been transformed in the four described steps consists only of expressions of the form $\vee$.

Proof: A basic formula of propositional logic (1.21) must obviously always have taken the value $\vee$: the reason is that *identical* subformulae were transformed in an *identical* manner in all transformation steps and whatever truth-value they may have received in steps 3 and 4, the total value obtained in step 4 must still have turned out to be $\vee$, since the initial formula was 'identically true', in the sense of propositional logic.

If, in addition, the truth-values obtained in the transformation of the formulae $\mathfrak{A}$ and $\mathfrak{A} \supset \mathfrak{B}$ were always $\vee$, then the expressions resulting from formula $\mathfrak{B}$ must all eventually have become $\vee$. For suppose $\mathfrak{B}^*$ is any expression resulting from $\mathfrak{B}$ by step 1. Then among the expressions that have resulted from the formulae $\mathfrak{A}$ and $\mathfrak{A} \supset \mathfrak{B}$ (in step 1) there must clearly be at least one pair of expressions of the form $\mathfrak{A}^*$ and $\mathfrak{A}^* \supset \mathfrak{B}^*$. On the other hand, $\mathfrak{A}^*$ eventually takes the value $\vee$, so does $\mathfrak{A}^* \supset \mathfrak{B}^*$; and this can happen only if $\mathfrak{B}^*$ also assumes the value $\vee$ (since $\vee \supset \wedge$ results in $\wedge$). (Here it is essential that identical parts of different expressions were transformed in *identical* ways in steps 2 to 4.)

In step 1, a basic formula of the form $(\forall \mathfrak{x}_\nu\, \mathfrak{A}) \supset \mathrm{Subst}\, \mathfrak{A} \begin{pmatrix} \mathfrak{x}_\nu \\ \mathfrak{a}_\nu \end{pmatrix}$ resulted in expressions of the form $(\forall \mathfrak{x}_\nu\, \mathfrak{A}^*) \supset \mathrm{Subst}\, \mathfrak{A}^* \begin{pmatrix} \mathfrak{x}_\nu \\ \gamma_\nu^{(\sigma)} \end{pmatrix}$; $1 \leq \sigma \leq \rho_\nu$.

In step 2, these became

$$\left[\mathrm{Subst}\, \mathfrak{A}^{**} \begin{pmatrix} \mathfrak{x}_\nu \\ \gamma_\nu^{(1)} \end{pmatrix} \& \ldots \& \mathrm{Subst}\, \mathfrak{A}^{**} \begin{pmatrix} \mathfrak{x}_\nu \\ \gamma_\nu^{(\rho_\nu)} \end{pmatrix} \right] \supset \mathrm{Subst}\, \mathfrak{A}^{**} \begin{pmatrix} \mathfrak{x}_\nu \\ \gamma_\nu^{(\sigma)} \end{pmatrix}.$$

Since one component of the square bracket must be equal to the expression on the right of the implication symbol, the calculation (in steps 3 and 4) must, in every case, have yielded the value $\vee$. In the case of a basic formula for $\exists$ (1.22) our reasoning is analogous.

Now the rule for $\forall$ (1.22): In step 2, all expressions of the form $\mathfrak{A} \supset \forall \mathfrak{x}_\nu\, \mathrm{Subst}\, \mathfrak{B} \begin{pmatrix} \mathfrak{a}_\nu \\ \mathfrak{x}_\nu \end{pmatrix}$ became expressions of the form

$$\mathfrak{A}^* \supset \left[\text{Subst } \mathfrak{B}^* \begin{pmatrix} \mathfrak{a}_\nu \\ \gamma_\nu^{(1)} \end{pmatrix} \& \ldots \& \text{Subst } \mathfrak{B}^* \begin{pmatrix} \mathfrak{a}_\nu \\ \gamma_\nu^{(\rho_\nu)} \end{pmatrix} \right].$$

On the other hand, every single one of the expressions

$$\mathfrak{A}^* \supset \text{Subst } \mathfrak{B}^* \begin{pmatrix} \mathfrak{a}_\nu \\ \gamma_\nu^{(\sigma)} \end{pmatrix} \qquad \sigma = 1, \ldots, \rho_\nu$$

must obviously have occurred among the expressions which resulted from the formula $\mathfrak{A} \supset \mathfrak{B}$ by steps 1 and 2. If all of these expressions had received the value $\vee$ in steps 3 and 4, then the above expression would also have taken that value. (For either $\mathfrak{A}^*$ became $\wedge$, in which case our assertion follows, or $\mathfrak{A}^*$ became $\vee$, in which case all of the expressions Subst $\mathfrak{B}^* \begin{pmatrix} \mathfrak{a}_\nu \\ \gamma_\nu^{(\sigma)} \end{pmatrix}$ must also have received the value $\vee$, so did therefore their conjunction.) Applications of the rule for $\exists$ are dealt with analogously.

In the case of a basic formula of the kind introduced in 1.23, the first reduction step resulted in expressions of the form

$$[\forall \mathfrak{x}_\nu \, (\gamma_{\nu+1}^{(\sigma)} \mathfrak{x}_\nu \supset \subset \gamma_{\nu+1}^{(\tau)} \mathfrak{x}_\nu)] \supset \gamma_{\nu+1}^{(\sigma)} = \gamma_{\nu+1}^{(\tau)}$$
$$(1 \leq \sigma \leq \rho_{\nu+1}; \quad 1 \leq \tau \leq \rho_{\nu+1}).$$

In steps 3 and 4, the square bracket can have become $\vee$ if all of the conjuncts $\gamma_{\nu+1}^{(\sigma)} \gamma_\nu^{(\mu)} \supset \subset \gamma_{\nu+1}^{(\tau)} \gamma_\nu^{(\mu)}$, for $\mu = 1, \ldots, \rho_\nu$, arising from step 2, took the value $\vee$; this in turn was possible only if σ was the same number as τ. In that case $\gamma_{\nu+1}^{(\sigma)} = \gamma_{\nu+1}^{(\tau)}$, i.e., (1.23) $\forall \mathfrak{y}_{\nu+2}(\mathfrak{y}_{\nu+2} \gamma_{\nu+1}^{(\sigma)} \supset \mathfrak{y}_{\nu+2} \gamma_{\nu+1}^{(\tau)})$ obviously yielded the value $\vee$. The fourth step therefore certainly resulted in the value $\vee$ in each case. (Since $\vee \supset \vee$ yields $\vee$, and if the square bracket took the value $\wedge$, then the total value is $\vee$; in any case.)

In the case of a basic formula that is an instance of the 'axiom of comprehension' (1.24), steps 1 and 2 lead to expressions of the form:

$$\left[\left(\gamma_{\nu+1}^{(1)} \gamma_\nu^{(1)} \supset \subset \text{Subst } \mathfrak{A}^* \begin{pmatrix} \mathfrak{a}_\nu \\ \gamma_\nu^{(1)} \end{pmatrix} \right) \& \right.$$
$$\left. \ldots \& \left(\gamma_{\nu+1}^{(1)} \gamma_\nu^{(\rho_\nu)} \supset \subset \text{Subst } \mathfrak{A}^* \begin{pmatrix} \mathfrak{a}_\nu \\ \gamma_\nu^{(\rho_\nu)} \end{pmatrix} \right) \right] \vee$$
$$\ldots \vee \left[\left(\gamma_{\nu+1}^{(\rho_\nu+1)} \gamma_\nu^{(1)} \supset \subset \text{Subst } \mathfrak{A}^* \begin{pmatrix} \mathfrak{a}_\nu \\ \gamma_\nu^{(1)} \end{pmatrix} \right) \& \right.$$
$$\left. \ldots \& \left(\gamma_{\nu+1}^{(\rho_\nu+1)} \gamma_\nu^{(\rho_\nu)} \supset \subset \text{Subst } \mathfrak{A}^* \begin{pmatrix} \mathfrak{a}_\nu \\ \gamma_\nu^{(\rho_\nu)} \end{pmatrix} \right) \right].$$

§ 2, THE CONSISTENCY PROOF

Suppose now that steps 3 and 4 have given rise to certain truth-values for the expressions Subst $\mathfrak{A}^* \begin{pmatrix} \mathfrak{a}_\nu \\ \gamma_\nu^{(\mu)} \end{pmatrix}$: Whatever these expressions were, there is always precisely *one* expression among the square brackets in which the truth-values to be substituted in step 3 for the expressions $\gamma_{\nu+1}^{(\sigma)}\gamma_\nu^{(\tau)}$ (for $\tau = 1, \ldots, \rho_\nu$) standing on the left, agree entirely with the truth-values to be substituted for the expressions Subst $\mathfrak{A}^* \begin{pmatrix} \mathfrak{a}_\nu \\ \gamma_\nu^{(\tau)} \end{pmatrix}$ standing on their right. This follows from the fact that at 2.23 the truth-values for $\gamma\gamma$-expressions were arranged in such a way that to every possible distribution of truth-values over the numbers from 1 to ρ_ν there corresponds exactly one $\gamma_{\nu+1}$ whose combinations with the γ_ν's received precisely these truth-values.

This means that the calculation of the truth-value of the square bracket concerned must have yielded the value $\vee$; the entire disjunction, therefore, also takes this value.

In the case of a formula that was formed by means of an application of the 'axiom of choice', step 1 leaves the structure unchanged, except for a $\gamma_{\nu+2}^{(\mu)}$ in the place of $\mathfrak{a}_{\nu+2}$. Let us consider the appearance of such an expression after step 2, in which all $\forall$-expressions were replaced by certain conjunctions and all $\exists$-expressions by certain disjunctions. If we call to mind the *informal sense* of the individual parts of which this formula is composed, then it follows easily that the total expression is 'true', (since the axiom of choice is trivially valid for finite domains) and this means formally that the formula must receive the value $\vee$. More precisely: If the value were $\wedge$, then the left component of the entire implication would have to have the value $\vee$ and the right component the value $\wedge$. From the former it follows that for every $\gamma_{\nu+1}^{(\sigma)}$ ($\sigma = 1, \ldots, \rho_{\nu+1}$) the implication

$$\gamma_{\nu+2}^{(\mu)}\gamma_{\nu+1}^{(\sigma)} \supset (\gamma_{\nu+1}^{(\sigma)}\gamma_\nu^{(1)} \vee \ldots \vee \gamma_{\nu+1}^{(\sigma)}\gamma_\nu^{(\rho_\nu)})$$

has the value $\vee$ and this means that if $\gamma_{\nu+2}^{(\mu)}\gamma_{\nu+1}^{(\sigma)}$ has the value $\vee$, then at least one of the expressions $\gamma_{\nu+1}^{(\sigma)}\gamma_\nu^{(\tau)}$ ($\tau = 1, \ldots, \rho_\nu$) has the value $\vee$. It follows analogously that the second part of the left component of the entire implication also has the value $\vee$ and so does its informal interpretation in terms of the γ's, i.e., for two distinct $\gamma_{\nu+1}$, for both of which $\gamma_{\nu+2}^{(\mu)}\gamma_{\nu+1}$ takes the value $\vee$, there exists no γ_ν which, in combination with either one of these two, yields the value $\vee$. Now we form the 'choice set', i.e., we choose for every $\gamma_{\nu+1}^{(\sigma)}$ for which $\gamma_{\nu+2}^{(\mu)}\gamma_{\nu+1}^{(\sigma)}$ has the value $\vee$, any one of the $\gamma_\nu^{(\tau)}$ for which the $\gamma_{\nu+1}^{(\sigma)}\gamma_\nu^{(\tau)}$ has the value $\vee$, and consider that $\gamma_{\nu+1}^{(\lambda)}$ which yields the value $\vee$ when combined with the chosen $\gamma_\nu^{(\tau)}$'s, and which does not yield this value in combination with any other $\gamma_\nu^{(\tau)}$. Such a $\gamma_{\nu+1}^{(\lambda)}$ must exist (2.23). By considering that component of the disjunction obtained from $\exists \mathfrak{w}_{\nu+1} \ldots$ which corresponds to this 'choice set', we can now see quite easily that it must have received the value $\vee$; the entire expression cannot, therefore, have taken the value $\wedge$, but only the value $\vee$.

We have therefore shown that by the four transformation steps the given derivation has been transformed entirely into expressions of the form $\vee$. From this the *consistency* follows at once, since a formula of the form $\mathfrak{A}\ \&\ \neg\ \mathfrak{A}$ would obviously have had to take the value $\wedge$ in the course of the transformations.

2.3. Concluding remarks. The arguments involved in the proof follow almost automatically, if we call to mind the *informal sense* of the γ's (2.1, 2.2) as well as that of the formulae and rules of inference (§ 1). It is then not difficult to see that the entire proof can easily be extended to the case where *an arbitrary finite domain of objects* is stipulated. Here the number of possible sets for each type remains after all finite. From this it follows that the simple theory of types also remains consistent if axioms are included which assert the existence of a certain finite number of objects. If the availability of *infinitely many* objects ('axiom of infinity') is demanded, however, then a radically different situation results which is still unresolved at the present time.

The consistency proof which has here been carried out is obviously completely 'finitist' in the sense of Hilbert's proof theory; it contains only the most elementary kinds of inferences and concepts.

6. THE CONCEPT OF INFINITY IN MATHEMATICS

The great controversy which has flared up in recent decades in connection with the foundations of mathematics is above all a controversy about the nature of infinity in mathematics. In the following, I shall try to characterize in as nontechnical a way as possible the precise problems that are here involved.

I shall first give a *classification of mathematics* into three distinct levels according to the degree to which the notion of 'infinite' is used in the various branches of mathematics. The first and lowest level is represented by elementary number theory, i.e., by the theory of numbers that does not make use of techniques from analysis. The infinite occurs here in its simplest form. An infinite sequence of objects, in this case the natural numbers, is involved. Several other branches of mathematics are logically equivalent to elementary number theory, viz., all those theories whose objects can be put into one-to-one correspondence with the natural numbers and which are therefore 'denumerable'. Almost the whole of algebra belongs here – the rational numbers, the algebraic numbers, also polynomials, can after all be proved to be denumerable – so does combinatorial topology, for example, i.e., that part of topology which deals only with objects whose properties are describable by finitely many data. The well-known four-colour problem belongs here. All these theories are, logically speaking, enrirely equivalent. It is therefore sufficient to deal with elementary number theory only; the theorems and proofs in the remaining theories can be reinterpreted as number-theoretical theorems and proofs by a correlation of their objects with the natural numbers. To the four-colour problem, for example, there corresponds indeed an equivalent number-theoretical problem, although our special interest in it derives of course solely from its intuitive topological formulation.

The second level of mathematics is represented by *analysis*. As far as the application of the concept of infinity is concerned, the essentially new feature here is the fact that now even *individual objects* of the theory may themselves be infinite sets. The real numbers, i.e., the objects of analysis, are after all defined as infinite sets, as a rule as infinite sequences of rational numbers. In this connection it makes no difference whether the particular definition chosen is that by nested intervals, or Dedekind cuts, or by some other means. The whole theory of complex functions also belongs at this level; nothing essentially new is here added. The third level of application of the concept of infinity, finally, is encountered in general set theory. Admitted as objects here are not only the natural numbers and other finitely describable quantities, as the first level, as well as infinite sets of these, as at the second level, but, in addition, infinite sets of infinite sets and again sets of such sets, etc., in the utmost conceivable generality.

The given classification subsumes every branch of mathematics. As far as geometry is concerned, for example, it no longer presents any special problems today in connection with the concept of infinity. What might appear to be such problems either belong to physics or occur in an equivalent form in analysis; the different geometries can after all always be interpreted in terms of logically equivalent models from analysis.

There are essentially two fundamentally different interpretations of the nature of infinity in mathematics, and I shall now go on to describe them. I shall call them the '*actualist*' ((*an sich*)) interpretation and the '*constructivist*' ((*konstruktiv*)) interpretation of infinity. The former is the interpretation of classical mathematics as we have all learned it at university. Several mathematicians have adopted the constructivist view – although not always to the same extent – among them Kronecker, Poincaré, Brouwer and Weyl. These names alone indicate that we are dealing with a direction of opinion that must indeed be taken seriously. I shall try to bring out the essence of the constructivist view vis-à-vis the actualist interpretation; in the short time available this can be done only imperfectly, especially since it must be kept in mind that by its very familiarity, the actualist interpretation has become second nature to us and that it is not easy to adopt, for once, a quite different way of thinking.

I shall begin with the antinomies of set theory. Here we have a situation in which actualist considerations have led to an absurdity which could not have resulted from the constructivist interpretation of the matter. For on the basis of the quite general concept of a set indicated earlier it is also possible to form, for example, the concept of the 'set of all sets'; this is a correctly defined set. Yet contradictions quite understandably follow from it: The set of all sets must after all contain itself as an element and in a certain sense – easily made precise – it must therefore be larger than itself, and this can obviously not be so. Upon closer examination it becomes easily apparent how the absurdity comes about : Strictly speaking, the 'set of all sets' must not itself be considered as belonging to the sets; it is a subsequent formation, as it were, which produces an entirely new collection from a given totality of sets. This is in fact the constructivist view of the situation: New sets may, as a matter of principle, be formed only constructively one by one, on the basis of already constructed sets. According to the actualist view, on the other hand, all sets are defined in advance by the abstract concept of a set and are therefore already available '*as such*' ((*an sich*)), quite independently of how individual sets may be selected from them by means of special constructions. This view had led to the antinomy.

If we were to try to express the essence of the constructivist view in as general a principle as possible, we would formulate it about as follows: 'Something *infinite* must never be regarded as *completed*, but only as something *becoming*, which can be built up constructively further and further.' I recall Gauss's well-known dictum that 'the use of an infinite quantity as something completed is never permissible in mathematics'.

If this principle of interpreting the infinite constructively is accepted, then differences vis-à-vis the actualist interpretation of classical mathematics manifest themselves not only in the theory of sets, but already in the realm of elementary number theory. I shall now discuss these differences in greater detail. In elementary number theory, we encounter the infinite only in its simplest form, viz,. in the form of the infinite sequence of the natural numbers. According to the actualist interpretation, we may regard this sequence as a *completed* infinite totality, whereas the constructivist interpretation allows us to say only this: We *can* progress further and further in the number sequence and always construct new numbers, but we must not speak of a completed totality. A proposition such as 'all natural numbers have the property $\mathfrak{P}$', for example, has in each case a somewhat different sense. According to the actualist interpretation, it says: The property $\mathfrak{P}$ holds for any number that may somehow be singled out from the complete totality of numbers. According to the constructivist interpretation we may say only this: Regardless of how far we progress in the formation of new numbers, the property $\mathfrak{P}$ continues to hold for these new numbers.

In practice, this difference in interpretation is here, however, immaterial. A proposition about all natural numbers is normally proved by complete induction, and this inference certainly appears to be in harmony also with the constructivist interpretation; particularly since complete induction is after all based on the idea of our *progressing* in the number sequence. The situation is different in the case of *existential propositions*. The proposition 'there exists a naural number with the property $\mathfrak{P}$' says, according to the actualist interpretation: 'Somewhere in the completed totality of the natural numbers there occurs such a number.' According to the constructivist interpretation, such an assertion is of course without sense. But this does not mean that under this interpretation existential propositions must be rejected outright. If a definite number n, for which the property $\mathfrak{P}$ holds, can actually be specified, then even under this interpretation can we speak of the existence of such a number; in reality, the existential proposition now no longer refers to the infinite totality of numbers; it would after all suffice to speak only of the numbers from 1 to n. The existence proofs that occur in practice are indeed mostly such that an example can actually be given. However, proofs are also possible where this is not the case, viz., *indirect existence proofs*: It is assumed that there is no number for which the property $\mathfrak{P}$ holds. If this assumption leads to a contradiction, it is inferred that a number for which the property $\mathfrak{P}$ holds exists after all. It may then happen that an effective procedure for actually producing such a number is altogether unobtainable. From the constructivist point of view, such a proof must consequently be rejected. Another technique of proof which likewise becomes unacceptable from this point of view and which is usually quoted in this connection, is the application of the 'law of the excluded middle' to propositions about infinitely many objects. According to the constructivist interpretation, for example, we cannot even say: 'A property $\mathfrak{P}$ holds for all natural numbers or it does not hold for all natural numbers.' The rejection of the law of the excluded middle seems particularly paradoxical, at first, but it is only a necessary consequence of the principle of interpretating the infinite potentially. After all, this law is based on the idea of the completed number sequence. This must not be interpreted to mean that the constructivists regard this law as altogether *false*; from their point of view it is more correct to regard it as being *without sense*. It thus makes no sense whatever to even speak of the totality of numbers as something completed, precisely because 'in reality', the number sequence is never completed, all that is given is an indefinitely extendable process of progression.

In practice these forms of inference, which are nonadmissible according to the constructivist interpretation, hardly ever occur in elementary number theory. The situation is different in analysis and set theory. Here the differences between the two interpretations are essentially the same as those described for the natural numbers; I shall therefore not discuss them further. In the case of analysis and set theory, however, the significance of the difference is considerably greater with the result that from the constructivist point of view *extensive* parts of analysis and almost all of set theory cannot be accepted.

In this connection, attention should be drawn to the fact that the delimitation between what is constructively permissible and what is not cannot be defined unequivocally in certain borderline cases and that the opinions of the different mathematicians representing this point of view are not identical. Yet these differences are not important enough to the picture as a whole to warrant a more detailed discussion. Words like 'intuitionist' (Brouwer) and 'finitist' (Hilbert) denote such somewhat different constructivist points of view.

Now the cardinal question becomes this: Which of the two interpretations is actually correct? Both are defended. On the one hand, we have the intuitionists under the leadership of Brouwer with the totally radical thesis that all of mathematics which is incompatible with the constructivist point of view must be discarded. On the other hand, the majority of mathematicians are understandably reluctant to make such a sacrifice. The antinomies, so they say, are indeed founded on inadmissible formation of concepts; but such concepts can be avoided by a proper delimitation; the whole of analysis and a *fortiori* number theory, so they claim, is entirely unobjectionable. Unfortunately, the delimitation of the inadmissible inferences can be carried out in basically different ways without their necessarily leading to a definite common point, and I must say that to me the clearest and most consequential delimitation seems to be that given by the principle of interpreting the infinite constructively.

We should nevertheless be reluctant to discard the extensive nonconstructive part of analysis which has, among other things, certainly stood the test in a variety of applications in physics. *Hilbert* sees in his *proof theory* a means of resolving these difficulties. This theory is intended to clarify as far as possible the mutual relationship between the two interpretations of the infinite by means of a purely *mathematical investigation.*

How can this be done? The first and formost task is to establish the *consistency* of mathematics, as far as such consistency exists. Here we after all have the strongest argument of the constructivists: The actualist interpretation has led to contradictions in set theory; who knows whether one day contradictions could not also occur in analysis. This objection could be considered as met by a consistency proof for analysis. It is in fact quite conceivable that the consistency of a methematical theory can be proved with exact mathematical techniques. In order to see this we call to mind the fact that a proposition asserting consistency is formalizable as a mathematical assertion; it says: There exists no *proof* within the theory which leads to a contradiction. The 'proofs' in a theory can be made the objects of a mathematical investigation, viz., 'proof theory', just as the natural numbers, for example, are made the objects of number theory. For this purpose it is customary to *formalize* the proofs, i.e., we replace the linguistic expressions in the proofs by definite symbols and combinations of symbols – to the inferences there now correspond certain formal rearrangements of combinations of symbols – so that we finally obtain as counterparts of the proofs certain figures composed of symbols. These figures are now susceptible of mathematical investigation in the same way as geometric figures. In order to be able to give a precise formal delimitation of the concept of a 'proof in a theory' it is of course essential, in particular, that a delimitation of the *forms of inference* occurring in the proof can be given in advance.

In practice, the number of forms of inference which are used in mathematics is fortunately relatively small.

If a consistency proof is then carried out, certain forms of inference must of course themselves be used for this proof. The correctness of *these* inferences must be presupposed from the outset, otherwise the whole proof would of course be circular. There can be no 'absolute' consistency proof. What kinds of inference must be presupposed as correct follows at once from our earlier considerations: The inferences must be compatible with the *constructivist* point of view. The reliability of the constructivist point of view is presupposed and not questioned. The objective then is to prove the consistency of the actualist interpretation by means of constructive inferences.

Recently I succeeded in carrying out such a proof for elementary number theory, i.e., for the first of the three given levels of the concepts of infinity[67]. Corresponding proofs must still be worked out for analysis and, finally, set theory in so far as the latter is actually consistent. In the course of this proof-theoretical investigation we can expect to gain insight into how far we can go without encountering antinomies and to find answers to further related questions.

What would be the relationship between the two interpretations of the infinite, if the consistency proof were complete? Even then different opinions could be held. One possibility would be to regard the consistency proof as still insufficiently secure by raising doubts about the constructive inferences used in the proof. I do not regard the danger of this objection as particularly great. *Something* will always have been gained if the reliability of the forms of inference of mathematics has been shown to depend on a minimum of fairly indisputable inferences; to do more is simply impossible; and I am certain that this foundation is considerably more secure than that provided by the actualist interpretation.

More important is a different objecten raised by the intuitionists: Even if the consistency were to have been proved, the propositions of actualist mathematics would still remain *without sense* and would therefore have to be rejected now as ever. An indirectly proved existential proposition, for instance, is claimed to be without sense; the reason is that the assertion of an existence is granted a real sense only if an example can actually be given. How can this be countered? It will have to be admitted that an indirectly proved existential proposition has a different, weaker sense than one that has been proved constructively; but a certain 'sense' is retained nevertheless. Furthermore, even if we do not concede an immediate sense to non-constructively proved propositions, the possibility still remains of using them to prove simple propositions such as directly verifiable numerical equations, which are certainly not without constructive sense; such propositions must then be true by virtue of the consistency proof and it might be the case that a direct constructive proof for the same proposition is more laborious or altogether unobtainable. This would seem to afford the actualist forms of inference at least a practical value which even the constructivists would have to acknowledge. This whole question of 'sense' does not seem to me at the moment to be ready for a final settlement. It is particularly from proof-theoretical research that significant contributions towards an answering of this question can be expected. A certain residual part will in the end always remain a matter of opinion. The objection against the sense of actualist propositions must in any case not be taken too lightly; it is not entirely without merit. I believe that in general set theory, for example, a careful proof-theoretical investigation will eventually confirm the view that all cardinalities exceeding the denumerable ones are in a very definite sense only fictitious entities ((*leerer Schein*)) and that it would be wise to do without these concepts.

After these general considerations I will now discuss in detail some of the difficulties arising in consistency proofs; I shall have to speak, in particular, of Gödel's theorem and of the significance of the transfinite ordinal numbers for consistency proofs.

Gödel has proved the important theorem: 'The consistency of a mathematical theory which contains elementary number theory cannot be proved – given that the theory is really consistent – with the techniques of proof of that theory itself.' At first sight it would seem that even the possibility of a consistency proof has thus become illusory since such a proof is intended to involve fewer techniques than those contained in the theory which is to be proved consistent. It remains quite conceivable, however, that the consistency of elementary number theory, for example, can be proved with techniques which are constructive on the one hand and do not involve the actualist aspects of elementary number theory, but which, on the other hand, still transcend the framework of elementary number theory. In my proof this technique is the rule of 'transfinite induction' applied to certain 'transfinite ordinal numbers'. I shall indicate briefly what is meant by this and how these concepts are connected with the consistency proof.

The concept of the 'transfinite ordinal numbers' goes back to G. Cantor and really belongs to set theory. We shall, however, require only a very limited portion of the ordinal numbers developed in that theory – a 'segment of the second number class' in the terminology of set theory – a segment which can be built up strictly constructively and which has nothing in common with the disputable aspects of the actualist interpretation which are especially blatant in set theory and which must be avoided in the consistency proof.

The transfinite ordinal numbers are constructed in the following way: First comes the sequence of the natural numbers: 1, 2, 3, etc. Then a new number ω is introduced, which is defined to rank behind all natural numbers. ω is followed by $\omega+1$, then $\omega+2$, $\omega+3$, etc. Behind all numbers of the form $\omega+n$ follows $\omega \cdot 2$, then $\omega \cdot 2+1$, $\omega \cdot 2+2$, etc., after these $\omega \cdot 3$, then $\omega \cdot 3+1$, $\omega \cdot 3+2$, etc., etc. Behind all numbers of the form $\omega \cdot n+n$ follows the number ω^2, then again ω^2+1, ω^2+2, ..., $\omega^2+\omega$, $\omega^2+\omega+1$, ..., $\omega^2+\omega \cdot 2$, ... $\omega^2+\omega \cdot 3$, ..., $\omega^2 \cdot 2$, ..., $\omega^2 \cdot 3$, ..., $\omega^2 \cdot 4$, etc., finally ω^3, and we can go on in this way to form ω^4, ..., ω^5, ..., finally ω^ω, and still further numbers, if desired. The entire procedure – which I have only sketched here – may at first seem somewhat bewildering. Yet basically only two operations are involved whose repeated application automatically generates all these numbers:

1) given an already existing number, we can form its successor (addition of 1);

2) given an infinite sequence of numbers, we can form a new number ranking behind the whole sequence (formation of a limit).

The concern that this procedure is nonconstructive since the actualist conception of the completed sequence of the natural numbers already seems to enter into the formation of ω, turns out to be unfounded. The concept of infinity can here definitely be interpreted potentially by saying, for example:

Regardless of how far we may go in constructively forming new natural numbers, the number ω stands in the order relation $n < \omega$ to any such natural number n. And the infinite sequences that arise in the formation of the other ordinal numbers should be interpreted in precisely the same way.

Now to the concept of 'transfinite induction': This induction is nothing more than the extension of the rule of complete induction from the natural numbers to the transfinite ordinal numbers. Complete induction may, as is well known, be formulated as follows: If a proposition holds for the number 1, and if it has been proved that its validity for all numbers preceding the number n entails its validity for n, then the proposition holds for all natural numbers. If we here replace 'natural number' by 'transfinite ordinal number', we have the rule of trnasfinite induction. We can easily convince ourselves of the correctness of this rule for initial segments of the transfinite number sequence as follows: Suppose the proposition holds for the number 1, and that it has been proved further that if the proposition holds for all numbers preceding a certain ordinal number it also holds for that ordinal number. Then we argue thus: The proposition holds for the number 1 hence also for the number 2, thus also for 3, etc., hence for all natural numbers. Consequently it also holds for the number ω, precisely because it holds for all its predecessors. For the same reason it holds for the number $\omega+1$, thus also for $\omega+2$, etc., finally for $\omega \cdot 2$; and, correspondingly, we show its validity further for $\omega \cdot 3$, $\omega \cdot 4$, etc., finally also for ω^2. Continuing in this way, we can convince ourselves of the validity of the rule of transfinite induction by ascending step by step in the sequence of transfinite ordinal numbers. As the numbers become larger, the situation admittedly begins to look rather complicated, but the principle always remains the same.

I shall explain now how the concepts of the transfinite ordinal numbers and the rule of transfinite induction enter into the consistency proof. The connection is quite natural and simple. In carrying out a consistency proof for elementary number theory we must consider all conceivable number-theoretical proofs and must show that in a certain sense, to be formally defined, each individual proof yields a 'correct' result, in particular, no contradiction. The 'correctness' of a proof depends on the correctness of certain other simpler proofs contained in it as special cases or constituent parts. This fact has motivated the arrangement of proofs in linear *order* in such a way that those proofs on whose correctness the correctness of another proof depends precede the latter proof in the sequence. This arrangement of the proofs is brought about by correlating with each proof a certain transfinite ordinal number; the proofs preceding a given proof are precisely those proofs whose ordinal numbers precede the ordinal number of the given proof in the sequence of ordinal numbers. At first, we might think that the *natural numbers* suffice as the ordinal numbers for such a classification. In actual fact, however, we need the transfinite ordinal numbers for the following reason: It may happen that the correctness of a proof depends on the correctness of infinitely many simpler proofs. An example: Suppose that in the proof a proposition is proved for *all* natural numbers by complete induction. In that case the correctness of the proof obviously depends on the correctness of every single one of the infinitely many individual proofs obtained by specializing to a particular natural number. Here a natural number is insufficient as an ordinal number for the proof, since each natural number is preceded by only finitely many other numbers in the natural ordering. We therefore need the transfinite ordinal numbers in order to represent the natural ordering of the proofs according to their complexity.

Now it also becomes clear why it is the rule of transfinite induction that is needed as the crucial rule for the consistency proof; this rule is used to prove the 'correctness' of each individual proof. Proof number 1 is after all trivially correct; and once the correctness of all proofs preceding a particular proof in the sequence has been established, the proof in question is also correct precisely because the ordering was chosen in such a way that the correctness of a proof depends on the correctness of certain earlier proofs. From this we can now obviously infer the correctness of all proofs by means of a transfinite induction, and we have thus proved, in particular, the desired consistency.

It turns out that this transfinite induction is precisely that inference in the consistency proof which necessarily, in agreement with Gödel's theorem, cannot itself be shown to be correct by means of techniques of elementary number theory.

The correctness of transfinite induction is actually established by a special argument of the kind used earlier up to the number ω^2. But even for elementary number theory we require a considerably larger segment of the transfinite numbers, viz.: In the same way in which, in outline, I defined ω^ω above, we obtain $\omega^{(\omega^\omega)}$, then $\omega^{(\omega^{(\omega^\omega)})}$, etc., by extending the procedure correspondingly; behind all these numbers follows the number ε_0, the first ε-number'. This number represents the upper limit of that segment of the transfinite ordinal numbers which is required for the consistency proof of elementary number theory, if that theory is formally delimited in the usual way.

I expect – although this is only a conjecture, for the time being – that the consistency of analysis – and of set theory, as far as possible – will become provable in the same way; in each case, we may of course have to advance a considerably greater distance into the sequence of numbers of the second number class. Altogether, the following picture would thus seem to emerge: The increase in complexity of the concept of infinity at the three levels of mathematics described at the beginning of this paper – elementary number theory, analysis, and set theory – is accompanied by a corresponding extension of the sequence of transfinite ordinal numbers; in the same way in which the number ε_0 forms the upper limit for elementary number theory, we would have a specific number of the second number class as the upper limit for analysis, and another number as the upper limit for a formally delimited set theory – in so far as such a set theory makes sense. But we should not overestimate the absolute significance of such limit numbers: Even in elementary number theory it happens that in order to solve certain associated problems, still further forms of inference would have to be included and this would widen the framework of elementary number theory further; this means that still higher ordinal numbers may be required for the consistency proof. There can be no upper bound for this; Gödel has shown that every formally delimited system of this kind is incomplete in the sense that certain associated problems can be solved only by including further techniques. This makes actually no difference for the consistency proof; we need only to extend the proof further with each inclusion of new techniques.

7. THE PRESENT STATE OF RESEARCH INTO THE FOUNDATIONS OF MATHEMATICS

§ 1. The different points of view concerning the question of the antinomies and the concept of infinity

The *antinomies of set theory* were discovered about forty years ago and, up to the present time, no final explanation for this matter has been found. *Research into the foundations of mathematics* has derived considerable impetus from this problem. The precariousness of certain fundamentals of mathematics which has clearly revealed itself at this point has prompted precisely some of the most prominent mathematicians – *Brouwer*, *Hilbert*, and *Weyl*, to mention but a few – to concern themselves with these questions, which are otherwise mostly foreign to the practising mathematician, who is frequently inclined to renounce them as incongurous with mathematical thinking because of the uncertainty and the multiplicity of opinions which their connection with *philosophy* evokes.

Several attempts have been made to find a '*solution*' *to the antinomies*, i.e., to point out clearly where 'the fallacy' lies. These attempts have not led to a conclusive result, and there is no reason to expect *such* a solution in the future. The situation is rather such that it is impossible to speak of a uniquely identifiable error in our thinking. All that can be said with certainty is that the appearance of the antinomies is connected with the *concept of infinity*. For no contradictions can occur, by human standards, in purely finite mathematics, as long as it is correctly constructed. Certain analogues of the antinomies in finite situations are founded in manifest inaccuracies in the formation of concepts.

In order to find a way out from the unpleasant state created by the antinomies, various directions have been chosen. The simplest procedure, to begin with, is to *draw a line between* permissible and nonpermissible forms of inference in mathematics, so that the inferences leading to the antinomies turn out to be nonpermissible. There is a whole multitude of such attempts;

in some cases the proposed delimitation is claimed to be natural on the basis of certain rationalizations, in other cases such justifications are not even attempted. Examples are axiomatic set theory and the system of the 'Principia Mathematica'.

This procedure is actually quite useful *in practice*, but not adequate *in principle*. First of all, the delimitation is rather *arbitrary* precisely because 'the error' cannot be definitely specified. *Second*, we are immediately faced with the question whether *contradictions* could not one day occur also in the realm of the permissible forms of inference. Several arguments can certainly be advanced which make it probable that the antinomies have been successfully precluded; yet this certainty is not particularly great. Indeed, it seems not entirely unreasonable to me to suppose that *contradictions* might possibly be concealed even in classical *analysis*. The fact that, so far, none have been discovered means very little when we consider that, in practice, mathematicians always work with a comparatively limited part of the logically possible complexities of mathematical constructs.

The most consequential form of delimitation is that represented by the '*intuitionist*' point of view, which has been formulated primarily by *Brouwer* and *Weyl*. This point of view can probably be understood most easily in terms of the following *fundamental principle*: Infinity must not be interpreted in a way that suggests that infinite sets *actually* exist '*as such*' and are in some sense only discovered by mathematicians – an interpretation which I shall briefly call the 'actualist interpretation' of infinity – but merely in the sense that an infinite totality can be built up *constructively* step by step starting from finite quantities, and here infinity must never be regarded as *completed*, but merely as an expression for the *possibility* of an unbounded extension of the finite.

This principle has undoubtedly much in its favour and even before the discovery of the antinomies efforts had been made in a similar direction. Once we subscribe to this principle the antinomies disappear since they obviously involve an illicit use of the actualist interpretation of infinite sets. On the other hand, the constructivist principle leads necessarily to the limitations imposed by the intuitionists on the forms of inference customary in mathematics today[68]. In order to give an example, let us consider the most important case of this kind in practice, that of *indirect existence proofs*:

According to the classical interpretation, the existence of a natural number, a number with the property $\mathfrak{P}$, for example, can be proved *indirectly* by assuming that no number possesses the property $\mathfrak{P}$, and by then deriving a contradiction from this assumption. Such a proof must be rejected from the constructivist point of view. Here an assumption about the infinite totality of all natural numbers is made; this is without sense – from the constructivist point of view – since this totality can never be given as a completed totality, because it can be conceived only as an uncompleted sequence which is extendable indefinitely. Even from this point of view, however, the existence of a natural number with the property $\mathfrak{P}$ can indeed be proved, provided that such a number can be directly *stated*, or that a procedure for its calculation can be produced. After all, the concept of the totality of all numbers here no longer enters into the proof.

An easily readable survey of the intuitionist point of view has recently been published by *Heyting*[69].

I believe that we must grant the intuitionists that they have drawn the most uncompromising conclusions from the unpleasantness caused by the antinomies. Serious objections can nevertheless be raised against a radical intuitionism which dismisses categorically as without sense everything in mathematics which does not correspond to the constructivist point of view. I shall discuss this point in greater detail in § 4. Here, I should like to mention only the following: If this point of view is adopted, then the whole of classical analysis is reduced to a field of rubble. Many, particularly a number of basic theorems, become invalid or have to be rephrased and proved in a different way. Added to this must be the fact that the formulations become mostly *more cumbersome* and the proofs *more tedious*. Existence proofs, for example, such as that of the 'fundamental theorem of algebra', must now be rephrased in such a way that a procedure for the *calculation* of the number whose existence is asserted is given, and special cases in which this cannot be done must be discarded.

We could certainly not shirk even the greatest sacrifice if it were really necessary. But is such a sacrifice actually necessary?

This brings me to *Hilbert's view* of the matter. Hilbert set up the programme of extricating the whole of classical mathematics as far as possible from what has become its critical state by proving its *consistency* along exact mathematical lines.

The realization if this programme is unfortunately in large measure still outstanding. It has turned out that the difficulties involved in such consistency proofs are greater than had at first been expected. (Cf. § 2, Gödel's theorem.) In 1936, I published a proof for the consistency of *elementary number theory*[70]; earlier partial results date back to Ackermann, von Neumann and Herbrand; but the most important proof of all in practice, that for *analysis*, is still outstanding.

In order to carry out a consistency proof, we naturally already require certain techniques of proof whose reliability must be *presupposed* and can no longer be justified along these lines. An absolute consistency proof, i.e., a proof which is free from presuppositions is of course impossible. The question therefore arises what techniques of proof can be taken as a foundation in this sense. The answer follows from what has been said earlir: We should be able to use those techniques of proof in which infinity is applied only in a constructive sense and avoid strictly everything that is based on the actualist interpretation of infinity and which is therefore of a suspect nature. This restriction amounts to about the same as what Hilbert calls the '*finitist point of view*'. As far as consistency proofs are concerned, however, it seems that somewhat stronger methods are required than those originally invisaged by Hilbert and which he had thought of as forming the 'finitist techniques of proof'. The required methods remain nevertheless in harmony with the *constructivist interpretation* of infinity; this is the point that matters and which distinguishes them fundamentally from the disputable techniques of proof.

A foremost characteristic of *Hilbert's* point of view seems to me to be the endeavour to withdraw the problem of the foundations of mathematics from *philosophy* and to tackle it as far as in any way possible with methods proper to mathematics. The problem cannot, of course, be solved entirely without extramathematical presuppositions. Hilbert's plan limits these to a *minimum*: It requires that we recognize the fundamental difference between the constructivist and the actualist interpretations of infinity and that we understand why the reasoning along constructivist lines enjoys a considerably greater measure of reliability, so that it can be chosen as a sufficiently secure basis for the consistency proof of those parts of mathematics that make use of the actualist interpretation of infinity.

I shall continue to avoid all *philosophical issues* whose discussion has no bearing on practical mathematics, and which frequently tend to blur the nature of the problem unnecessarily and make it seem difficult.

We must still mention briefly the so-called '*logicism*', usually listed beside *intuitionism* and *Hilbert's point of view* as the third important approach to the question of the foundations of mathematics. Its tenets are founded in certain philosophical attitudes which, in accordance with what was said above, will not be discussed here. In connection with the problem of the antinomies and infinity, which is more important for practical mathematics than any other, the logicists have adopted an essentially noncommittal or undecided position; and their contribution to the solution of this problem is therefore negligible since they are basically interested in different questions, e.g., the definition of the concept of a number.

§ 2. Exact foundational research in mathematics: axiomatics, metalogic, metamathematics. The theorems of Gödel and Skolem

In the following, I shall discuss some of the more recent findings and, in particular, some of the especially important earlier results obtained in the exact foundational research in mathematics, i.e., in that branch of mathematics in which the foundations of mathematics are investigated. The objects of inquiry are, for example, *axiom systems* for mathematical theories– a topic of ancient vintage – with special emphasis, in recent times, on the *logical forms of inference* and, generally, the *methods of proof* of mathematics.

In the last decades, a large number of scholars from all countries have concerned themselves with these questions and have obtained a multitude of results. It seems fair to say that in *Germany* metalogical and metamathematical research is, at the present time, carried out regularly only in Münster; abroad we must mention primarily *America* and *Poland* as the main centres of this type of research in mathematics.

A main task of metamathematics is the development of the *consistency proofs* required for the realization of *Hilbert's* programme. Other major problems are: The *decision problem*, i.e., the problem of finding a procedure for a given theory which enables us to decide of every conceivable assertion in that theory whether it is true or false; further, the question of *completeness*, i.e., the question of whether a specific system of axioms and forms of inference for a specific theory is *complete*, in other words, whether the truth or falsity of every conceivable assertion of that theory can be proved by means of these forms of inference.

Highly pertinent to these fundamental problems are several important theorems which *Gödel* proved about eight years ago[71], and which have attracted much attention and have sometimes also been misinterpreted.

To begin with, we have the theorem about *consistency proofs* which says that the consistency of a mathematical theory in which elementary number theory is contained, and which really is consistent, cannot be proved with the techniques of proof of that theory itself and naturally therefore not with any part of those techniques. This theorem has frequently been taken as conclusive proof that Hilbert's programme is unrealizable. This view is based on the conviction – and there seemed to be some evidence in its favour

– that the 'finitist' or 'constructivist' forms of inference by which the consistency proofs were to be carried out, merely represent part of the precisely formalizable forms of inference occurring in *elementary number theory*. If this were the case then, according to Gödel's theorem, these forms of inference would no longer be sufficient to prove the consistency of number theory. I am of the opinion, however, that there are forms of inference which are still in harmony with the constructivist interpretation of infinity and which, nevertheless, transcend the framework of formalized number theory, indeed, by their very nature, these techniques can presumably be extended beyond the framework of *any* formally delimited theory. I have already stated the relevant inferences in # 4, as far as they were required for the consistency proof of elementary number theory[72]. They are closely connected with the 'transfinite induction' occurring in set theory, but this does not mean that they are subject to the same contingencies as that rule; on the contrary, they are proved *constructively in a way entirely independent* of set theory. – Quite unscathed by these facts, Gödel's theorem retains of course great significance as a very valuable result which renders a great service, particularly in connection with the discovery of consistency proofs, by telling us specifically which techniques do not lead to the desired end.

Another of Gödel's theorems concerns the *decision problem*, particularly as it applies to the so-called 'predicate logic'. It says that there are theorems of this system which cannot be decided by certain very strong mathematical techniques.

This theorem has recently been considerably strenghtened by *Church* in this respect that by introducing a very general notion of 'procedure' he was able to show that there can be *no general decision procedure* for predicate logic *at all* and that thus the decision problem is therefore not generally solvable[73]. The situation is actually such that if the decision problem were solved for predicate logic, then the truth or falsity of Fermat's last theorem, for example, and similar number-theoretical problems could, in principle, simply be *calculated*, and it seems fair to say in advance that it is not very probable that such a decision procedure can ever be found. Nevertheless, it is of course very valuable to have this suspicion confirmed by an explicit proof. Church's proof is actually based on the assumption that the notion of a 'calculation procedure' which he has introduced is the most general possible. If someone were to succeed in finding yet another kind of calculation procedure, then it would be conceivable that a general decision procedure could after all be obtained. It is fairly clear, however, that the concept introduced by Church is so general that it is practically impossible to think of any kind of procedure that might not fall under this concept. Also in favour of Church's formulation points the fact that even by taking quite different starting points, we always end up with the same concept, or one that is equivalent to it.

A third result by Gödel concerns the *completeness problem*. It says that every formally delimited consistent mathematical theory is *incomplete* in the sense that number-theoretical theorems can be stated which are *true*, but which are *not provable* with the techniques of that theory. This is undoubtably a very interesting, but certainly not an alarming result. We can paraphrase it by saying that for number theory no adequate system of forms of inference can be specified for once and for all, but that, on the contrary, new theorems can always be found whose proof requires new forms of inference. That this is so may not have been anticipated in the beginning, but it is certainly not implausible.

The theorem reveals of course a certain weakness of the *axiomatic method*.

Since *consistency proofs* generally apply only to delimited systems of techniques of proof, these proofs must obviously also be *extended* when an extension of the techniques of proof takes place.

It is remarkable that in the whole of existing mathematics only very few easily classifiable and constantly recurring forms of inference are used, so that an extension of these methods may be desirable in theory, but is insignificant in practice. In fact, the nonprovable number-theoretical theorems presented by Gödel were in each case specially constructed for this purpose and are of no practical importance; with one notable exception, however: The assertion of the *consistency* of a theory, which also belongs to the theorems that are not provable within the theory. For this reason the consistency proof must indeed make use of new forms of inference which, in this case, must moreover be of a constructive nature; this has already been discussed.

I shall now mention some results which concern the *theory of sets*.

In an attempt to salvage set theory from the damaging blow of the antinomies, certain restrictive conditions have been formulated by which the contradictions are eliminated. For this purpose several *axiom systems* for set theory have been developed; the most well-known is the system of *Zermelo* and *Fraenkel*. For part of this system, the so-called 'general set theory', *Ackermann* recently carried out a consistency proof, or rather, traced the consistency of this system back to the consistency of elementary number theory[74]. 'General set theory' results, if we omit from the full axiom system the 'axiom of infinity' which asserts the existence of infinitely many objects of the theory. Ackermann's proof is based on the fact that for this part of set theory a *model consisting of natural numbers* can be constructed, a result which has been known for some time. If the axiom of infinity is included, the same kind of construction does not seem possible since this introduces precisely the existence of nondenumerable cardinalities into set theory.

In this connection we must, however, mention a theorem which is at first sight rather surprising and which leads to interesting consequences. It was first formulated by *Skolem* about fifteen years ago and he called it the 'theorem concerning the relativity of the concept of a set'[75]. In contrast with the metamathematical theorems mentioned earlier, Skolem's theorem is no longer compatible with the *constructivist interpretation* of mathematics and must be considered as falling under the *actualist interpretation* if for no other reason than that it deals with nondenumerable cardinalities. (This fact in no way detracts from the significance of the theorem, which lies precisely in its application to actualist mathematics.) Skolem's theorem says: If a model of arbitrarily high cardinality exists at all for an axiom system of a particular type, then there exists a *denumerable* model which also satisfies the axiom system. – All axiom systems that have been used up to now belong to this type or can, at any rate, be converted to it; nor is it apparent how an axiom system could be formulated which did not fall within the range of Skolem's concept.

If this theorem is applied to any axiom system of *set theory*, it follows that if the system is satisfiable at all, which we naturally wish to assume, it is already satisfiable by a *denumerable* model. It seems fair to say that this result is no an exactly pleasant one for axiomatic set theory. It says, after all, that all nondenumerable cardinalities with which set theory is concerned are in a certain sense only fictitious entities to this extent that, without altering the validity of any theorem, such sets may simply be replaced by certain *denumerable* sets.

At first sight, this discovery appears certain to lead to *contradictions*. In axiomatic set theory it is proved, for example, that the set of all real numbers is nondenumerable. More precisely, the following theorem is actually proved: There exists no one-to-one correspondence between the natural numbers and the real numbers. Let us, therefore, examine the denumerable model for set theory which, by Skolem's result, exists. This model contains objects which represent the natural numbers of the axiom system, others which represent the real numbers, and still others which represent the *correspondences* which can be set up between these numbers on the basis of the axiom system; here each type comprises at most denumerably many objects. Nevertheless, the mentioned theorem remains valid for this model for the very reason that among the *correspondences* available in the model there is, in fact, none which correlates the denumerably many representatives of the 'natural numbers' one-to-one with the denumerably many representatives of the 'real numbers'. Such a correspondence certainly exists in principle, but it is not contained among the correspondences occurring in the model.

Perhaps this rather intricate state of affairs will become somewhat clearer if it is examined from a different angle and here I shall, at the same time, confine myself to the continuum of the real numbers as the prototype of a 'non-denumerable' cardinality: For the sake of argument, we adopt the actualist position and maintain that the continuum is actually given *a priori*, as the set of all arbitrary infinite decimal fractions, for example. By *Cantor's* method we can show the nondenumerability of this system. On the other hand, we can say the following: Every *axiom system* for analysis which is constructive is in a certain sense *insufficient* for the purpose of encompassing completely this conceived continuum. Skolem's theorem after all entails that by specifying a particular axiom system, this continuum can be *replaced* by a denumerable model which satisfies in the same way all properties of the continuum determined *by the axiom system*. According to this point of view, Skolem's result does not, so to speak, exhibit a weakness of the continuum or of higher cardinalities, but rather a weakness in man's ability to comprehend these cardinalities.

How abstract cardinal set theory can be extricated from the Scylla of the antinomies and the Charybdis of Skolem's theorem of relativity, indeed, whether it can be extricated at all, only the future will decide.

It should be explicitly mentioned that other parts of set theory (point sets, the second number class) are affected only to a lesser degree by these difficulties and will definitely always retain a certain significance.

If *Skolem's* theorem is compared with *Gödel's* incompleteness theorem, it can be said that both theorems illustrate certain imperfections inherent in formally delimited axiom systems (which now also include the permissible *techniques of proof*). What might appear, at first sight, to be a rather surprising state of affairs arising from Gödel's theorem, viz., that even by specifying the most complicated axiom systems of analysis etc. there remain *number-theoretical* theorems which are unprovable, is to some extent explained by *Skolem's* theorem: According to this theorem the most complicated axiom systems are basically reducible to a denumerable model and hence also to the natural numbers; the theorems of these systems can therefore be reinterpreted collectively as *number-theoretical* theorems; all these systems are therefore basically *number theory*.

Another result by *Skolem*[76] clearly illustrates the imperfection of the axiomatic method in connection with number theory; it says: If any axiom system of the above quite general type is given for the natural numbers, then these numbers may be replaced by a model which is *nonisomorphic to the natural number sequence* and which also satisfies the axiom system.

§ 3. The continuum

In this and the following paragraphs, I shall now examine somewhat more closely the differences between the *actualist interpretation* and the *constructivist interpretation* of *analysis*, which is the most important branch of practical mathematics. In this paragraph, I shall, in particular, contrast the two approaches to the definition of *real numbers* and *real functions* and, in § 4, propose a possible *reconciliation between the various points of view*.

The concept of an *irrational number* may, as is well known, be obtained roughly as follows: The interval from 0 to 1 is divided into two parts, each part again into two parts, etc.; in this way finer and finer subdivisions are progressively obtained. A sequence of such intervals in which each term forms part of the term preceding it tends to contract more and more to a point as the process is continued. This is where the actualists take the plunge into the class of completed infinities by declaring that an *infinitely* long sequence of this kind is a '*real number*'.

From this interpretation some unusual consequences follow, which, apart from the general disputability of the actualist interpretation arising from the antinomies, could be advanced as additional arguments against this view: On the one hand, it may be proved in the usual way that these real numbers form a *nondenumerable* set. On the other hand, however, all theorems, all definitions, and all proofs which can ever be formulated or carried out are *denumerable* since they can always be characterized by finitely many symbols. This leads to the conclusion that there are real numbers which can in no way be individually defined and valid theorems which are unutterable and which no one will ever be able to prove. If *Skolem's theorem of relativity* is invoked, it follows further that the whole of conventional analysis remains quite generally valid in all of its parts if it is interpreted in a certain *denumerable* model.

Here we might well be tempted to say: If 'the nondenumerable continuum' in this way succeeds in completely eluding our comprehension, is there then any point at all in speaking of it as something *real*? In § 4, I shall show how, in a restricted sense, we can nevertheless answer this question in the affirmative.

At first it will be examined what the constructivist point of view has to offer as a *replacement* for the actualist concept of irrational numbers.

The sequence of divisions of intervals may be *begun* as before. The notion of the *completed* infinite sequence of intervals, however, must be rejected as being without sense. The infinite is after all to be regarded only as a *possibility, as an expression for the unboundness of the finite*. We can therefore reasonably say: It is possible to push this subdivision further and further. However, no irrational number is obtained in this way; at each stage of the subdivision we are left with a collection of eventually rather proximate *rational* numbers. It is in this sense that *Kronecker* maintained that 'there are no irrational numbers at all'[77]. Even a *constructivist*, however, does not have to be this narrow-minded; there are possibilities of going further. An irrational number can after all be regarded as given if a *rule* is available which permits the computation of a sequence of intervals of the mentioned kind *arbitrarily far*. (Here it is convenient, in order to avoid certain formal difficulties, to base the argument on *double* dyadic intervals, i.e., those which result from pairing two neighbouring dyadic intervals of the same subdivision.)

Such a rule is easily stated, for example, for $\sqrt{2}$, generally for $\sqrt[n]{m}$, but also for transcendental numbers such as π and e, in fact, quite generally, for practically all numbers which are needed in analysis as *individually defined* numbers; viz., in all cases in which the number concerned can actually be *calculated* to any desired degree of accuracy.

In order to remain faithful to the constructivist point of view, however, the 'numbers' defined in this way must be handled with care. The temptation must be resisted of regarding such a number as a completed infinitely long dyadic fraction; what is given in this sense is not really the entire number, but only the rule for its progressive realization; the rule itself is *finite* and merely a suitable *representative*, in certain contexts, of an infinitely long number of which it can be said, now as ever, that *it actually does not really exist*.

In his paper entitled 'Das Kontinuum', *Weyl* has attempted to construct an analysis on the basis of this kind of number concept[78]. (In this paper, Weyl has not exploited the full consequences of the constructivist attitude toward the natural number sequence; but this was later remedied[78].) One of the difficulties arising in this connection is that of deciding what techniques should be permitted in the computation and hence the definition of numbers.

Initially, Weyl carried out a definite delimitation of techniques; usually, however, intuitionists avoid such a delimitation altogether, a point of view which is not entirely inappropriate since a universally valid delimitation involves fundamental difficulties analogous to those existing for delimited axiom and inference systems, already discussed above, viz: That such delimitations are always open to and in need of further *extensions*. This actually represents no serious shortcoming; in cases of immediate practical importance it is perfectly clear how the concept of 'calculability' is intended.

Church, as already mentioned in § 2, made this concept of calculability more precise; independently of Church, *Turing* formulated an equivalent concept and also applied it, in particular, to the calculability of real numbers[79]. The greater precision consists here in the formulation of a *unified concept embracing* all 'calculation procedures'; this makes it feasible to carry out an *impossibility* proof of the kind mentioned in § 2; even at this level of precision, however, it cannot be decided of every procedure falling under this concept whether it is a 'calculation' procedure or not.

An extension of the constructivist concept of a real number was devised by *Brouwer* with his 'free choice sequences'[78]. These sequences are the logical outcome of an attempt to introduce the concept of a *function* of real numbers. In actualist mathematics this concept, is, as is well-known, defined simply as a relation which correlates with every arbitrary real number a second real number as its functional value. The concept of a *completed infinity* is here involved threefold: first in the two real numbers, and second in the universal abstract 'correlation'. This concept is therefore of no use to the constructivist. One of the ways open to him is to define a function as a *rule* which correlates with every rule defining a real number a second rule defining another real number. It is easily seen, however, that the following less restrictive version, which is closer to the actualist function concept, is still entirely compatible with constructivist principles: In place of the concept of an individual real number given by a rule, the procedure of subdividing intervals is once again taken as a starting point, by defining a function specifically as a *rule* with the following properties: As the sequence of intervals of the above described kind is chosen *in some way*, the functional rule correlates with a certain finite initial segment of this sequence a first interval of the 'functional value', and after having continued the sequence up to a certain further point, a *second* such segment, etc. The correlated intervals are therefore once again designed to form a 'nest of intervals'. – *In short*: In each case a desired finite number of initial places of the functional value should be *calculable* by the function rule from a sufficiently large number of initial places of the argument value.

That this function concept is still considerably narrower than the actualist one can already be seen from the fact that such a 'function' is always a *continuous* function. *Brouwer* proves moreover the *uniform continuity* of these functions and, in doing so, he makes a rather extensive use, for a constructivist, of 'transfinite' induction[80].

The argument values occurring in connection with this function concept are what Brouwer calls *'free choice sequences'*, viz., sequences of intervals in which successive terms can in each case be *freely chosen* – subject only to the restriction imposed by the fundamental conditions for nested intervals.

Even this number concept must be handled with care; it really has no independent meaning, but only a meaning within a proper context. After all, the concept of a *completed* infinite sequence is still entirely without sense; free choice sequences may thus be used only in contexts in which a finite initial segment of them or, at most, the *possibility* of their arbitrary extension, is involved. This is guaranteed in the case of the stated definition of a function.

By means of *Brouwer's function concept* the most frequently used functions in analysis can now be given suitable constructivist definitions without difficulty. Most of them are after all such that their functional values can be calculated more and more accurately as the argument value is progressively narrowed down.

Considerable *differences* between intuitionist and classical analysis nevertheless manifest themselves in the further development of the theory, especially in connection with *existence theorems*, as already mentioned in § 1. Constructivists must, after all, insist that a *calculation rule* is specified for the number whose existence is asserted; actualist existence proofs often do not meet this requirement.

Intuitionist analysis thus becomes much *more complicated* than classical analysis. This may already have been noticed in connection with the definitions of the fundamental concepts. Constructivists require, for example, *different* concepts of real numbers for different uses, whereas a single simple concept suffices in actualist analysis.

Nevertheless, a *constructivist consistency proof* for classical analysis is urgently needed because of the fundamental disputability of the actualist interpretation. I anticipate that such a proof will most likely be obtained by a further extension of the same techniques which made the consistency proof for number theory possible. It might be supposed that by being *non-*

denumerable, the continuum introduces a fundamentally new difficulty. To this it can be replied, for example, that every formally delimited system of analysis – and only for such unequivocally fixed systems the consistency needs to be proved – is already satisfied by a *denumerable* model, according to Skolem's theorem of relativity, so that even for the question of the *consistency* the non-denumerability presents only an *apparent* problem.

§ 4. The possibility of reconciling the different points of view

I am of the opinion that once the consistency proof for analysis has been successfully carried out, there should be no obstacle in the way of an agreement among the representatives of the different tendencies – i.e., among the constructivists or intuitionists on the one hand, and Hilbert's supporters, as well as the representatives of a purely actualist interpretation, on the other – *to retain classical analysis* in its existing form. At present, however, the situation is such that the radical constructivists do not agree with this conclusion, and it is here that the actual fundamental difference of opinion between *Brouwer* and *Hilbert* reveals itself. That is to say, the intuitionists regard all theorems of mathematics that are based on the actualist interpretation of infinity as *without sense* and all associated forms of inference as the components of a futile game with symbols without meaning.

In the preceding paragraphs, I have mentioned several facts which lend some support to the latter view. On the other hand, the tremendous wealth of successful applications of classical analysis in *physics*, to mention only *one* aspect of greatest significance, weighs heavily against this conclusion. In the following, I shall try to make clear how someone, even upon acknowledging the fundamental thesis of constructivism, can still reach the conclusion that actualist analysis should be retained and continued.

Hilbert himself has here shown the way: viz., by the method of *ideal elements*[81].

I.e.: propositions which talk about the infinite in the sense of the actualist interpretation are regarded as 'ideal propositions', as propositions which do not really mean at all what the words in them purport to mean, but which can be of greatest value in rounding off a theory, in facilitating its proofs, and in making the formulation of its results more straightforward. In projective geometry, for example, *ideal points* are introduced for the same reason, with the advantage that many theorems are simplified which would otherwise be plagued by exceptions. In the bargain, we must of course accept the fact that in some cases the *sense* of a theorem is now no longer the usual one. The following is asserted, for example: 'Two straight lines always have a point in common'. If the straight lines happen to be parallel, however, then they plainly have *no* point in common in reality. The procedure is harmless, because it has been precisely specified what, in such exceptional cases, is to be understood by the notion of a 'point', which has now a *wider* sense.

We might consider still another example which, in its relation to physics, seems to provide even more striking analogies to the relationship between constructivist mathematics and actualist mathematics:

I am thinking of the occasional attempt to construct a 'natural geometry', i.e., a geometry which is better suited to physical experience than the usual (Euclidean) geometry, for example[82]. In this natural geometry, the theorem 'precisely one straight line passes through two distinct points' holds only if the points are not lying too close together. For if they are lying very close together, then *several* adjacent straight lines can obviously be drawn through the two points. The *draftsman* must take these considerations into account; in *pure geometry, however,* this is not done because here the points are *idealized*. The *extended* points of experience are replaced by the *ideal*, unextended, 'points' of theoretical mathematics which, in reality, have no existence. That this procedure is benefical is borne out by its success: It results in a mathematical theory which is of a much simpler and considerably smoother form than that of natural geometry, which is continually concerned with unpleasant exceptions.

The relationship between actualist mathematics and constructivist mathematics is quite analogous: Actualist mathematics idealizes, for example, the notion of 'existence' by saying: A number *exists* if its existence can be proved by means of a proof in which the logical deductions are applied to *completed infinite* totalities in the same form in which they are valid for finite totalities; entirely as if these infinite totalities were actually present quantities. In this way the concept of existence therefore inherits the advantages and the disadvantages of an ideal element: The *advantage* is, above all, that a considerable simplification and elegance of the theory is achieved – since intuitionist existence proofs are, as mentioned, mostly very complicated and plagued by unpleasant exceptions–, the *disadvantage*, however, is that this ideal concept of existence is no longer applicable to the same degree to physical reality as, for example, the constructive concept of existence.

As an *example*, let us consider the equation $a \cdot x = b$ over the real numbers. According to the actualist interpretation, the following may simply be asserted: The equation has a root, as long as a is not equal to 0. The intuitionist, however, says: The equation has a root if it has been determined that a is different from 0. It may happen, however, that from the way a is given it cannot be anticipated whether a is equal to 0 nor whether a is different from 0.

In this case the question of the existence of a root remains open. It must certainly be admitted that this interpretation corresponds more to the position of the *physicist* who may have to determine the coefficient a from an *experiment* which is not precise enough to establish with certainty a difference between a and 0.

The question now arises: What use are elegant bodies of knowledge and particularly *simple* theorems if they are not applicable to physical reality in their literal sense? Would it not be preferable in that case to adopt a procedure which is more *laborious* and which yields *more complicated* results, but which has the advantage of making these results immediately meaningful in reality? The answer lies in the *success* of the former procedure: Again consider the example of geometry. The great achievements of mathematics in the advancement of physical knowledge stem precisely from this method of *idealizing* what is physically given and thereby *simplifying* its investigation. In any application of the general results to reality, their special status due to this idealization must, of course, be kept in mind and a corresponding *reinterpretation* must be carried out. This is where *applied* mathematics has its realm of activity.

For the sake of comparison, I quote from *Heyting* and *Weyl*:

Heyting, the intuitionist, says in one place[83]:

'From the formalist standpoint the aim of physics can be characterized as the mastering of nature. If this aim can be achieved by formal methods' – i.e., by actualist mathematics – 'then no argument is tenable against them'.

In the dispute between *Brouwer* and *Hilbert*[84], *Weyl* formulates his position as follows:

'In studying mathematics for its own sake one should follow *Brouwer* and confine oneself to discernible truths into which the infinite enters only as an open field of possibilities; there can be no motive for exceeding these bounds. In the natural sciences, however, a sphere is touched upon which is no longer penetrable by an appeal to visible self-evidence, in any case; here cognition necessarily assumes a symbolic form. For this reason it is no longer necessary, as mathematics is drawn into the process of a theoretical reconstruction of the world by physics, to be able to isolate mathematics into a realm of the intuitively certain: On this higher plane, from which the whole of science appears as a unit, I agree with *Hilbert*'.

I am under the impression that certain fundamental *intuitionist* concepts, e.g., the concept of existence or that of a real number, are strictly speaking already 'ideal elements'. Yet this may remain debatable; it is difficult to discuss and not that important. In any case, it would not mean that the application of such concepts also requires a *consistency proof*; they are after all applied only in such a way that their precise constructive sense always remains apparent (cf. § 3). The same is true of the 'ideal points' in projective geometry; the situation is *different* in the case of the ideal concepts of actualist mathematics which – looked at from the constructivist point of view – do not involve any inherent 'sense' at all, but which, in spite of this, are used as if, by their very wording, they were endowed with such a sense.

While the constructivists, on the one hand, are thus conceding a purpose to actualist mathematics, it seems reasonable that the *constructivist point of view* should, on the other hand, be given a greater role in mathematics than it has at present. In foundational research it is already customary to carry the proofs out along constructivist lines whenever possible, not only because of their greater indisputability – this is not always the aim of a proof – but also because of the greater tangible content of the result. For it is clear that a *constructivist* existence proof means more than an indirect actualist proof. Particularly in elementary *number theory* and, generally, in all theories dealing only with *finitely* describable objects, it is natural to take as a basis the constructivist point of view[85]. In the past this was done quite automatically, in any case; the genuinely naive reasoning in which no special attention is paid at all to the methods of proof, is by nature chiefly *constructive*, i.e., it shuns the 'infinite'. In these areas, moreover, the application of transfinite actualist forms of inference serves hardly any practical purpose. Not so in the realm of the *continuum*, in *analysis* and *geometry*: Here the actualist approach celebrates its triumphs; here the constructivist approach is inferior in practise.

In conclusion, it can therefore be said: The *constructivist* ('intuitionist', 'finitist') mathematics constitutes an *important realm* within the whole of mathematics because of its great self-evidence and the particular significance of its results. Yet no compelling reasons exist why all parts of analysis that are based on the actualist interpretation should be *radically rejected*; on the contrary, they are afforded a *great significance in their own right*, above all in view of their physical applications.

Whether the continuum should ultimately be regarded as a mere *fiction*, as an ideal construction, or whether it should be insisted upon that it possesses a *reality* independent of our methods of construction, in the sense of the actualist interpretation, is a purely theoretical question whose answer will probably remain a matter of taste; for *practical* mathematics it has hardly any further significance.

8. NEW VERSION OF THE CONSISTENCY PROOF FOR ELEMENTARY NUMBER THEORY

In the following I shall present a new version of the consistency proof contained in section IV of #4; only this time the main emphasis will be placed on developing the *fundamental ideas* and on making every single step of the proof as *lucid* as possible. For this purpose I shall in places dispense with the explicit exposition of certain details, viz., in those places where this is unimportant for the understanding of the context as a whole and where it can furthermore be supplied by the reader himself without much difficulty.

Sections I and III of # 4 contain considerations the knowledge of which need not be presupposed for an appreciation of the logic of the consistency proof, even though they are indispensable for the understanding of its *purpose*. In section II, I developed a quite detailed formalization of elementary number theory which preserves a close affinity with mathematical practice. This formalization is of great value now as ever; although a complete formal system could have been written down from the start, it seems to me that by doing so an essential part of the context as a whole would have fallen by the wayside.

Added to this must be the fact that the formal representation of the forms of inference (# 4, § 5), directly based, as it was, on mathematical practice, with the characteristic concept of the 'sequent', already proves quite suitable for metamathematical investigations, in fact, judging by my own experience, it is better suited to most purposes than the methods of representation generally customary to date.

Nevertheless it cannot be said that the 'most natural' logical calculus, simply because it corresponds most closely to real reasoning, is also the most suitable calculus for proof-theoretical investigations. For the consistency proof, in particular, a somewhat different version has proved to be even more suitable and will therefore be adopted in this paper. I am referring to that formalization of the logical forms of inference which I had already developed in # 3 as the '*LK*-calculus'[86]. A knowledge of that paper is not however presupposed. I shall merely require a few basic concepts from section II of the earlier consistency proof and shall, in each case, give the appropiate reference.

The constructivist proof of the 'theorem of transfinite induction' (up to ε_0), article 15.4 of #4, is retained unchanged as the conclusion of the consistency proof and will not be revised for the time being; cf. the concluding remarks at the end of the present paper.

§ 1. New formalization of number-theoretical proofs

I shall formulate the concepts involved and in each case include some explanatory remarks.

1.1. '*Formula*'.

The definition of a *formula* is adopted from the earlier paper (article 3.2), but with the following simplification:

Only 1 is used as a numeral. Functions are not admitted (cf. however the concluding remarks) with the exception of a single one, the successor function, which is denoted by a prime: $\mathfrak{a}'$ has the same informal meaning as $\mathfrak{a}+1$. By means of this function symbol the natural numbers can now be represented 1, 1′, 1″, 1‴ etc. *Terms* are therefore now always of the form 1 or 1′ or 1″ etc. or $\mathfrak{a}$ or $\mathfrak{a}'$ or $\mathfrak{a}''$ etc., where $\mathfrak{a}$ stands for an arbitrary free variable. The former we call *numerical terms* and they therefore correspond to the earlier numerals; the latter *variable terms*. Predicate symbols are admitted as before according to need; it is required only (#4, article 13.3) that they are decidably defined i.e., that it can be decided of every definite natural number whether the predicate does or does not hold. On the basis of these concepts of term and predicate the old definition of a formula (article 3.23) is now preserved, but the logical connective $\supset$ will no longer be used. This represents no significant restriction since it is well known that $\supset$ can be replaced by & and $\neg$, or $\vee$ and $\neg$. In addition we could still eliminate $\vee$ and $\exists$, as was done in #4 (§ 12); but this is unnecessary since by being in exact correspondence with & and $\forall$, these connectives cause no difficulties whatever in the '*LK*-calculus'.

Example of a formula:

$$\forall x \, (x > 1' \, \& \, \exists y \, (y''' = a))$$

where a is a free variable and x and y are bound variables.

Three simple auxiliary concepts will still be needed below:

A *prime formula* is a formula containing no logical connectives.
Example: $a''' = 1''$.

The *terminal connective* of a formula which is not a prime formula is that logical connective which is adjoined last in the construction of that formula (according to article 3.23 of #4).

The *degree* of a formula is the total number of logical connectives occurring in it.

Examples: A prime formula has degree 0. The formula $\forall x\,(x > 1'\ \&\ \exists y\,(y''' = a))$ has degree 3 and its terminal connective is the $\forall$.

1.2. '*Sequent*'.

A *sequent* is an expression of the form

$$\mathfrak{A}_1, \mathfrak{A}_2, \ldots, \mathfrak{A}_\mu \rightarrow \mathfrak{B}_1, \mathfrak{B}_2, \ldots, \mathfrak{B}_\nu,$$

where arbitrary formulae may take the place of

$$\mathfrak{A}_1, \mathfrak{A}_2, \ldots, \mathfrak{A}_\mu, \mathfrak{B}_1, \mathfrak{B}_2, \ldots, \mathfrak{B}_\nu.$$

The $\mathfrak{A}$'s are called the *antecedent formulae*, the $\mathfrak{B}$'s the *succedent formulae* of the sequent. Both the antecedent and the succedent may be empty.

Suppose that it is known of each antecedent and succedent formula of a sequent without free variables whether it is 'true' or 'false'. Then the sequent is 'false' if all of its antecedent formulae are true and all of its succedent formulae are false. (Moreover, a sequent which has neither antecedent nor succedent formulae is also false.) In every other case the sequent is 'true'.

Explanatory Remarks. We shall make use of the definition of 'true' and 'false' only in connection with the concept of a 'basic sequent' and here the $\mathfrak{A}$ and $\mathfrak{B}$ will be prime formulae without free variables and therefore immediately decidable. In general the concept of the 'truth' of a formula is of course not formally defined at all. The definition can nevertheless serve quite generally to explain the informal sense of a sequent, but it should still be added that a sequent with free variables is considered to be true if and only if every arbitrary substitution of numerals for free variables yields a true sequent. The informal meaning of a sequent without free variables can be expressed briefly as follows: 'If the assumptions '$\mathfrak{A}_1$', ..., '$\mathfrak{A}_\mu$' hold, then at least one of the propositions '$\mathfrak{B}_1$', ..., '$\mathfrak{B}_\nu$' holds.'

In #4, I had introduced the concept of a sequent with only *one* succedent formula for the immediate purpose of providing a natural representation of mathematical proofs (§ 5). Considerations of this kind will in fact also lead to the new symmetric concept of a sequent in situations where the aim is a particularly natural representation of the *distinction of cases* (cf. § 4 of #4, and 5.26 in particular). For, a ∨-elimination can now be represented simply as follows: From $\to \mathfrak{A} \vee \mathfrak{B}$ we infer $\to \mathfrak{A}, \mathfrak{B}$, which reads: 'One of the two possibilities, $\mathfrak{A}$ or $\mathfrak{B}$, holds.' It must be admitted that this new concept of a sequent in general already constitutes a departure from the 'natural' and that its introduction is primarily justified by the considerable *formal advantages* exhibited by the representation of the *forms of inference* following below which this concept makes possible.

It should still be pointed out that the informal sense of a sequent is to be considered to coincide with the given definition in those cases in which the sequent possesses no antecedent formulae or no succedent formulae: if there are no antecedent formulae, the sequent expresses the fact that at least one of the propositions '$\mathfrak{B}_1$', ..., '$\mathfrak{B}_\nu$' holds, this time independently of any assumptions. If there are no succedent formulae, the sequent expresss the fact that on the basis of the assumptions $\mathfrak{A}_1, \ldots, \mathfrak{A}_\mu$ *no possibility* remains open, i.e.: the assumptions are incompatible, they lead to a contradiction. A sequent without antecedent and succedent formulae, the 'empty sequent', therefore indicates that without any assumptions at all a contradiction results, i.e.: if this sequent is derivable in a system then that system itself is inconsistent.

Example of a sequent.

$$\forall x\,(x' > 1) \to a > 1 \vee a = 1,\ 1' > 1,\ 1'' = b,$$

where a and b are free variables and x a bound variable.

1.3. '*Inference figure*'.

An *inference figure* (the formal counterpart of an *inference*) consists of a *line of inference*, a *lower sequent*, written below the line, and *upper sequents* (one or more), written above the line. The lower sequent here stands for the conclusion of the inference which has been drawn from the premisses represented by the upper sequents.

The only inference figures admitted into our formalism are those obtainable from one of the following twenty *inference figure schemata* by a substitution of the following kind:

The variables $\mathfrak{A}, \mathfrak{B}, \mathfrak{D}$, and $\mathfrak{E}$ may be replaced by arbitrary formulae; the formulae $\forall \mathfrak{x}\, \mathfrak{F}(\mathfrak{x})$ and $\exists \mathfrak{x}\, \mathfrak{F}(\mathfrak{x})$ by arbitrary formulae of the same form; the formulae $\mathfrak{F}(\mathfrak{a})$ and $\mathfrak{F}(\mathfrak{t})$ by the formulae resulting from $\mathfrak{F}(\mathfrak{x})$ by the substitution of an arbitrary free variable $\mathfrak{a}$ and an arbitrary term $\mathfrak{t}$, resp., for the bound variable $\mathfrak{x}$. The letters Γ, Δ, Θ and Λ may be replaced by arbitrary, possibly empty sequences of formulae, separated by commas.

The following *restriction on variables* is to be observed: the free variable designated by $\mathfrak{a}$ – which we call the *eigenvariable* of the inference figure concerned – may not occur in the *lower sequent* of this inference figure.

The Inference Figure Schemata.

1.31. Schemata for *structural inference figures*:

$$\text{Thinning:} \quad \frac{\Gamma \to \Theta}{\mathfrak{D}, \Gamma \to \Theta} \qquad \frac{\Gamma \to \Theta}{\Gamma \to \Theta, \mathfrak{D}}$$

$$\text{Contraction:} \quad \frac{\mathfrak{D}, \mathfrak{D}, \Gamma \to \Theta}{\mathfrak{D}, \Gamma \to \Theta} \qquad \frac{\Gamma \to \Theta, \mathfrak{D}, \mathfrak{D}}{\Gamma \to \Theta, \mathfrak{D}}$$

$$\text{Interchange:} \quad \frac{\Delta, \mathfrak{D}, \mathfrak{E}, \Gamma \to \Theta}{\Delta, \mathfrak{E}, \mathfrak{D}, \Gamma \to \Theta} \qquad \frac{\Gamma \to \Theta, \mathfrak{D}, \mathfrak{E}, \Lambda}{\Gamma \to \Theta, \mathfrak{E}, \mathfrak{D}, \Lambda}$$

$$\text{Cut:} \quad \frac{\Gamma \to \Theta, \mathfrak{D} \qquad \mathfrak{D}, \Delta \to \Lambda}{\Gamma, \Delta \to \Theta, \Lambda}.$$

The two formulae in the last schema designated by $\mathfrak{D}$ are called *cut formulae*, their *degree* the degree of the cut.

1.32. Schemata for *operational inference figures*:

$$\&: \begin{cases} \dfrac{\Gamma \to \Theta, \mathfrak{A} \quad \Gamma \to \Theta, \mathfrak{B}}{\Gamma \to \Theta, \mathfrak{A} \& \mathfrak{B}} \end{cases} \qquad \dfrac{\mathfrak{A}, \Gamma \to \Theta}{\mathfrak{A} \& \mathfrak{B}, \Gamma \to \Theta} \qquad \dfrac{\mathfrak{B}, \Gamma \to \Theta}{\mathfrak{A} \& \mathfrak{B}, \Gamma \to \Theta}$$

$$\vee: \begin{cases} \dfrac{\mathfrak{A}, \Gamma \to \Theta \quad \mathfrak{B}, \Gamma \to \Theta}{\mathfrak{A} \vee \mathfrak{B}, \Gamma \to \Theta} \end{cases} \qquad \dfrac{\Gamma \to \Theta, \mathfrak{A}}{\Gamma \to \Theta, \mathfrak{A} \vee \mathfrak{B}} \qquad \dfrac{\Gamma \to \Theta, \mathfrak{B}}{\Gamma \to \Theta, \mathfrak{A} \vee \mathfrak{B}}$$

$$\forall: \begin{cases} \dfrac{\Gamma \to \Theta, \mathfrak{F}(\mathfrak{a})}{\Gamma \to \Theta, \forall \mathfrak{x}\, \mathfrak{F}(\mathfrak{x})} \end{cases} \qquad \dfrac{\mathfrak{F}(t), \Gamma \to \Theta}{\forall \mathfrak{x}\, \mathfrak{F}(\mathfrak{x}), \Gamma \to \Theta}$$

$$\exists: \begin{cases} \dfrac{\mathfrak{F}(\mathfrak{a}), \Gamma \to \Theta}{\exists \mathfrak{x}\, \mathfrak{F}(\mathfrak{x}), \Gamma \to \Theta} \end{cases} \qquad \dfrac{\Gamma \to \Theta, \mathfrak{F}(t)}{\Gamma \to \Theta, \exists \mathfrak{x}\, \mathfrak{F}(\mathfrak{x})}$$

$$\neg: \begin{cases} \dfrac{\mathfrak{A}, \Gamma \to \Theta}{\Gamma \to \Theta, \neg \mathfrak{A}} \end{cases} \qquad \dfrac{\Gamma \to \Theta, \mathfrak{A}}{\neg \mathfrak{A}, \Gamma \to \Theta}.$$

That formula in the schema which contains the *logical connective* is called the *principal formula* of the operational inference figure concerned.

1.33. Schema for *CJ – inference figures* (the formal counterparts of complete inductions):

$$\frac{\mathfrak{F}(\mathfrak{a}), \Gamma \to \Theta, \mathfrak{F}(\mathfrak{a}')}{\mathfrak{F}(1), \Gamma \to \Theta, \mathfrak{F}(t)}.$$

The degree of the CJ-inference figure is the degree of that formula in the schema which is designated by $\mathfrak{F}(1)$ – and which is, of course, the same as that of $\mathfrak{F}(\mathfrak{a})$, $\mathfrak{F}(\mathfrak{a}')$ and $\mathfrak{F}(\mathfrak{t})$.

Example of an inference figure:

$$\frac{\to a' = 1', 1 < 1'' \,\&\, a = 1''}{\to a' = 1', \exists z (1 < z \,\&\, a = 1'')},$$

where a is a free and z a bound variable.

Explanatory remarks about the inference figure schemata will follow below in connection with the concept of a derivation.

1.4. '*Basic sequents*'.

We shall distinguish basic '*logical*' and '*mathematical*' sequents.

A *basic logical sequent* is a sequent of the form $\mathfrak{D} \to \mathfrak{D}$, where an arbitrary formula may stand for $\mathfrak{D}$.

A *basic mathematical sequent* is a sequent consisting entirely of prime formulae and becoming a 'true' sequent with every arbitrary substitution of numerical terms for possible occurrences of free variables.

In accordance with our assumption of the decidability of all predicates, the 'truth' of a prime formula without free variables is always verifiable. Whether or not a sequent with free variables is a basic mathematical sequent is of course not generally decidable; nor is this actually essential.

Examples of basic sequents:

$$\forall x \, \exists y \, (x'' = a \,\&\, y > x) \to \forall x \, \exists y \, (x'' = a \,\&\, y > x)$$
$$a = b, b = c \to a = c$$
$$a < 1 \to$$
$$\to 1' > 1$$
$$a = b \to a = b$$
$$\to a' > a$$
$$\to 1''' = 1 \,(\text{mod } 1'').$$

1.5. '*Derivation*'.

A *derivation* is a figure in tree-form consisting of a number of sequents (at least one) with one lowest sequent, the *endsequent*, and certain *uppermost sequents* which must be basic sequents; the connection between the uppermost sequents and the endsequent is established by inference figures.

It should be intuitively obvious how this is meant; yet let me paraphrase it again as follows: suppose that an endsequent is given. This sequent is either already an uppermost sequent – in which case it alone constitutes at once the entire derivation – or it is the lower sequent of an inference figure. Every upper sequent of this inference figure is in turn either an uppermost sequent of the derivation or the lower sequent of a further inference figure, etc.

The reader should always visualize a derivation quite intuitively as a tree-like structure, then the transformations on a derivation to be performed in § 3 become most easily intelligible.

Example of a derivation:

$$\frac{\rightarrow 1 = 1 \quad \dfrac{a = a \rightarrow a' = a'}{1 = 1 \rightarrow b = b}\ CJ\text{-inference figure}}{\dfrac{\rightarrow b = b}{\rightarrow \forall x(x = x)}}\ \text{cut} \qquad \dfrac{\dfrac{1''' = 1''' \rightarrow 1''' = 1'''}{\forall x\,(x = x) \rightarrow 1''' = 1'''}}{\rightarrow 1''' = 1'''}\ \begin{array}{l}\forall\text{-inference}\\ \text{figures}\\ \text{cut.}\end{array}$$

For a further example, cf. 1.6.

Another auxiliary concept which will be needed later:

A '*path*' in a derivation is, briefly speaking, a sequence of sequents which we must run through in descending from a given uppermost sequent to the endsequent. At each step, we pass from an upper sequent of an inference figure to the lower sequent of that inference figure.

It is furthermore immediately obvious what is meant by the following: a sequent in the derivation stands *above* or *below* another sequent occurring in the same path (i.e., not only immediately above or below it, but any number of steps apart). It is understood that wherever the notion of 'above' or 'below' is used, the sequents concerned belong to a common path; otherwise the concept is without sense.

1.6. *Explanatory remarks about the new formalization of number-theoretical proofs.*

As a result of the revised concept of a derivation a formalization of number-theoretical proofs is given which distinguishes itself from my earlier 'natural' version mainly *in two points*. *First*: the rules of inference belonging to the logical connectives, i.e. the 'introduction' and 'elimination' of a logical connective, have now been reformulated throughout in such a way that the *lower sequent* always contains the 'principal formula', whereas the upper sequents contain the associated side formulae. To the earlier 'introduction' of a logical connective now corresponds the occurrence of that connective in a *succedent formula* of the lower sequent, to the 'elimination' of a logical connective corresponds the occurrence of that connective in an *antecedent formula* of the lower sequent. The reader should convince himself of the equivalence of the old and new versions by examining, for example, the ∀-rules of inference (disregarding, for the time being, the occurrence of several succedent formulae). The 'cut' and the basic logical sequents must be used in the proof of equivalence. Cf. the derivation with the ∀-introduction on the left and the subsequent '∀-elimination', given as an example in 1.5.

This part of the conversion from the former to the new rules of inference amounts to an abandoning of the *natural succession of the propositions* in number-theoretical proofs and to the introduction, in its place, of an artificial arrangement of the propositions along special lines with the result that in operational inferences the *simpler* proposition now always comes *first* and is *followed* by the more complex proposition, viz., the proposition with the additional connective. This rearrangement proves of practical value for the consistency proof because of the essential role which the concept of the *complexity* of a derivation and, with it, the complexity of a particular formula (which increases as the *degree* of the formula increases) plays in the consistency proof.

The *second* important distiction vis-à-vis the old concept of a derivation consists in the symmetrization of the sequents by the admission of arbitrarily many succedent formulae. This makes it unfortunately more difficult to grasp the informal sense of the various inference schemata and to persuade oneself of their ·correctness'. To overcome this difficulty the reader should first conceive of the presence of only one succedent formula and should then convince himself that the inference remains correct even if several succedent formulae occur and also if no succedent formula occurs. As the reader becomes more familiar with this concept of a derivation, he should be able to realize that transformations of derivations and other proof-theoretical investigations can be carried out particularly simply and elegantly with this concept. The decisive advantages are these:

There exists complete symmetry between & and ∨, ∀ and ∃. All of the connectives &, ∨, ∀, ∃ and ¬ have, to a large extent, equal status in the system; no connective ranks notably above any other connective. The special position of the negation, in particular, which constituted a troublesome exception in the natural calculus (cf. articles 4.56 and 5.26 of ≠ 4), has been completely removed in a seemingly magical way. The manner in which this observation is expressed is undoubtedly justified since I myself was completely surprised by this property of the '*LK*-calculus' when first formulating that calculus. The 'law of the excluded middle' and the 'elimination of the double negation' are implicit in the new inference schemata – the reader may convince himself of this by deriving both of them whithin the new calculus – but they have become completely harmless and no longer play the least special role in the consistency proof that follows.

If we think of the $\Gamma, \Delta, \Theta, \Lambda$ as absent from the inference figure schemata, we see that the schemata are of the greatest simplicity and likeness in the sense that none of them any longer contains anything that is not absolutely essential; the $\Gamma, \Delta, \Theta, \Lambda$ constitute an appendage which signifies merely that *additional* antecedent and succedent formulae are carried forward unchanged from the upper sequent to the lower sequent.

The new formulation of the concept of the 'basic mathematical sequent' still requires an explanation. In # 4, this concept was defined differently (articles 5.23 and 10.14). It turns out however that the former basic sequents are *derivable* in the new system. An example which typifies the general aspects of the situation may make this clear:

The following 'basic mathematical sequent' in the old sense

$$\rightarrow \forall x \, \forall y \, \neg (x = y \, \& \, \neg y = x)$$

is derivable thus:

$$\cfrac{\cfrac{\cfrac{a = b \rightarrow b = a}{a = b \, \& \, \neg b = a \rightarrow b = a}}{\cfrac{\neg b = a, \; a = b \, \& \, \neg b = a \rightarrow}{\cfrac{a = b \, \& \, \neg b = a, \; a = b \, \& \, \neg b = a \rightarrow}{\cfrac{a = b \, \& \, \neg b = a \rightarrow}{\begin{array}{l} \rightarrow \neg (a = b \, \& \, \neg b = a) \\ \rightarrow \forall y \, \neg (a = y \, \& \, \neg y = a) \\ \rightarrow \forall x \, \forall y \, \neg (x = y \, \& \, \neg y = x) \end{array}}}}}}{}$$

All usual 'basic mathematical sequents' in the old sense are derivable in the same way from informally synonymous basic mathematical sequents in the new sense[87]. The fact that the new concept of a derivation is actually *equivalent* with that of the earlier paper – apart from the restriction which results from the initially limited admission of *functions* in the new system – can be verified without great difficulty from the observations made above, and I shall discuss it no further[88].

§ 2. Outline of the consistency proof

It is to be shown that every derivation is *consistent*[89]; this may be paraphrased by saying that no derivation has an *empty endsequent*. For from a contradiction, $\to \mathfrak{A}$ and $\to \neg \mathfrak{A}$, we can first of all derive the sequents $\to \neg \mathfrak{A}$ and $\neg \mathfrak{A} \to$, and from them, by means of a cut, the empty sequent. (Conversely, from the empty sequent every arbitrary sequent can be derived by 'thinnings'.)

It makes sense that we should begin by proving the consistency of simple derivations, then of *more complex ones*, using the consistency of the simpler derivations, and so forth. We thus proceed 'inductively'. It is furthermore not implausible that this procedure repeatedly requires the examination of an already *infinite* sequence of derivations before a more complex class can be tackled; for example, first all derivations consisting of only one sequent, then all derivations consisting of two sequents, etc. This actually means that we are applying a '*transfinite induction*'. In practice the pattern of this analysis is of course considerably more involved than in the case of the given example.

The proof is carried out in three stages:

1. The consistency of an arbitrary derivation is reduced to the consistency of all 'simpler' derivations. This is done by defining an – unambiguous – *reduction step* for arbitrary 'contradictive derivations', i.e. derivations with *the empty sequent as endsequent*; this step transforms such a derivation into a 'simpler' derivation with the same endsequent. The definition of this reduction step forms the contents of § 3.

2. Then a *transfinite ordinal number is correlated* with every derivation and it is shown that in a reduction step the contradictive derivation concerned is turned into a derivation with a *smaller ordinal number*. In this way the so far only loosely determined concept of 'simplicity' receives its precise sense: the larger the *ordinal number* of a derivation, the greater is its '*complexity*' in the context of this consistency proof. This forms the contents of § 4.

3. From this the consistency of all derivations then obviously follows by 'transfinite induction'. The inference of transfinite induction which, at first, is a rather 'disputable' inference, may not be presupposed in the consistency proof nor proved as in set theory. This inference requires rather a separate justification by means of indisputable 'constructive' forms of inference. At the end of § 4, the reader is at this time referred to # 4 in this connection.

§ 3. A reduction step on a contradictive derivation

3.1. *Underlying ideas.*
Suppose that a derivation is given whose endsequent is the empty sequent.

This derivation is to be transformed into a (in some sense) *simpler* derivation with the same endsequent. What is here meant by 'simpler' can at present only be stated roughly and will be made precise later through the ordinal numbers.

What are the considerations that make us suspect at all that, given a proof for a contradiction, there already exists an even *simpler* way of proving such a contradiction? By a contradiction is meant a proposition of a quite simple structure, for example '$1 = 2$'. If such a *simple* proposition can be proved by means of a complex proof, it is reasonable to suspect that the proof can be *simplified*. The following argument might conceivably be used: Somewhere in the proof there must after all occur a proposition of *maximal complexity*. In that case it must be assumed that this 'complexity extremum' (formally, this might be a formula of maximal *degree* occurring in the derivation) must somehow be 'reducible'. The only way in which this proposition could in general enter into the proof is by an 'introduction' of its terminal connective and a subsequent 'elimination' of the same connective. But if a connective is first introduced and then again eliminated it can be left out altogether by direct passage from the preceding sub-propositions to the corresponding succeeding sub-propositions[90].

This is the basic idea underlying the 'operational reduction' to be outlined below. In actual fact, however, the situation will turn out not to be quite as simple as assumed in the argument just sketched. One of the difficulties that may arise is the occurrence of a *complete induction* in the proof; viz., in the case where the proposition with the maximal number of conectives in question is not directly proved by an 'introduction' inference, but rather by a complete induction. This requires a further special kind of reduction step which will be called a '*CJ*-reduction'. The form of this reduction step is extremely simple and precisely what we would expect: if the term t in the schema of the *CJ*-inference figure is a *numerical term*, thus denoting a *fixed number*, the complete induction can naturally be replaced by a number of ordinary inferences – in our formalization a number of 'cuts'. This constitutes the '*CJ*-reduction'.

If a *CJ*-inference figure occurs in the derivation whose t is a *variable* term – and this is in fact normally the case – then this figure cannot of course be reduced *immediately* in this way. The reduction procedure may be arranged in such a way that with successive reduction steps more and more variable terms are gradually replaced by numerical terms so that eventually even initially irreducible *CJ*-inference figures become in turn reducible. This remark is incidental. Here we are actually concerned only with the definition of *one single reduction step* so that regardless of the nature of the given contradictive derivation, at least *one place* can be found in it to which a reduction can be applied.

Let us suppose therefore that there is no place in the derivation in which a *CJ*-reduction can be carried out. Then, as will be shown in detail below, an 'operational reduction' is always feasible. On the other hand, it cannot be expected that a formula of highest degree in the entire derivation is always amenable to a reduction. As mentioned before, this formula may have been introduced by a *CJ*-inference figure and this figure can contain a variable t. It is nonetheless possible in each case to locate a formula in the derivation which represents a 'relative extremum', viz., a formula which is introduced by the introduction of its terminal connective and whose further use in the derivation then consists in the elimination of that connective, and which is therefore reducible. Why such a formula must exist is best seen within the context of the proof following below (3.43).

A special phenomenon should still be mentioned: it may happen that the formula which is intended to form the starting point of the operational reduction is used again in the derivation not only once but several times. (An example: suppose that the formula has the form $\forall \mathfrak{x}\, \mathfrak{F}(\mathfrak{x})$, and from it are inferred $\mathfrak{F}(1)$ and $\mathfrak{F}(1''')$ or in another place perhaps even $\forall \mathfrak{x}\, \mathfrak{F}(\mathfrak{x}) \vee \mathfrak{A}$.) In the general case all that can be achieved is that in one place of application the formula is used in the form of an elimination of its terminal connective. About the remaining places nothing can be said. In this general case the formula can therefore not be reduced away completely; we can merely bring about a simplification *of this one place of application* which, *at this point*, makes the passage via the formula redundant. The occurrence of this formula in the remaining places, remains unaffected. It turns out that this is sufficient.

These preliminaries have been carried out against the background of the '*natural proof*' with the natural succession of the individual propositions. For their application to our formalism developed in § 1, a corresponding translation must be made: To the 'introduction' of a connective here corresponds its occurrence in a succedent formula of the lower sequent, to the 'elimination' of that connective its occurrence in an antecedent formula of the lower sequent of the operational inference figure. All other details will follow from the precise formal development now to be carried out; the preliminaries ought not and cannot of course do more than indicate to the reader in a superficial way the main ideas of the procedure and in doing so facilitate the understanding of the actual presentation.

3.2. *Elimination of redundant free variables in preparation for the reduction step. – The 'ending'.*

We begin with the definition of the 'reduction step on a contradictive derivation' by stipulating: before the reduction step proper the following simple transformation must be carried out:

All free variables in the derivation are to be replaced by the numeral 1; except that the eigenvariable (1.3) of an inference figure is retained in all derivation sequents occurring above the lower sequent of the inference figure concerned.

What is the effect of this preliminary step? Actually, a free variable *normally* serves as eigenvariable of an inference figure and may here occur only above the lower sequent of this inference figure; its occurrence in the lower sequent itself is even expressly forbidden by the restriction on variables (1.3). Wherever else free variables may occur they are completely *redundant* and can equally well be replaced by 1. It is fairly obvious that this leaves the derivation correct. The empty endsequent remains of course unchanged.

We furthermore require a simple auxiliary concept – the *ending of a derivation* – which is defined thus: the ending consists of all those derivation sequents that are encountered if we ascend each individual path (1.5) from the endsequent and stop as soon as we arrive at the line of inference of an operational inference figure. Thus the lower sequent of this inference figure in each case still belongs to the ending, but its upper sequents do so no longer. If a path crosses no line of inference of an operational inference figure at all, then it is of course completely included in the ending.

Among the inference figures, the ending obviously contains only structural and *CJ*-inference figures.

We now distinguish two cases:
1. The ending of our contradictive derivation contains at least one *CJ*-inference figure. In that case a *CJ-reduction* is carried out, cf. 3.3.
2. The ending contains no *CJ*-inference figure. In that case an *operational reduction* is carried out (3.5) after a further preparatory step (3.4).

3.3. *The CJ-reduction.*

If the ending of the given contradictive derivation contains at least one *CJ*-inference figure after the stated preparatory step, then the reduction step proper consists of the transformation of the derivation described next.

We select a *CJ*-inference figure in the ending which is such that it occurs *above* no other *CJ*-inference figure. (i.e.: the derivational path which goes through the lower sequent of the selected *CJ*-inference figure must not cross the line of inference of any *CJ*-inference figure between that sequent and

§ 3, A REDUCTION STEP ON A CONTRADICTIVE DERIVATION

the endsequent.) In order to make the reduction step unambiguous, an appropriate procedure for the unique determination of the *CJ*-inference figure to be selected must still be given; there is a simple way in which this can be done.

The *CJ*-inference figure has the form:

$$\frac{\mathfrak{F}(\mathfrak{a}), \Gamma \to \Theta, \mathfrak{F}(\mathfrak{a}')}{\mathfrak{F}(1), \Gamma \to \Theta, \mathfrak{F}(\mathfrak{n})},$$

where $\mathfrak{n}$ designates a *numerical term*. For, by virtue of the preparations made no variable term could here possibly occur; in fact the lower sequent cannot contain a single free variable: after the preparatory step, free variables can occur only above inference figures with one eigenvariable and no such figure occurs below our *CJ*-inference figure. Indeed, the section of the derivational path between the lower sequent and the endsequent of this figure from here on crosses only lines of inference of *structural inference figures*.

This *CJ*-inference figure is now replaced by a system of structural inference figures of the following kind:

$$\frac{\dfrac{\mathfrak{F}(1), \Gamma \to \Theta, \mathfrak{F}(1') \quad \mathfrak{F}(1'), \Gamma \to \Theta, \mathfrak{F}(1'')}{\dfrac{\mathfrak{F}(1), \Gamma, \Gamma \to \Theta, \Theta, \mathfrak{F}(1'')}{\dfrac{\mathfrak{F}(1), \Gamma \to \Theta, \mathfrak{F}(1'') \quad \mathfrak{F}(1''), \Gamma \to \Theta, \mathfrak{F}(1''')}{\dfrac{\mathfrak{F}(1), \Gamma, \Gamma \to \Theta, \Theta, \mathfrak{F}(1''')}{\mathfrak{F}(1), \Gamma \to \Theta, \mathfrak{F}(1''')}}}}{\mathfrak{F}(1), \Gamma \to \Theta, \mathfrak{F}(\mathfrak{n})} \quad \text{etc.}$$

cut; possibly several interchanges and contractions; cut; possibly several interchanges and contractions.

Above the sequents $\mathfrak{F}(1), \Gamma \to \Theta, \mathfrak{F}(1')$ and $\mathfrak{F}(1'), \Gamma \to \Theta, \mathfrak{F}(1'')$ etc., we write in each case that section of the derivation which precedes $\mathfrak{F}(\mathfrak{a}), \Gamma \to \Theta, \mathfrak{F}(\mathfrak{a}')$, where we replace the free variable $\mathfrak{a}$ in the entire section – except in the case where it at the same time happens to be the eigenvariable of an inference figure occurring in that section, in all sequents occurring above the lower sequent of that inference figure – by the numerical terms 1 or 1' or 1'' etc. From the sequent $\mathfrak{F}(1), \Gamma \to \mathfrak{F}(\mathfrak{n})$ downwards, the ending is finally continued by adjoining the unchanged remainder of the old derivation. To put it precisely: all derivational paths which did not go through this sequent have been preserved unchanged and those which did go through it remain unchanged from the endsequent up to this point.

If $\mathfrak{n}$ is equal to 1, then the reduction proceeds somewhat differently: in that case the lower sequent of the *CJ*-inference figure runs $\mathfrak{F}(1), \Gamma \to \Theta, \mathfrak{F}(1)$, This sequent is derived from the *basic logical sequent* $\mathfrak{F}(1) \to \mathfrak{F}(1)$ by thinnings and interchanges, as required. Whatever preceded this lower sequent in the derivation is omitted; everything else is retained unchanged, as in the general case.

It is easily seen that in the *CJ*-reduction step the given contradictive derivation is in all parts transformed into another correct contradictive derivation. All we need to realize here in essence is that the replacement of $\mathfrak{a}$ by a numerical term turns every inference figure into another correct inference figure.

Comments about the nature of the reduction step should no longer be required; as stated at 3.1, its informal significance is exceedingly simple: a complete induction up to a definite number is replaced by a corresponding number of ordinary inferences.

3.4. *Preliminaries and preparatory step for an operational reduction.*

We must now deal with the case whesre the contradictive derivation contains no *CJ*-inference figure in its ending after the preparatory step 3.2.

The 'operational reduction' to be carried out in this case is preceded by a further preparatory step (3.42) whose purpose it is to eliminate all possible occurrences of *thinnings* and *basic logical sequents* from the ending, since these would otherwise give rise to bothersome exceptions in the actual operational reduction.

For this purpose, and also for the sake of its further use, we must first examine *the structure of the ending* more closely.

3.41. *The ending* of our derivation contains only *structural* inference figures. Its uppermost sequents are the uppermost sequents of the entire derivation, or the lower sequents of operational inference figures. The ending contains no free variables (since it contains no inference figures with eigenvariables). This is all quite obvious.

We now introduce two simple auxiliary concepts:

Identical sequent formulae in the upper sequents and the lower sequent of a structural inference figure *corresponding to one another* according to the inference figure schema will be called '*clustered*'.

Clustered are, for example, the three formulae designated by $\mathfrak{D}$ in the schema of a contraction; likewise the first of the formulae designated by Γ in the upper sequent with the first of the formulae designated by Γ in the lower sequent; the second formula of the upper sequents with the second formula of the lower sequent; etc.; the two cut formulae of a cut are clustered; etc.

§ 3, A REDUCTION STEP ON A CONTRADICTIVE DERIVATION

The totality of all formulae in the ending of the derivation obtained by starting with *a particular formula* and collecting all of its *clustered* formulae, then all formulae clustered with these, etc., is called a *cluster of formulae*; we can also say: *the cluster associated* with the relevant *initial* formula.

About the form of this cluster we can say the following:

With every cluster *is associated a cut* in the sense that its cut formulae belong to the cluster. This is so since every formula which occurs somewhere in the ending, as is evident from the structural inference figure schemata, is always clustered with a formula in the sequent standing immediately below it, except when it is a cut formula. Since the endsequent of our derivation is empty, we must at some point reach such a cut in tracing a cluster downwards towards the endsequent.

We now start with this cut and trace the location of the cluster upwards from the two cut formulae belonging to the cluster. With the following result: That portion of the cluster which is obtained by starting with the *left* cut formula – we call it *the left side of the cluster* – is in tree-form; a branching takes place if, in coming from below, we reach a contraction whose $\mathfrak{D}$ belongs to the cluster; a branch may *terminate* at some point if the $\mathfrak{D}$ of a thinning or the uppermost sequent of the ending is reached; in that case we speak of an *uppermost formula of the cluster*. All formulae of the left side of the cluster are *succedent formulae* of the sequents concerned. Exactly analogous remarks apply to the *right side of the cluster*, obtained by starting with the *right* cut formula; it too is in tree-form, etc., all its formulae are *antecedent formulae*. It follows further that no cut formulae other than the two formulae from which we started belong to the cluster; hence *the cut associated with* a cluster is uniquely determined and so are therefore the concepts of the left side and the right side of the cluster. No formulae of the cut other than the cut formulae belong to the cluster. All formulae belonging to the cluster occur above the lower sequent of the cut. (I.e.: *all* sequents containing cluster formulae occur above that sequent.) The left and right sides together therefore constitute the whole cluster.

The correctness of all these assertions is easily seen by tracing the cluster mentally from the cut formulae upwards and by visualizing with the help of the schemata of the structural inference figures the kinds of procedure which alone lead to new clustered formulae.

3.42. We can now turn to the *preparatory step for the operational reduction* which, as said earlier, is intended to accomplish *the elimination of all thinnings and basic logical sequents from the ending*. This can clearly be done. After all, a 'thinning' represents only a weakening of the informal sense of a sequent; if a contradiction can be derived from the weakened sequent, the same can obviously also be derived from the stronger upper sequent alone; and a basic logical sequent, being a pure tautology, is also dispensable in the context of mere structural transformations.

The procedure almost suggests itself. Let us begin with the thinnings: We select a thinning above which – in the ending – no other thinning occurs. We then simply cancel its lower sequent and then use the upper sequent in its place. In order to leave the derivation correct, we continue downwards and in the next lower sequent cancel the formula *clustered* with the formula $\mathfrak{D}$ in the thinning, as well as the formula clustered with the latter in the subsequent lower sequent, etc. Can this procedure lead to new difficulties? Actually, a contraction may arise in which a $\mathfrak{D}$ of the upper sequent is to be cancelled. All the better, the upper sequent becomes then *identical* with the lower sequent; the contraction becomes redundant and we have finished. There may be other occasions in the procedure in which the upper and lower sequents of an inference figure become *identical*; in that case we simply omit the inference figure and write the sequent down only once. If we encounter a *cut* in which the formula to be cancelled is a *cut formula*, we cancel the other upper sequent of the cut, together with whatever stands above it, and derive the lower sequent from the remaining upper sequent alone by thinnings and interchanges (as far as necessary).

The new thinnings which arise are again eliminated by the same procedure. That this procedure terminates, thus ridding the ending of thinnings completely, follows from the fact that with each reduction step we find ourselves *lower down* in the derivation (measured in terms of the total number of cuts up to the endsequent, for example).

We leave it to the reader to give an exact demonstration of the feasibility of the indicated procedure, as well as to formulate it unambiguously; this presents no essential difficulties.

Next we eliminate the *basic logical sequents*: In the ending such a sequent can now occur only as the upper sequent of a cut since no contractions and interchanges are applicable to it; the lower sequent of the cut is therefore, as is easily seen, *identical with the other* upper sequent. We therefore simply omit the cut and have thus finished.

As a result we finally obtain a contradictive derivation whose ending has the same properties as those stated above with the additional property of containing *no thinnings* and *no basic logical sequents* (as uppermost derivation sequents).

3.43. *Further preliminaries to the operational reduction.*

I now assert: There exists at least one cluster of formulae in the ending of our derivation, with at least one uppermost formula both on its left side and on its right side, which is *the principal formula of an operational inference figure.*

At this point the connection between our formal procedure and the fundamental ideas sketched in 3.1 becomes apparent: the concept of the *cluster of formulae* makes it possible for us to grasp in its entirety the collection of all occurrences of a 'proposition' in the 'proof' (i.e.: *formula in the derivation*). A *principal formula* as the uppermost formula *on the left side* corresponds to an *introduction instance* of the terminal connective of the proposition concerned; a principal formula *on the right side* – which is, after all, an *antecedent formula* – corresponds to a subsequent *elimination instance* of that *connective.* The *cut* associated with the cluster represents nothing more than the formal establishment of the connection between the two instances made necessary by the particular structure of our formalism. The fact that branchings of a cluster occur, corresponds to the difficulty discussed at the end of 3.1; branchings on the right side, for example, represent a *multiple application* of the proposition. The fact that branchings can appear both on the left and the right is due to the general symmetry of our formalism and renders it more difficult to carry over the fundamental ideas to each individual detail of the reduction. It suffices, however, if we have a reasonable conception of the fundamental ideas and continue to let ourselves be guided simply by formal analogies; this is precisely what I have done in formulating the consistency proof.

We must now prove the above assertion which can be interpreted as asserting *the existence of a suitable place for an operational reduction* in our derivation.

In this connection we first observe that our derivation must contain at least one operational inference figure. If this were not the case, the ending would represent the entire derivation. This would mean that a 'false' sequent has been derived from basic mathematical sequents which contain no free variables, and are therefore 'true' sequents, by means of the application of structural inference figures alone and without thinnings. At the same time the only formulae occurring in the whole derivation are prime formulae without free variables, thus *decidable* formulae, so that it can be decided of each sequent whether it is true or false. (A formula with logical connectives cannot occur because no such connective occurs in the basic sequents, and because none could have been introduced by the possible inference figures.) This would mean that at least one inference figure occurs whose lower sequent is 'false' whereas its upper sequents are 'true'. This is easily seen to be impossible.

In order to prove the above assertion, we now examine all those paths of the ending whose uppermost sequent is the lower sequent of an operational inference figure. We follow these paths from the top down and record whether in the sequents which we encounter a formula occurs which belongs to the same cluster as one of the *principal formulae* standing immediately above it (or whether it itself is a principal formula). This is so in the case of the uppermost sequents of our paths and, as we continue downwards in a path, this property is generally inherited. It is preserved trivially in passing through contractions and interchanges (by the definition of cluster). If we reach a cut in which two paths of the considered type meet, it may happen, however, that this property is not transferred to the lower sequent; but this can arise at most in the case where the cluster belonging to the *cut formulae* contains a principal formula on both sides. This is precisely the case specified in the *assertion*. Since the empty endsequent does not possess the mentioned property in any case, the assertion is proved as long as this case really is the only possible one in which the property under discussion *fails* to be passed on as we trace out the paths under examination. To this needs to be added only one more case, viz., the case in which, coming from above, a cut is encountered whose other upper sequent belongs to none of the paths examined and can therefore occur only in paths of the ending that are bordered above by *basic* mathematical *sequents*. This upper sequent can then contain only *prime formulae*, and the cut formulae are therefore also prime; indeed, a formula occurring in the traced upper sequent and belonging to the same cluster as a principal formula cannot be a cut formula since its degree is greater than 0, and it is therefore clustered with a formula with the same property in the *lower sequent*.

This concludes the proof of the existence of a cluster of formulae suitable for an operational reduction.

Now one last auxiliary concept that will be of central importance for the definition of the 'measure of complexity' of a derivation:

By the *level of a derivation sequent* we mean the highest degree of any cut or of a *CJ*-inference figure whose lower sequent stands below the sequent concerned. If there is no such inference figure, then the level is equal to 0.

Comments about the importance of this concept will follow below.

3.5. *The operational reduction.*

Now the operational reduction proper can be defined. Given is contradictive derivation whose ending includes at least one cluster of formulae

§ 3, A REDUCTION STEP A ON CONTRADICTIVE DERIVATION

containing on each side at least one principal formula of an operational inference figure. We select *such* a cluster of formulae and *from each* of its sides *one* uppermost formula of the kind mentioned. In order to make this step unambiguous, a certain procedure concerning the type of choice to be made must be specified; this is not difficult. We shall first deal with the case in which the terminal connective of the clustered formulae is a $\forall$. The remaining cases are dealt with almost in the same way and can be disposed of later in a few words. The derivation therefore looks like this:

$$
\begin{array}{cc}
\vdots\,(\mathfrak{a}) & \vdots \\
\dfrac{\Gamma_1 \to \Theta_1, \mathfrak{F}(\mathfrak{a})}{\Gamma_1 \to \Theta_1, \forall\mathfrak{x}\,\mathfrak{F}(\mathfrak{x})} \quad & \dfrac{\mathfrak{F}(\mathfrak{n}), \Gamma_2 \to \Theta_2}{\forall\mathfrak{x}\,\mathfrak{F}(\mathfrak{x}), \Gamma_2 \to \Theta_2} \quad \text{The two operational inference figures}
\end{array}
$$

$$
\text{level } \rho \quad \dfrac{\Gamma \to \Theta, \forall\mathfrak{x}\,\mathfrak{F}(\mathfrak{x}) \quad \forall\mathfrak{x}\,\mathfrak{F}(\mathfrak{x}), \Delta \to \Lambda}{\Gamma, \Delta \to \Theta, \Lambda} \quad \text{The cut associated with the cluster}
$$

level ρ
level $\sigma < \rho$ $\overline{}$ 'level line'
$$\Gamma_3 \to \Theta_3$$

$$\overline{} \to \quad \text{the empty sequent.}$$

Explanatory remarks:

The dots are intended to indicate that further paths may enter into the traced paths from both sides in arbitrary fashion. In addition, entire derivational sections of a certain kind may stand above the operational inference figures. The term $\mathfrak{n}$ can only be a *numerical term* since no inference figure with an eigenvariable can occur below it (3.2, 3.41). Suppose that $\Gamma_3 \to \Theta_3$ is the first sequent encountered, as we trace the path from $\Gamma, \Delta \to \Theta, \Lambda$ to the endsequent, which is of *a lower level* than the upper sequents of the cut belonging to the cluster. (Such a sequent must always exist since the level of the endsequent equals 0, while that of the upper sequents of the cut in question is at least 1, since the degree of the cut itself is at least 1.) It may happen that the sequent $\Gamma, \Delta \to \Theta, \Lambda$ is already the desired sequent; the above diagram must then be interpreted correspondingly. It may of course equally well happen that an upper sequent of the cut is itself already the lower sequent of the operational inference figure; and, finally, the sequent $\Gamma_3 \to \Theta_3$ may be identical with the endsequent; none of this makes any difference to the reduction.

The *reduction step* consists now of the transformation of the derivation into the form indicated by the following diagram:

§ 3, A REDUCTION STEP ON A CONTRADICTIVE DERIVATION

$$
\begin{array}{c}
\vdots(\mathfrak{n}) \\
\dfrac{\Gamma_1 \to \Theta_1, \mathfrak{F}(\mathfrak{n})}{\Gamma_1 \to \mathfrak{F}(\mathfrak{n}), \Theta_1, \mathfrak{x}\mathfrak{F}(\mathfrak{x})} \text{ Interchanges and thinning} \qquad \dfrac{\mathfrak{F}(\mathfrak{n}), \Gamma_2 \to \Theta_2}{\mathfrak{x}\mathfrak{F}(\mathfrak{x}), \Gamma_2 \to \Theta_2} \\
\dfrac{\Gamma \to \mathfrak{x}\mathfrak{F}(\mathfrak{x}), \Theta, \Lambda}{\Gamma, \Lambda \to \mathfrak{F}(\mathfrak{n}), \Theta, \Lambda} \\
\vdots \\
\dfrac{\Gamma_3 \to \mathfrak{F}(\mathfrak{n}), \Theta_3}{\Gamma_3 \to \Theta_3, \mathfrak{F}(\mathfrak{n})}
\end{array}
\qquad
\begin{array}{c}
\vdots(\alpha) \\
\dfrac{\Gamma_1 \to \Theta_1, \mathfrak{F}(\alpha)}{\Gamma_1 \to \Theta_1, \mathfrak{x}\mathfrak{F}(\mathfrak{x})} \qquad \dfrac{\mathfrak{F}(\mathfrak{n}), \Gamma_2 \to \Theta_2}{\mathfrak{x}\mathfrak{F}(\mathfrak{x}), \Gamma_2, \mathfrak{F}(\mathfrak{n}) \to \Theta_2} \text{ Interchanges and thinning} \\
\dfrac{\Gamma \to \mathfrak{x}\mathfrak{F}(\mathfrak{x}), \Theta}{\Gamma, \Delta, \mathfrak{F}(\mathfrak{n}) \to \Theta, \Lambda} \\
\vdots \\
\dfrac{\Gamma_3, \mathfrak{F}(\mathfrak{n}) \to \Theta_3}{\mathfrak{F}(\mathfrak{n}), \Gamma_3 \to \Theta_3} \text{ Interchanges, if necessary. New cut}
\end{array}
$$

$$
\dfrac{\Gamma_3, \Gamma_3 \to \Theta_3, \Theta_3}{\Gamma_3 \to \Theta_3} \text{ Contractions and interchanges, if necessary}
$$

$$
\vdots
$$

$\uparrow$ The endsequent.

How the diagram is intended should basically be obvious. The old derivation for $\Gamma_3 \to \Theta_3$ is written down twice side by side and the first instance is modified in such a way that *the left operational inference figure vanishes*; in the section of the derivation standing immediately above it, every occurrence of the free variable $\mathfrak{a}$ is replaced by the numerical term $\mathfrak{n}$ – except, again, where it happens to be used simultaneously as the eigenvariable of an occurring inference figure in the sequents standing above the lower sequent of that inference figure –; the formula $\forall \mathfrak{x} \mathfrak{F}(\mathfrak{x})$ is reintroduced nevertheless, but this time by a *thinning*; everything else is left exactly as it was before, with the single exception that in the path going through $\Gamma_1 \to \mathfrak{F}(\mathfrak{n}), \Theta_1, \forall \mathfrak{x} \mathfrak{F}(\mathfrak{x})$ the formula $\mathfrak{F}(\mathfrak{n})$ is carried along as an additional succedent formula. It can be seen at once by reference to the inference figure schemata that this leaves all inference figures correct; the same is true of the replacement of $\mathfrak{a}$ by $\mathfrak{n}$. The *second* copy of the old derivation of $\Gamma_3 \to \Theta_3$ is then modified is analogously. Here the *right* operational inference figure vanishes without necessitating the replacement of a variable; and the formula $\mathfrak{F}(\mathfrak{n})$ is carried along as an additional *antecedent* formula. From the two sequents $\Gamma_3 \to \mathfrak{F}(\mathfrak{n}), \Theta_3$ and $\Gamma_3, \mathfrak{F}(\mathfrak{n}) \to \Theta_3$, the old sequent $\Gamma_3 \to \Theta_3$ is then obtained by a *new cut*, together with the applications of interchanges and contractions, and the rest of the old derivation is then added unchanged.

The reader can convince himself without difficulty that the reduction step here defined turns the given derivation into another entirely correct derivation in the sense of our formalism.

Remarks about the significance of this reduction step.

Let us recall the fundamental ideas of the operational reduction (3.1) and compare them with the formal procedure just described. The two operational inference figures represent an introduction and elimination of the $\forall$ in $\forall \mathfrak{x} \mathfrak{F}(\mathfrak{x})$. According to the original fundamental idea, the two inference figures should have been omitted and the $\forall \mathfrak{x} \mathfrak{F}(\mathfrak{x})$ replaced by the 'simpler' $\mathfrak{F}(\mathfrak{n})$ – whose degree is smaller by 1 –; the place of the cut with the cut formulae $\forall \mathfrak{x} \mathfrak{F}(\mathfrak{x})$ should have been taken by a new cut with the cut formulae $\mathfrak{F}(\mathfrak{n})$. There remains, however, the difficulty already mentioned that the formula $\forall \mathfrak{x} \mathfrak{F}(\mathfrak{x})$ may have several application instances, even several introduction instances – i.e., the formula cluster may have *branchings on both sides* and contain several uppermost formulae. It is therefore actually necessary, both in connection with the cancellation of the *left* operational inference figure and that of the *right* operational inference figure, to retain also the old cut with $\forall \mathfrak{x} \mathfrak{F}(\mathfrak{x})$; a '*simplification*' has nevertheless been achieved *in each case* by the *omission* of an operational inference figure above this cut. (Although interchanges and a thinning have taken the place of this figure, these 'do not count' in the determination of the 'complexity' of the derivation. – The fact that $\forall \mathfrak{x}\, \mathfrak{F}(\mathfrak{x})$ is reintroduced by a thinning is motivated only by convenience since its reappearance further down in the derivation must be expected in any case, and since this is the most convenient way of obtaining the new form of the derivation from the old one.)

Further down in the new derivation then follows the 'new cut' with the cut formulae $\mathfrak{F}(\mathfrak{n})$. Precisely why has this cut been placed below the 'level line'? (Basically, it could have been introduced at any stage below the two $\forall \mathfrak{x}\, \mathfrak{F}(\mathfrak{x})$-cuts up to the end of the derivation; we would merely also have had to write down twice the section of the derivation from these cuts up to the new cut with $\mathfrak{F}(\mathfrak{n})$ as an additional antecedent or succedent formula and to leave the section below the new cut unchanged.)

This leads us to the purpose of the concept of a level in general. What actually matters *here* is that in the reduction a *'simplification'* of the derivation is achieved in a sense *to be made precise* in the next paragraph through the ordinal numbers. At first glance, to be sure, the new form of the derivation looks *more complex* than the old form: one and the same section of the derivation now occurs twice, although *in each case* in somewhat *simpler* form than before because of the omission of an operational inference figure. In defining a measure of complexity for derivations, it will therefore be easy to achieve that *each individual* section standing above the new cut is valued somewhat lower than the corresponding section of the old derivation. How is it to be accomplished, now that a new cut has been adjoined, that the *entire* section of the derivation up to $\Gamma_3 \to \Theta_3$ is valued lower than the old derivation up to the same sequent? The new cut has a *lower degree* than the old cut; it is this feature to which we must cling. The new cut is *thus* placed *below* the collection of *all* cuts whose degree is equal to that of the old cut, so that after the reduction the collection of cuts above any one of these cuts *of high degree* is *no larger* than before, but is at most the same or, in a certain sense, 'simplified'. On the other hand, the new cut and everything below it now extends over a larger collection than before. This is compensated for by the fact that all of these cuts are *of lower degree* than the old cut. Our success in achieving a lowering of the ordinal number of the derivation through the reduction depends merely on our exploiting these facts properly when assigning ordinal numbers below.

Especially great weight will thus have to be attached to the degree of a cut in this connection.

In this discussion it was tacitly assumed as normal that cuts of higher degree generally occur above cuts of lower degree in the derivation. Since, in reality, this need of course not be the case, the 'degree' is replaced by the concept of 'level', and all this means is that cuts of a lower degree above cuts of a higher degree are treated *as if they also* possessed the higher degree; once this has been done, the main ideas stated above carry over without difficulty.

In determining the level of arbitrary derivation sequents, the *CJ*-inference figures are furthermore treated like cuts, since in the course of their reduction they would be resolved into cuts of the same degree in any case.

The form of the reduction step for other connectives.

We must still specify how the reduction step is to be modified if the terminal connective of the cluster formulae is not a $\forall$, as in the case explicitly presented, but a $\&$, $\exists$, $\vee$ or $\neg$. The differences are only minor:

If the cluster formulae have the form $\mathfrak{A} \& \mathfrak{B}$, we imagine the above diagrams suitably modified; in place of $\forall \mathfrak{x} \, \mathfrak{F}(\mathfrak{x})$ stands $\mathfrak{A} \& \mathfrak{B}$, and the operational inference figures run thus:

$$\frac{\Gamma_1 \to \Theta_1, \mathfrak{A} \quad \Gamma_1 \to \Theta_1, \mathfrak{B}}{\Gamma_1 \to \Theta_1, \mathfrak{A} \& \mathfrak{B}} \text{ and } \frac{\mathfrak{A}, \Gamma_2 \to \Theta_3}{\mathfrak{A} \& \mathfrak{B}, \Gamma_2 \to \Theta_2} \text{ or } \frac{\mathfrak{B}, \Gamma_2 \to \Theta_2}{\mathfrak{A} \& \mathfrak{B}, \Gamma_2 \to \Theta_2}.$$

In the new derivation, $\mathfrak{A} \& \mathfrak{B}$ now takes the place of $\forall \mathfrak{x} \, \mathfrak{F}(\mathfrak{x})$, and in place of $\mathfrak{F}(\mathfrak{n})$ occurs $\mathfrak{A}$ or $\mathfrak{B}$, depending on which of the two possible forms the right operational inference figure (the '&-elimination') has had. In the place from which the left operational inference figure was omitted, we retain only the derivation of $\Gamma_1 \to \Theta_1, \mathfrak{A}$ or $\Gamma_1 \to \Theta_1, \mathfrak{B}$, resp., the other derivation being omitted. (This corresponds to the replacement of $\mathfrak{a}$ by $\mathfrak{n}$ in the $\forall$-case.) The rest of the procedure is exactly the same as above; even the indicated differences completely suggest themselves.

If the terminal connective of the cluster formulae is a $\exists$ or $\vee$, the reduction proceeds *completely symmetrically* to the cases $\forall$ and $\&$. Right and left are here interchanged.

If the cluster formulae finally have the form $\neg \mathfrak{A}$, nothing changes essentially: the formula $\mathfrak{F}(\mathfrak{n})$ in the new derivation then corresponds to the formula $\mathfrak{A}$, except that, as a consequence of the omission of the *left* operational inference figure, the latter formula occurs as an additional *antecedent* formula and, correspondingly, as a consequence of the omission of the *right* operational inference figure, as a *succedent* formula. In both cases the formula $\mathfrak{A}$ is carried forward up to the sequent $\Gamma_3 \to \Theta_3$, as usual; the only difference is that now the left and the right upper sequent of the '*new cut*', i.e., the complete derivational sections standing above it, must be *interchanged* with one another.

This completes the definition of a reduction step on a contradictive derivation.

§ 4. The ordinal numbers – concluding remarks

4.1. *The transfinite ordinal numbers below ε_0.*

I shall now define the ordinal numbers to be used. These will not be written as decimal fractions, as in $\#$ 4; this time I shall adopt the notation customary in set theory. (In spite of this all definitions and proofs given in the following paragraphs are entirely 'finitist' and are of an especially elementary nature in this respect, as were the corresponding sections in the earlier proof. Here we are not really concerned with a study of transfinite induction; cf. below.)

Recursive definition of the ordinal numbers, also of equality and the order relation ($<$) between them:

The system $\mathfrak{S}_0$ consists of the number 0. We define: $0 = 0$ and not $0 < 0$.

Suppose that the numbers of the system $\mathfrak{S}_\rho$ (where ρ is a natural number or 0) are already defined, as well as $=$, and the $<$-relation between them. An arbitrary number of the system $\mathfrak{S}_{\rho+1}$ then has the form

$$\omega^{\alpha_1} + \omega^{\alpha_2} + \cdots + \omega^{\alpha_\nu},$$

where the α's are numbers of the system $\mathfrak{S}_\rho$, with $\alpha_1 \geq \alpha_2 \geq \ldots \geq \alpha_\nu$; ν designates a natural number. The number 0 also belongs to the system $\mathfrak{S}_{\rho+1}$.

A $\mathfrak{S}_{\rho+1}$-number β is equal to a $\mathfrak{S}_{\rho+1}$-number γ if their representations coincide. A $\mathfrak{S}_{\rho+1}$-number β is smaller than (larger than) a $\mathfrak{S}_{\rho+1}$-number γ, if the first non-coinciding 'exponent' α in the representation of β is smaller than (larger than) the corresponding exponent in the representation of γ. If $\beta = \gamma + \cdots$, then $\beta > \gamma$. The ordinal number 0 is considered to be smaller than any other number. $\beta > \gamma$ means of course *the same as* $\gamma < \beta$.

This completes the definition. It is easily seen that each system includes all preceding systems, and that the relations of 'smaller than' and 'equal to' between two numbers are independent of the system to which these numbers are considered to belong. It also follows quite clearly that of a given expression it can always be *decided* whether it is an ordinal number or not, and that of two given ordinal numbers it can be decided (in a simple way) whether they are equal or which is the smaller one. (These concepts are therefore indeed 'finitist'.)

For our purposes, the symbols '0', '+', and 'ω', as well as the 'exponentiation', which occur in the representations of ordinal numbers, should be interpreted quite formally without our having to associate with them any particular sense such as that of regarding ω as 'an infinite number' or '+' as a symbol for 'addition'. Such visualizations are of use merely for the *understanding* of the context as a whole. Solely for the purpose *of comparing* the size of the individual systems, the following might still be said if we make use of concepts and results from set theory:

The system $\mathfrak{S}_1$ consists of the numbers: 0, ω^0, $\omega^0+\omega^0$, ... i.e., in the usual notation: 0, 1, 2, ...; that is to say, of 0 and the *natural numbers*.

The limit number of the system is ω.

$\mathfrak{S}_2$ contains all numbers below ω^ω, viz:

$$0, \omega^0, \omega^0+\omega^0, \ldots, \omega^{\omega^0}, \omega^{\omega^0}+\omega^0, \ldots, \omega^{\omega^0}+\omega^{\omega^0},$$
$$\omega^{\omega^0}+\omega^{\omega^0}+\omega^0, \ldots, \omega^{\omega^0+\omega^0}, \ldots, \omega^{\omega^0+\omega^0+\omega^0}, \ldots,$$

thus: 0, 1, 2, ..., ω, $\omega+1$, ..., $\omega \cdot 2$, $\omega \cdot 2+1$, ..., ω^2, ..., ω^3, ...; in general, all polynomials $\omega^{\nu_1} \cdot \mu_1 + \cdots + \omega^{\nu_\sigma} \cdot \mu_\sigma$; the ν's and μ's designate natural numbers or 0; $\nu_1 > \nu_2 > \cdots > \nu_\sigma$.

$\mathfrak{S}_3$ contains all numbers below ω^{ω^ω} (i.e., $\omega^{(\omega^\omega)}$; in the following, multiple exponentiations are to be interpreted correspondingly).

$\mathfrak{S}_4$ contains all numbers below $\omega^{\omega^{\omega^\omega}}$, etc.

The limit number of all systems taken together is the number ε_0, the 'first ε-number'.

We shall use the symbol 1 *as an abbreviation* for ω^0. We also need the concept of the '*natural sum*' of two (non-zero) ordinal numbers, which is defined as follows:[91]

Suppose that $\alpha = \omega^{\gamma_1}+\omega^{\gamma_2}+\cdots+\omega^{\gamma_\mu}$ and $\beta = \omega^{\delta_1}+\omega^{\delta_2}+\cdots+\omega^{\delta_\nu}$ ($\mu \geq 1$, $\nu \geq 1$). The 'natural sum $\alpha \# \beta$' is then obtained by arranging the $\mu+\nu$ terms ω^γ and ω^δ by size and joining them back together again by '+'-symbols, the largest term first, the smallest last, with equal terms of course side by side. In this way another correct ordinal number obviously results.

An example: If

$$\alpha = \omega^{\omega^1+1}+1 \quad \text{and} \quad \beta = \omega^{\omega^{\omega^1+1+1}+1}+\omega^{\omega^1+1}+\omega^1,$$

then

$$\alpha \# \beta = \omega^{\omega^{\omega^{1+1+1}+1}} + \omega^{\omega^{1+1}} + \omega^{\omega^{1+1}} + \omega^{1} + 1.$$

In all cases $\alpha \# \beta = \beta \# \alpha$. The natural sum of arbitrarily many ordinal numbers is independent of the order of the individual summations. $\alpha \# \beta > \alpha$. If $\alpha^* < \alpha$ then $\alpha^* \# \beta < \alpha \# \beta$. These facts are easily proved.

4.2. *The correlation of ordinal numbers with derivations.*

Suppose that an arbitrary derivation is given. *Its ordinal number* is calculated by passing downward from the uppermost sequents and assigning to each individual derivational sequent as well as to each line of inference an ordinal number (> 0) on the basis of the following stipulations:

Each uppermost sequent receives the ordinal number 1 (i.e., ω^0).

Suppose that the ordinal numbers of the upper sequents of an inference figure have already been determined. The ordinal number of the *line of inference* is then obtained as follows:

If *the inference figure is structural*, then the ordinal number of the upper sequent is adopted unchanged or, in the case of a *cut*, *the natural sum* of the ordinal numbers of the two upper sequents is formed.

If *the inference figure is operational*, then $+1$ is adjoined to the ordinal number of the upper sequent; if the figure has two upper sequents, the larger of the two ordinal numbers is selected and $+1$ is adjoined to it.

If a *CJ-inference figure* is finally encountered – whose upper sequent has the ordinal number $\omega^{\alpha_1} + \cdots + \omega^{\alpha_v}$ ($v \geq 1$), then ω^{α_1+1} is taken as the ordinal number of the line of inference. If $\alpha_1 = 0$, then this number is of course ω^1.

From the ordinal number of a line of inference – call it α – the ordinal number of the *lower sequent* of the inference figure concerned is obtained in the following way:

If the *level* of the lower sequent is the same as that of the upper sequent, then the ordinal number of the lower sequent is equal to α. If its level is lower by 1, then the ordinal number of the lower sequent is ω^α. If lower by 2, the ordinal number is ω^{ω^α}, if lower by 3: $\omega^{\omega^{\omega^\alpha}}$, etc.

The ordinal number which is finally obtained for the *endsequent* of the derivation is *the ordinal number of the derivation*.

The reader can easily convince himself that the operations mentioned really do yield ordinal numbers as defined above. For the time being I shall not comment on this method of correlating ordinal numbers; it is really quite simple; of special interest is only the evaluation of the *CJ*-inference figures and that of the different levels; in both cases this evaluation will be most easily understood through its effect later on.

4.3. *The decrease of the ordinal number in the course of a reduction step on a contradictive derivation.*

It still remains to be shown that in the course of a reduction step according to § 3, the ordinal number of an inconsistent derivation *decreases*. This no longer presents any special difficulties; all we need to do is *to examine* the correctness of this assertion carefully for each individual case.

The *preparatory step* 3.2. obviously leaves the ordinal number entirely unaffected. What is the situation in the case of a *CJ-reduction* (3.3)?

Suppose that the ordinal number of the upper sequent of the *CJ*-inference figure is $\omega^{\alpha_1} + \cdots + \omega^{\alpha_\nu}$ ($\nu \geqq 1$), that of the line of inference therefore ω^{α_1+1}. This is also the ordinal number of the lower sequent whose level cannot be lower than that of the upper sequent, since the cuts associated with the clusters to which $\mathfrak{F}(1)$ and $\mathfrak{F}(\mathfrak{n})$ belong, and which have the same degree as the *CJ*-inference figure, must still occur further down in the derivation. Let us now examine the figure which has replaced the *CJ*-inference figure in the reduction (first for $\mathfrak{n}$ not equal to 1). In the new derivation each one of its uppermost sequents obviously receives the same ordinal number $\omega^{\alpha_1} + \cdots + \omega^{\alpha_\nu}$. Furthermore, all sequents of the replacement figure have the same level, viz., that level which the two sequents of the *CJ*-inference figure had before. (The newly occurring cuts have of course the same degree as that of the *CJ*-inference figure.) The ordinal number of the lowest sequent of this figure is therefore obviously equal to the natural sum of all numbers $\omega^{\alpha_1} + \cdots + \omega^{\alpha_\nu}$. Consequently it begins: $\omega^{\alpha_1} + \cdots$. *It is therefore smaller than* ω^{α_1+1}, according to the definition of 'smaller than' for ordinal numbers.

From this it now follows easily that the ordinal number of the entire derivation has also been decreased. After all, from the *CJ*-inference figure downwards nothing has changed in the derivation; in fact, all levels have here also remained the same. The decrease which has occurred at one place is *preserved* in the calculation of the ordinal number further down to the endsequent; what is essential is that in proceeding downwards only *structural inference figures* are encountered and that the following holds: If $\alpha^* < \alpha$, then $\omega^{\alpha^*} < \omega^\alpha$ and $\alpha^* \# \beta < \alpha \# \beta$. (Suppose that α, α^* and $\beta \neq 0$.) Both requirements are satisfied at once by definition.

Now the purpose of ω^{α_1+1} in the evaluation of a *CJ*-inference figure also becomes clear: in the reduction the figure is broken up into a number of cuts; and in some sense the n-fold multiple of one and the same derivational section occurs. In order to achieve a decrease in the ordinal number, we must therefore choose as the ordinal number of the original derivational section up to the *CJ*-inference figure *the 'limit number' of all 'n-fold multiples'*

§ 4, THE ORDINAL NUMBERS – CONCLUDING REMARKS

of the ordinal number of the upper sequent, i.e., $\omega^{\alpha_1+1} = `\omega^{\alpha_1} \cdot \omega'$. (The expressions in ` ' serve of course only as illustrations; they are not even defined in this context.)

Now there remains only the case where $\mathfrak{n}$ equals 1: in the new derivation the sequent $\mathfrak{F}(1), \Gamma \to \Theta, \mathfrak{F}(1)$ receives the ordinal number 1. In the old derivation its ordinal number was at least equal to ω^1. Here we have an obvious decrease which is at the same time passed on to the ordinal number of the entire derivation.

This proves the decrease of the ordinal number of a contradiction derivation in a *CJ*-reduction. There still remains the case of the *operation reduction*. Here it must first be observed that the further *preparatory step* (3.42) cannot cause an increase in the ordinal number.

The proof of this fact presents certain difficulties in spite of the obvious external simplification of the derivation in this step. I shall sketch only briefly what kind of reasoning is here required – the reader interested only in bare essentials may skip this paragraph –:

The omission or adjunction of formulae and other transformations within structural inference figures except cuts have no influence whatever on the ordinal number. This is different in the case of the cancellation of a cut through the omission of an upper sequent together with everything standing above it. If we disregard for the time being the *change of levels* which this cancellation entails, then a decrease in the ordinal number results from the replacement of the natural sum of two numbers by only one of these two numbers. Added to this must be the fact that through the omission of a cut the *level* of a whole collection of sequents above this cut may be *reduced* to a greater or lesser extent (not only in the ending, but in the entire derivation). In order to recognize that this rather entangled transformation cannot effect an increase in the ordinal number of the entire derivation we argue thus: we imagine that we can fix the levels quite arbitrarily. We begin with the old derivation, omit the cut and, at first, leave all levels untouched. Then we gradually change these levels to the values which the transformed derivation really should have, according to the definition of level, by carrying out a succession of single steps of the following kind: the level of the upper sequents of *one* inference figure whose lower sequent has a lower level than the upper sequent is in each case diminished by 1. It is easily seen that the entire change of levels can in fact be made up of such operations. (We begin from below.) What exactly happens to the ordinal numbers in the course of a *single* such change of levels? Suppose that before the change the ordinal numbers of the upper sequents are α and β (if there is only one such number, we simply think of the second number below as not being present). After the change they then take the form ω^α and ω^β. (Except if one upper sequent is an uppermost sequent of the derivation, in which case its ordinal number was and remains equal to 1, and this simplifies the following discussion further.) Before the change, the ordinal number of the line of inference was thus either α, $\alpha \# \beta$, max $(\alpha+1, \beta+1)$, or ω^{α_1+1} (where $\alpha = \omega^{\alpha_1} + \cdots$, in the case of a *CJ*-inference figure), depending on the kind of inference figure involved. After the change, the ordinal number takes the form of either ω^α, $\omega^\alpha \# \omega^\beta$, max $(\omega^\alpha+1, \omega^\beta+1)$, or ω^{α_1+1}. Now to *the lower sequent*: If the difference in levels between it and the upper sequent was 1 before and is therefore now 0, then the change of levels has brought about a change in the ordinal number of this sequent from ω^α to ω^α, from $\omega^{\alpha \# \beta}$ to $\omega^\alpha \# \omega^\beta$, from max $(\omega^{\alpha+1}, \omega^{\beta+1})$ to max $(\omega^\alpha+1, \omega^\beta+1)$, or finally from $\omega^{\omega^{\alpha_1+1}}$ to $\omega^{\omega^{\alpha_1+\cdots+1}}$. *In each case the ordinal number has either remained the same or has become smaller*; this should be verified by the reader from case to case from the definition of 'smaller than'. If the difference in levels between the upper and lower sequents was greater than 1, nothing has essentially changed: in each case the numbers mentioned are augmented by an equal number of exponentiations with ω. This property of not becoming larger carries over to the ordinal number of *the entire derivation*, and this number can therefore increase neither in a single step of the described change of levels nor, quite generally, in the preparatory step for the operational reduction as a whole.

We now come to *the operational reduction proper* (3.5), in which we must demonstrate a *decrease* of the ordinal number. We shall again base our discussion upon the case presented in detail above (with $\forall$ as the connective to be reduced). The ordinal number of *each of the two lines of inference* standing immediately above the sequents $\Gamma_3 \to \mathfrak{F}(\mathfrak{n}), \Theta_3$ and $\Gamma_3, \mathfrak{F}(\mathfrak{n}) \to \Theta_3$ in the new derivation – which we denote by α_1 and α_2 – is smaller than the ordinal number α of the 'level line' in the old derivation. This is so since the derivational sections standing above the lines of inference essentially correspond to one another; all levels, in particular, are the same as those in the old derivation – the levels of the sequents standing immediately above the mentioned lines of inference are equal to ρ throughout –; in each case *only one operational inference figure has disappeared* and been replaced by structural inference figures which have no influence on the ordinal number. At this point a decrease in the ordinal number has therefore taken place which *is preserved* as we pass through the subsequent structural inference figures up to the mentioned lines of inference. Also: the sequent $\Gamma_3 \to \Theta_3$

has of course the same level σ in the new derivation as in the old one; $\sigma < \rho$. The sequent $\Gamma_3, \Gamma_3 \to \Theta_3, \Theta$ has of course the level σ. The level τ of the *upper sequents* of the 'new cut' satisfies $\rho > \tau \geq \sigma$. The inequality on the right is trivial; and that $\rho > \tau$ is recognized thus: by the definition of level, τ is the maximum of the two numbers σ and 'the degree of $\mathfrak{F}(\mathfrak{n})$'. If $\tau = \sigma$, then $\tau < \rho$, since $\sigma < \rho$. If τ equals the degree of $\mathfrak{F}(\mathfrak{n})$, then $\tau < \rho$, since the degree of $\mathfrak{F}(\mathfrak{n})$ is smaller than the degree of $\forall \mathfrak{x}\, \mathfrak{F}(\mathfrak{x})$, and since ρ is at least equal to the latter.

Let us first suppose that the differences between the levels ρ, τ and σ are *minimal*, i.e., that $\rho = \tau+1$ and $\tau = \sigma$. In this case our demonstration is completed as follows: In the old derivation the level line had the ordinal number α, the sequent $\Gamma_3 \to \Theta_3$ therefore the ordinal number ω^α. In the new derivation the lines of inference corresponding to this level line have the ordinal numbers α_1 and α_2, both are smaller than α, and the upper sequents of the new cut therefore have the ordinal numbers ω^{α_1} and ω^{α_2}; the sequent $\Gamma_3 \to \Theta_3$ receives the ordinal number $\omega^{\alpha_1}+\omega^{\alpha_2}$. (Without loss of generality we may assume that $\alpha_1 \geq \alpha_2$.) *The latter number is obviously smaller then ω^α*; and we have thus finished. For, on the basis of an already repeatedly applied argument this decrease carries over to the ordinal number of the endsequent and therefore to the derivation as a whole. (Below $\Gamma_3 \to \Theta_3$ nothing has of course changed.) If the distances between the levels ρ, τ and σ are *greater*, our argument is not essentially different. The place of the inequality

$$\omega^\alpha > \omega^{\alpha_1}+\omega^{\alpha_2} \quad (\text{where } \alpha > \alpha_1 \geq \alpha_2)$$

is then simply taken by the inequality

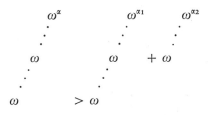

and the latter inequality is also easily seen to be valid.

It now becomes apparent how through the method of definition of the ordinal numbers, together with the concept of level, the difficulties associated with the apparent increase in complexity of a derivation as a result of the operational reduction have been overcome. The main idea is: in the reduction the same derivational section occurs *twice*, although both times somewhat simplified. In the general case, however, $\alpha < \alpha_1+\alpha_2$, where α_1

and α_2 are assumed to be smaller than α. For the exponential expression, however, it holds that $\omega^\alpha > \omega^{\alpha_1} + \omega^{\alpha_2}$. (Just as in the case of the natural numbers, we can put any number $\geqq 3$ for ω.)

The 'simplification' of the figure as a whole has thus been achieved, as long as it is always possible to insert an exponentiation; and this is made feasible by the fact that the degree of the new cut is *smaller* than the degree of the old $\forall \mathfrak{x} \, \mathfrak{F}(\mathfrak{x})$-cut. It was for the purpose of exploiting this fact that the general concept of level was introduced and applied in the correlation of ordinal numbers.

The cases where the connective to be reduced is a &, $\exists$, $\vee$, or $\neg$, are so similar that a special discussion of them becomes superfluous.

The decrease of the ordinal number of a contradictive derivation in the reduction step has thus been proved.

4.4. Concluding remarks.

If we had *not* admitted *CJ-inference figures* into our formalism, it would be possible to make do with *the natural numbers as ordinal numbers*. In order to realize this, the reader should omit 4.1 and in 4.2 replace ω by 3 throughout and 'natural sum' simply by 'sum'. Sums and powers are to be understood in the way customary for the natural numbers. 4.3 then remains valid throughout, as is easily verified; the *CJ*-reduction must here of course be left out. The consistency proof could then be concluded by an ordinary complete induction instead of a transfinite induction.

As soon as we admit *CJ*-inference figures, thus obtaining our full formalism, the following remarkable *connection* between the magnitude of the *ordinal number* of a derivation and the highest *degree* of the formulae occurring in the derivation holds: the ordinal number of a derivation in which only formulae of degree 0 occur is smaller than ω^ω (i.e., ω^{ω^1} in our notation). If the highest degree of a formula equals 1, then its ordinal number is smaller than ω^{ω^ω} if the degree equals 2, then the ordinal number is smaller than $\omega^{\omega^{\omega^\omega}}$, etc. This is not difficult to prove.

These theorems are of course meaningful only relative to our special correlation of ordinal numbers. Yet it is reasonable to assume that by and large this correlation is already fairly *optimal*, i.e., that we could *not* make do with *essentially lower* ordinal numbers. In particular, the totality of all our derivations cannot be handled by means of ordinal numbers all of which lie below a number which is smaller than ε_0. For transfinite induction up to such a number is itself provable in our formalism; a consistency proof carried out by means of this induction would therefore contradict Gödel's theorem (given, of course, that the other techniques of proof used, especially the correlation of ordinal numbers, have not assumed forms that are non-representable in our number-theoretical formalism). By the same roundabout argument we can presumably also show that certain subclasses of derivations cannot be handled by ordinal numbers below certain numbers of the form $\omega^{\cdot^{\cdot^{\omega}}}$. It is quite likely that one day a direct approach to such impossibility proofs will be found.

If we include *arbitrary functions* in our formalism, then the consistency proof remains valid with minor modifications: all that needs to be shown is that at some point in the reduction, following the first preparatory step, for example, all terms without free variables can be evaluated and replaced by their numerical values. It is presupposed that all functions can be effectively calculated for all given numerical values. There still arise certain formal difficulties from the fact that although a term may be calculable, a corresponding term in another place in the same inference figure may still contain a variable (cf. article 14.22 of # 4); these difficulties do not affect the main ideas involved.

In its purpose and intention, *section V of # 4* continues to apply to this new version of the consistency proof. I have not given a new proof of the 'reducibility' of arbitrary derivable sequents; nor do I attach any special importance to this. (I had previously advanced it as an argument against radical intuitionism – article 17.3 –, but it is not particularly essential for this purpose.)

Transfinite Induction.

I have not given a new proof of the transfinite induction which concludes the consistency proof, since I intend to discuss the question involved at this point separately at some later date. For the conclusion of the present proof, the earlier proof of the 'theorem of transfinite induction' (articles 15.4 and 15.1) is therefore to be adopted for the time being. For this purpose the new ordinal numbers must be made to correspond to the decimal fractions used in the earlier paper; this presents no special difficulties. (Both systems are after all of the same 'order type ε_0'.)

The transfinite induction occupies quite a special position within the consistency proof. Whereas all other forms of inference used are of a rather elementary kind, from the point of view of being 'finitist' – this applies to the new proof as much as it does to old one – this cannot be maintained of the transfinite induction. Here we therefore have *a task of a different kind*: we are not merely required *to prove* transfinite induction – this is not particularly difficult and possible in various ways – but rather to prove it on a *finitist* basis, i.e., to establish clearly that it is a form of inference which is in harmony with the principle of the *constructivist interpretation* of infinity; an undertaking which is no longer purely *mathematical*, but which nevertheless forms part of a consistency proof.

We might be inclined to doubt the finitist character of the 'transfinite' induction, even if only because of its suspect name. In its defense it should here merely be pointed out that most somehow constructivistically orientated authors place special emphasis on *building up constructively* (up to ω^ω, for example) *an initial segment* of the transfinite number sequence (within the 'second number class'). And in the consistency proof, and in possible future extensions of it, we are certainly dealing only with an initial part, a 'segment' of the second number class, even though this is an already comparatively extensive segment, and must probably be extended considerably further for a consistency proof for analysis. I fail to see, however, at what 'point' that which is constructively indisputable is supposed to end, and where a further extension of transfinite induction is therefore thought to become disputable. I think, rather, that the reliability of the transfinite numbers required for the consistency proof compares with that of the first initial segments, say up to ω^2, in the same way as the reliability of a numerical calculation extending over a hundred pages with that of a calculation of a few lines: it is merely a considerably *vaster* undertaking to convince oneself of this certainty from beginning to end. A detailed discussion of these matters (which seem to me now to have been discussed somewhat too sketchily in $\neq$ 4, article 16.11) will as said before, follow at a later date.

9. PROVABILITY AND NONPROVABILITY OF RESTRICTED TRANSFINITE INDUCTION IN ELEMENTARY NUMBER THEORY

The impossibility of proving transfinite induction up to the ordinal number ε_0 with elementary number-theoretical techniques may be inferred indirectly from the following two facts:

1. Gödel's theorem: The consistency of elementary number theory cannot be proved with the techniques of that theory[92].

2. The consistency of elementary number theory has been proved by applying transfinite induction up to ε_0, together with exclusively elementary number-theoretical techniques[93].

In the following I shall give a direct proof for the nonprovability of transfinite induction up to ε_0 in elementary number theory. The procedure of proof will, in addition, make it possible to show that still further restricted forms of transfinite induction to numbers *below* ε_0 are not provable in certain subsystems of the number-theoretical formalism.

On the other hand, it is known that transfinite induction up to any ordinal number *below* ε_0 is provable in elementary number theory[94]. In § 2, I shall indicate how such proofs are *formalized* in the number-theoretical formalism.

I shall make frequent use of # 8, which will simply be referred to as '*New*'.

§ 1. *TJ*-derivations

We begin by defining the concept of an elementary number-theoretical proof for the validity of transfinite induction up to a definite ordinal number below ε_0. The formalized version of such a proof will briefly be called a '*TJ*-derivation'.

For this purpose we must extend the concept of a number-theoretical derivation as it was formalized, for example, in my earlier papers, in two ways: In the first place we must adjoin the transfinite ordinal numbers to the natural numbers as further objects, as well as certain associated functions, predicates and mathematical axioms.

Second, a formalized version of the rule of transfinite induction must be stated, and it must be explained what is to be understood by a *derivation for this rule of inference*.

This will now be done formally by extending *New*, § 1. It will be stated explicitly, in each case, what is to be adopted from that paper.

All ordinal numbers below ε_0 can be represented uniquely by means of 0, ω, addition, and ω-exponentiation in the way described in *New*, 4.1. Expressions of this form will be called 'numerical terms'. (They designate the objects of our theory; the natural numbers are special cases of such terms.)

Numerical variables, briefly called '*variables*', are subdivided into *free* and *bound* variables. They may be designated by arbitrary symbols, provided that these symbols have not yet been used in another sense; in each case it must, however, be specified whether such a symbol is to designate a free or a bound variable.

An arbitrary number of *function symbols* and *predicate symbols* may occur; but we require that these functions and predicates are *decidably defined*, i.e., that a procedure is given at the same time which enables us to calculate for every given combination of numerical terms, from the number of argument places of the function or predicate, what is the functional value – again in the form of a numerical term – or, in the case of a predicate, whether the predicate does or does not hold for the given numbers.

In the following, we shall make use of the following specific predicate and function symbols (α and β indicate the argument places):

Predicate symbols: $\alpha = \beta$, $\alpha < \beta$, $\alpha > \beta$, $\alpha \leq \beta$, $\alpha \geq \beta$. (Decidably defined in *New*, 4.1).

Function symbols: First, $\alpha+\beta$, $\alpha \cdot \beta$ and ω^α. These (as well as the predicate symbols) are considered to have the meanings customary in set theory; to define these symbols decidably for the numbers below ε_0 presents no essential difficulties; I shall assume that such definitions have been given and shall discuss this point no further[95].

The occurrence of sums and powers of ω in the numerical terms must be shown to be in harmony with the definition of these functions; again this presents no significant difficulty.

We furthermore use the single-place function symbol ω_α with the following meaning: $\omega_0 = 0$, $\omega_1 = 1$ (as in *New* 4.1, 1 is to be considered as an abbreviation for ω^0, and ω as an abbreviation for ω^1, wherever it occurs without an exponent or subscript), $\omega_2 = \omega$, $\omega_3 = \omega^\omega$ etc. (2 and 3 are, of course, abbreviations for $1+1$ and $1+1+1$) in general, $\omega_{\alpha+1} = \omega^{\omega_\alpha}$, where $\alpha < \omega$. If $\alpha \geq \omega$, we define $\omega_\alpha = 0$.

Finally, we require two somewhat more complicated functions of the following kind:

If $\alpha < \beta+\omega^\gamma$ and $\gamma > 0$ (were α, β, γ are ordinal numbers of our domain) then there exist two numbers $fu_1(\alpha, \beta, \gamma)$ and $fu_2(\alpha, \beta, \gamma)$ such that $fu_1 < \gamma$, $fu_2 < \omega$ and $\alpha \leq \beta+\omega^{fu_1} \cdot fu_2$. These can be decidably defined as follows: Written as a numerical term, $\beta+\omega^\gamma$ has the form $\omega^{\delta_1}+\omega^{\delta_2}+\cdots+\omega^{\delta_\nu}$ ($\delta_1 \geq \delta_2 \cdots \geq \delta_\nu$, where ν is a natural number).

Here $\delta_\nu = \gamma$, according to the arithmetical rules for addition. If $\alpha \leq \beta$, then we simply assign to fu_1 and fu_2 the value 0. Yet if $\alpha > \beta$, then α, written as a numerical value, has the form

$$\omega^{\delta_1}+\cdots+\omega^{\delta_{\nu-1}}+\omega^\kappa+\ldots+\omega^\kappa+\omega^\lambda+\cdots, \quad \text{where } \gamma > \kappa > \lambda.$$

(If ν equals 1, then the terms occurring before the first ω^κ are simply disregarded. At least one ω^κ-term must occur. ω^λ and the other places on the right may be empty.) The summand ω^κ is considered to occur μ times, where μ is a natural number.

We now put $fu_1 = \kappa$, $fu_2 = \mu+1$, with the restrictions satisfied, since

$$\beta+\omega^{fu_1} \cdot fu_2 \geq \omega^{\delta_1}+\cdots+\omega^{\delta_{\nu-1}}+\omega^\kappa \cdot (\mu+1) > \alpha.$$

(If $\alpha \geq \beta+\omega^\gamma$ or $\gamma = 0$, then, fu_1 and fu_2 are assumed to be 0.)

Recursive definition of a *'term'* (in the customary way): Numerical terms and free variables are terms. If the argument places of a function symbol are filled by arbitrary terms, then another term results.

Recursive definition of a *'formula'*:

If the argument places of a predicate symbol are filled by arbitrary terms, then a formula results. Such a formula is called a *'prime formula'*.

A formula also results if a term is placed in the argument place of the one-place 'predicate variable' $\mathscr{E}$. (The symbol $\mathscr{E}$ is required for the formulation of the TJ-derivations, cf. below. It is not counted among the 'variables', these are intended to refer to numerical variables only.)

If $\mathfrak{A}$ and $\mathfrak{B}$ are formulae, then $\mathfrak{A} \& \mathfrak{B}$, $\mathfrak{A} \vee \mathfrak{B}$, $\mathfrak{A} \supset \mathfrak{B}$ and $\neg \mathfrak{A}$ are also formulae. From a formula another formula results if every occurrence of a free variable is replaced throughout by a bound variable $\mathfrak{x}$ not yet occurring in the formula and if $\forall \mathfrak{x}$ or $\exists \mathfrak{x}$ are at the same time placed in front of the resulting expression. (Brackets are used as usual in order to display unambiguously the scope of a logical or functional symbol.)

The definition of a *'sequent'* is to be adopted verbatim from *New*, 1.2.

The concept of an *'inference figure'* is also adopted from *New*, 1.3, with the following modifications:

Since we now also wish to use the logical connective $\supset$ (in § 2), we introduce the following additional inference figure schema:

$$\frac{\Gamma, \mathfrak{A} \to \mathfrak{B}}{\Gamma \to \mathfrak{A} \supset \mathfrak{B}}.$$

The schema for *CJ*-inference figures also requires a modification because of the inclusion of the transfinite numbers; it shall now run:

$$\frac{\mathfrak{a} < \omega, \mathfrak{F}(\mathfrak{a}), \Gamma \to \Theta, \mathfrak{F}(\mathfrak{a}+1)}{\mathfrak{t} < \omega, \mathfrak{F}(0), \Gamma \to \Theta, \mathfrak{F}(\mathfrak{t})},$$

with the following rule of replacement: For $\mathfrak{F}(\mathfrak{a})$ we may put an arbitrary formula containing a free variable $\mathfrak{a}$. Then $\mathfrak{F}(\mathfrak{a}+1)$, $\mathfrak{F}(0)$, or $\mathfrak{F}(\mathfrak{t})$, resp., are to be replaced by those formulae which result from $\mathfrak{F}(\mathfrak{a})$ if the free variable $\mathfrak{a}$ is replaced throughout by $\mathfrak{a}+1$ or 0 or an arbitrary term $\mathfrak{t}$. The remaining parts of the schema are to be understood in the obvious way. As usual, the eigenvariable $\mathfrak{a}$ must not occur in the lower sequent.

As '*basic sequents*' we shall use the following:

Basic logical sequents are sequents of the form $\mathfrak{D} \to \mathfrak{D}$, where $\mathfrak{D}$ stands for an arbitrary formula, as well as (in § 2) $\mathfrak{A} \supset \mathfrak{B}, \mathfrak{A} \to \mathfrak{B}$ (where $\mathfrak{A}$ and $\mathfrak{B}$ are arbitrary formulae).

Basic mathematical sequents are, as in *New*, 1.4, sequents consisting of prime formulae which become 'true' sequents in the sense of the decidable definition of functions and the decidability of predicates, together with the definition of the truth of a sequent (*New*, 1.2), with every arbitrary substitution of numerical terms for possible occurrences of free variables. (Such sequents may therefore not contain the symbol $\mathscr{E}$.)

Here, as in my earlier papers, I shall dispense with the statement of definite basic mathematical sequents (as mathematical 'axioms'). I shall nevertheless make use of a number of such axioms in § 2; that these are 'true' will be assumed as proved, since this involves no essential difficulties.

We furthermore introduce a third kind of basic sequent, viz., '*basic equality sequents*', to be formed according to the following schema:

$$\mathfrak{s} = \mathfrak{t}, \mathfrak{F}(\mathfrak{s}) \to \mathfrak{F}(\mathfrak{t})$$

together with the replacement rule: $\mathfrak{s}$ and $\mathfrak{t}$ may be replaced by arbitrary terms, $\mathfrak{F}(\mathfrak{s})$ by an arbitrary formula containing (at least) one occurrence of the term $\mathfrak{s}$, $\mathfrak{F}(\mathfrak{t})$ by a formula which results from $\mathfrak{F}(\mathfrak{s})$ by the replacement of *one occurrence* of $\mathfrak{s}$ by $\mathfrak{t}$.

(N.B.: The basic equality sequents would be superfluous if the calculus did not contain the predicate variable $\mathscr{E}$, since the latter sequents would then be derivable from basic mathematical sequents.)

The concept of a '*derivation in the ordinary sense*' is adopted verbatim from *New*, 1.5, with the following addition: The symbol $\mathscr{E}$ may not occur in a derivation of this kind.

TJ-derivations.

In order to make the concept of a *TJ*-derivation precise, I shall begin by formulating a 'schema for *TJ*-inference figures', similar to the schema for *CJ*-inference figures:

$$\frac{\mathfrak{a} > 0,\ \forall\mathfrak{x}\,[\mathfrak{x} < \mathfrak{a} \supset \mathfrak{F}(\mathfrak{x})],\ \Gamma \to \Theta,\ \mathfrak{F}(\mathfrak{a})}{\mathfrak{F}(0),\ \Gamma \to \Theta,\ \mathfrak{F}(\mathfrak{\hat{s}})}.$$

The replacement rule is analogous to the rule for the *CJ*-inference figures; $\mathfrak{\hat{s}}$ must be replaced by a *numerical term*.

This schema formalizes 'transfinite induction up to the ordinal number represented by the numerical term $\mathfrak{\hat{s}}$'. (The only use that will be made of this schema is that of elucidating the definition that follows.)

Now the definition:

A '*TJ-derivation up to $\mathfrak{\hat{s}}$*' (where $\mathfrak{\hat{s}}$ is a numerical term) is a derivation in the ordinary sense, but with the following modifications of that concept: The predicate variable $\mathscr{E}$ occurs in the derivation (which means, of course, that it may occur in certain derivation-sequent formulae by virtue of the definition of a formula); the endsequent of the derivation runs

$$\mathscr{E}(0) \to \mathscr{E}(\mathfrak{\hat{s}});$$

where, in addition to basic sequents, sequents of the following form, called '*TJ*-upper sequents', are admitted as uppermost sequents:

$$\mathfrak{r} > 0,\ \forall\mathfrak{x}\,[\mathfrak{x} < \mathfrak{r} \supset \mathscr{E}(\mathfrak{x})] \to \mathscr{E}(\mathfrak{r}),$$

here $\mathfrak{r}$ stands for an arbitrary term and $\mathfrak{x}$ for an arbitrary bound variable.

It ought to be fairly clear that this concept really reflects that which is informally understood by a proof for the validity of transfinite induction up to the number designated by $\mathfrak{\hat{s}}$. The following connection exists, in particular, with the stated 'schema for *TJ*-inference figures': If in addition to the kinds of inference figures admissible in a 'derivation in the ordinary sense' a '*TJ*-inference figure', according to the above schema, were to occur, and if there were also available a '*TJ*-derivation up to $\mathfrak{\hat{s}}$', then this *TJ*-derivation could be 'substituted' for the occurrence of the *TJ*-inference figure in such a way that altogether a correct 'derivation in the ordinary sense' would result, now without a *TJ*-inference figure.

This substitution would simply have to be carried out in such a way that all expressions of the form $\mathscr{E}(\mathfrak{v})$ in the '*TJ*-derivation up to $\mathfrak{z}$' are replaced by $\mathfrak{F}(\mathfrak{v})$ (here $\mathfrak{v}$ designates any arbitrary expression standing in the argument place of $\mathscr{E}$), that the sequences Γ and Θ of formulae are adjoined to the *TJ*-upper sequents, and that Γ and Θ are then 'carried forward', so that they finally also occur in the endsequent of the *TJ*-derivation; that sequent has then already the form $\mathfrak{F}(0), \Gamma \to \Theta, \mathfrak{F}(\mathfrak{z})$, and we can adjoin to it that part of the original derivation which followed the *TJ*-inference figure; finally, we must place above the *TJ*-sequents that part of the derivation which occurred above the upper sequent of the *TJ*-inference figure, with every occurrence of the free variable $\mathfrak{a}$ replaced by the term $\mathfrak{r}$ occurring in the appropriate *TJ*-upper sequent. (Possible conflicts in the labelling of variables can be avoided by trivial redesignations.)

This consideration shows at the same time that it is appropriate to regard a '*TJ*-derivation' in our sense as a derivation still belonging to *elementary number theory*, in spite of the fact that it contains a 'predicate variable' ('formula variable', 'set variable') $\mathscr{E}$: This variable occurs only as a free variable, and the *TJ*-derivation becomes an elementary number-theoretical derivation in the ordinary sense as soon as it is *specialized* to a definite 'number-theoretical' predicate $\mathfrak{F}$.

As far as the extension of 'elementary number theory' by the inclusion of transfinite ordinal numbers, together with the specification of a *designatory* schema for them, is concerned, incidentally, we note that it is basically of the same kind as the inclusion of negative numbers, fractions, etc., and that it introduces in no way elements incompatible with elementary number theory[96].

§ 2. Characterization of *TJ*-derivations

I shall now state a procedure for the systematic construction of *TJ*-inference figures up to $\omega_0 = 0, \omega_1 = 1, \omega_2 = \omega, \omega_3 = \omega^\omega$; in general up to $\omega_\mathfrak{n}$, where $\mathfrak{n}$ designates an arbitrary natural number.

The letters $\alpha, \beta, \gamma, \delta$ will be used as free variables, the letters ξ, η as bound variables.

I shall not list all inference figures and basic sequents individually, but enough of them to enable the reader to supply those missing; all inference figures and basic sequents that are needed below, as will as those explicitly written down, will be labelled in the following way by means of special symbols according to the individual schemata to which they belong:

L: Basic logical sequent; *M*: Basic mathematical sequent; *G*: Basic equality sequent; *J*: *TJ*-upper sequent; *S*: One or several structural inference figures; *CJ*: *CJ*-inference figure. For the operational inference figures I shall use the combinations of symbols $\&-I$, $\&-E$, $\vee-I$, $\vee-E$, etc., where I ('introduction' of the connective) designates that kind of inference figure in whose schema the connective occurs in a succedent formula of the lower sequent, whereas E ('elimination' of the connective, in the sense of the 'natural' succession of inferences) denotes that kind of inference figure in whose schema the connective occurs in an antecedent formula of the lower sequent. (It should be noted that for the $\supset$-elimination, we now use the second kind of the basic logical sequents.)

The 'truth' of the occurring basic mathematical sequents is, as said earlier, assumed as known.

2.1. Characterization of a *TJ*-derivation up to 0:

$$\mathscr{E}(0) \to \mathscr{E}(0) \quad L.$$

2.2. *Transformation of a TJ-derivation up to ω_n into a TJ-derivation up to $\omega_{n+1} = \omega^{\omega_n}$.* (Here $\mathfrak{n}$ designated a natural number or 0.)

Suppose that a *TJ*-derivation up to ω_n is given.

We may assume, without loss of generality, that this derivation does not contain the variables introduced below. (This can be achieved by a trivial relabelling, in any case.)

We first replace the symbol $\mathscr{E}$ in the entire derivation as follows: $\mathscr{E}(\mathfrak{v})$ is in each case turned into $\forall \xi [\mathscr{E}^*(\xi) \supset \mathscr{E}^*(\xi + \omega^{\mathfrak{v}})]$, where $\mathfrak{v}$ in each case designates the expression standing in the argument place of $\mathscr{E}$, and where $\mathscr{E}^*(\mathfrak{w})$ is an abbreviation for $\forall \eta [\eta \leq \mathfrak{w} \supset \mathscr{E}(\eta)]$, with $\mathfrak{w}$ arbitrary.

It is easily seen that all inference figures, all basic logical sequents and all basic equality sequents remain correct, also of course the basic mathematical sequents, which remain entirely unaffected by this modification.

TJ-upper sequents, however, are turned into sequents of another kind and the endsequent also undergoes an essential change. In those places the derivation must be augmented in the way now to be described.

2.21. After the replacement, the *endsequent* obviously runs:

$$\forall \xi [\mathscr{E}^*(\xi) \supset \mathscr{E}^*(\xi+\omega^0)] \to \forall \xi [\mathscr{E}^*(\xi) \supset \mathscr{E}^*(\xi+\omega^{\omega_n})].$$

We adjoin the following new conclusion:

$$\frac{\forall \eta [\eta > \alpha +1 \supset \mathscr{E}(\eta)] \rightarrow \mathscr{E}(\alpha+1)}{\alpha+1 > 0, \ \forall \eta [\eta > \alpha+1 \supset \mathscr{E}(\eta)] \rightarrow \mathscr{E}(\alpha+1)} \ J$$

$$\frac{\ }{\alpha+1 > 0} \ M$$

$$\frac{\mathscr{E}^*(\alpha), \ \beta > \alpha+1 \rightarrow \mathscr{E}(\beta)}{\mathscr{E}^*(\alpha) \rightarrow \forall \eta [\eta > \alpha+1 \supset \mathscr{E}(\eta)]} \ L$$

$$\frac{\beta \leqq \alpha \supset \mathscr{E}(\beta), \ \beta \leqq \alpha \rightarrow \mathscr{E}(\beta)}{\mathscr{E}^*(\alpha), \ \beta \leqq \alpha \rightarrow \mathscr{E}(\beta)} \ \forall\text{-}E$$

$$\beta < \alpha+1 \rightarrow \beta \leqq \alpha \quad M, \ \vee\text{-}I, \ S, \ \vee\text{-}I, \ S$$

$$\gamma \leqq \alpha+1 \rightarrow \gamma \leqq \alpha \vee \alpha+1 = \gamma$$

$$\frac{\mathscr{E}^*(\alpha), \ \gamma \leqq \alpha+1 \rightarrow \mathscr{E}(\gamma)}{\mathscr{E}^*(\alpha) \rightarrow \mathscr{E}^*(\alpha+1)} \ \forall\text{-}I, \ \wedge\text{-}I, \ \supset\text{-}I$$

$$\gamma \leqq \alpha \vee \alpha+1 = \gamma, \ \mathscr{E}^*(\alpha) \rightarrow \mathscr{E}(\gamma) \quad \forall\text{-}E \text{ (as above)}$$

$$\frac{\mathscr{E}^*(\alpha), \ \alpha+1 = \gamma \rightarrow \mathscr{E}(\gamma)}{\mathscr{E}^*(\alpha) \rightarrow \mathscr{E}(\alpha+1)} \ G, \ S$$

$$\vdash \forall \xi [\mathscr{E}^*(\xi) \supset \mathscr{E}^*(\xi+1)] \quad \supset\text{-}I, \ \wedge\text{-}I$$

$$\frac{\mathscr{E}^*(0) \rightarrow \mathscr{E}^* (0+\omega^{\omega_n})}{\vdash \mathscr{E}^*(0) \supset \mathscr{E}^*(0+\omega^{\omega_n})} \ S$$

The endsequent of the old derivation after the replacement:

$$\vdash \forall \xi [\mathscr{E}^*(\xi) \supset \mathscr{E}^*(\xi + \omega^{\omega_n})] \quad L, \ \forall\text{-}E, \ S$$

$$\vdash \overline{\forall \xi [\mathscr{E}^*(\xi) \supset \mathscr{E}^*(\xi + \omega^{\omega_n})]} \quad S$$

$$\frac{\mathscr{E}^0(0) \rightarrow \mathscr{E}^0(0)}{\mathscr{E}^0(0) \rightarrow \mathscr{E}^*(\alpha)} \ \alpha = 0, \ \mathscr{E}^0(0) \rightarrow \mathscr{E}(\alpha) \quad G$$

$$\frac{\ }{\alpha \leqq 0} \ \alpha = 0, \ \alpha \leqq 0 \quad G, \ M, \ S$$

$$\mathscr{E}^0(0) \rightarrow \mathscr{E}^*(0)$$

$$\mathscr{E}^0(0) \rightarrow \mathscr{E}^*(\omega_{n+1})$$

$$\mathscr{E}^0(0), \ \omega_{n+1} \leqq \omega_{n+1} \rightarrow \mathscr{E}(\omega_{n+1})$$

$$\mathscr{E}^0(0) \rightarrow \mathscr{E}(\omega_{n+1}) \ \omega_{n+1} \leqq \omega_{n+1} \quad M$$

§ 2, CHARACTERIZATION OF TJ-DERIVATIONS

We have therefore obtained the desired endsequent of the new derivation.
2.22. After the replacement, one of the *TJ-upper sequents* obviously runs:

$$\mathfrak{r} > 0, \forall \mathfrak{x} \, [\mathfrak{x} < \mathfrak{r} \supset \forall \xi [\mathscr{E}^*(\xi) \supset \mathscr{E}^*(\xi + \omega^{\mathfrak{x}})]] \to \forall \xi [\mathscr{E}^*(\xi) \supset \mathscr{E}^*(\xi + \omega^{\mathfrak{r}})].$$

This sequent is derived as follows:

Suppose that $\mathfrak{C}$ is an abbreviation for its second antecedent formula. The meanings of $\mathfrak{s}$ and $\mathfrak{t}$ (both of which designate certain *terms*) will be explained later.

$$\begin{array}{c}
L, \forall\text{–}E \\
\dfrac{\mathfrak{C}, \mathfrak{s} < \mathfrak{r} \quad \to \forall \xi[\mathscr{E}^*(\xi) \supset \mathscr{E}^*(\xi+\omega^{\mathfrak{s}})]}{\mathfrak{C}, \mathfrak{s} < \mathfrak{r}, \mathscr{E}^*(\gamma+\omega^{\mathfrak{s}} \cdot \alpha) \to \mathscr{E}^*(\gamma+\omega^{\mathfrak{s}} \cdot \alpha+\omega^{\mathfrak{s}})} \; L, \forall\text{–}E, S \\
\dfrac{}{\mathfrak{C}, \mathfrak{s} < \mathfrak{r}, \mathscr{E}^*(\gamma+\omega^{\mathfrak{s}} \cdot \alpha) \to \mathscr{E}^*(\gamma+\omega^{\mathfrak{s}} \cdot (\alpha+1))} \; G, M, S \\
\dfrac{}{\alpha < \omega, \mathfrak{C}, \mathfrak{s} < \mathfrak{r}, \mathscr{E}^*(\gamma+\omega^{\mathfrak{s}} \cdot \alpha) \to \mathscr{E}^*(\gamma+\omega^{\mathfrak{s}} \cdot (\alpha+1))} \; S \\
\dfrac{}{\mathfrak{t} < \omega, \mathfrak{C}, \mathfrak{s} < \mathfrak{r}, \mathscr{E}^*(\gamma+\omega^{\mathfrak{s}} \cdot 0) \to \mathscr{E}^*(\gamma+\omega^{\mathfrak{s}} \cdot \mathfrak{t})} \; CJ, S \\
\dfrac{}{\mathfrak{t} < \omega, \mathfrak{C}, \mathfrak{s} < \mathfrak{r}, \mathscr{E}^*(\gamma) \quad\quad \to \mathscr{E}^*(\gamma+\omega^{\mathfrak{s}} \cdot \mathfrak{t})} \; G, M, S \\
\dfrac{}{\mathfrak{t} < \omega, \mathfrak{C}, \mathfrak{s} < \mathfrak{r}, \mathscr{E}^*(\gamma), \beta \leq \gamma+\omega^{\mathfrak{s}} \cdot \mathfrak{t} \to \mathscr{E}(\beta)} \; L, \forall\text{–}E, S.
\end{array}$$

The symbols $\mathfrak{s}$ and $\mathfrak{t}$ are given the following meanings: $\mathfrak{s}$ is an abbreviation for $fu_1(\beta, \gamma, \mathfrak{r})$, $\mathfrak{t}$ for $fu_2(\beta, \gamma, \mathfrak{r})$; fu_1 and fu_2 have the meanings stated in § 1; accordingly, the following basic mathematical sequents are valid:

$$\mathfrak{r} > 0, \beta < \gamma+\omega^{\mathfrak{r}} \to \beta \leq \gamma+\omega^{\mathfrak{s}} \cdot \mathfrak{t}, \text{ as well as } \mathfrak{r} > 0, \beta < \gamma+\omega^{\mathfrak{r}} \to \mathfrak{s} < \mathfrak{r}$$

$$\text{and } \mathfrak{r} > 0, \beta < \gamma+\omega^{\mathfrak{r}} \to \mathfrak{t} < \omega.$$

From these and the earlier derived sequents we obtain the following by three cuts and further structural inference figures:

$$\dfrac{\mathfrak{r} > 0, \mathfrak{C}, \mathscr{E}^*(\gamma), \beta < \gamma+\omega^{\mathfrak{r}} \quad \to \mathscr{E}(\beta)}{\text{further: } \mathfrak{r} > 0, \mathfrak{C}, \zeta^*(\gamma) \to \forall \xi[\xi < \gamma+\omega^{\mathfrak{r}} \supset \mathscr{E}(\xi)]} \; \supset\text{–}I, \forall\text{–}I$$

By cutting this sequent with the lower sequent of the following derivation:

$$\dfrac{\overset{M}{\to \gamma+\omega^{\mathfrak{r}} > 0} \quad \overset{J}{\gamma+\omega^{\mathfrak{r}} > 0, \forall \xi[\xi < \gamma+\omega^{\mathfrak{r}} \supset \mathscr{E}(\xi)] \to \mathscr{E}(\gamma+\omega^{\mathfrak{r}})}}{\forall \xi[\xi < \gamma+\omega^{\mathfrak{r}} \supset \mathscr{E}(\xi)] \to \mathscr{E}(\gamma+\omega^{\mathfrak{r}})} \; S,$$

we obtain, *on the one hand*, the sequent:

$$\dfrac{\mathfrak{r} > 0, \mathfrak{C}, \mathscr{E}^*(\gamma) \quad \to \mathscr{E}(\gamma+\omega^{\mathfrak{r}})}{\text{further: } \mathfrak{r} > 0, \mathfrak{C}, \mathscr{E}^*(\gamma), \delta = \gamma+\omega^{\mathfrak{r}} \to \mathscr{E}(\delta)} \; G, S.$$

On the other hand, by means of L, $\forall$–E, and S, the above sequent – whose derivation must therefore be written down a second time – yields:

$$\mathfrak{r} > 0, \mathfrak{C}, \mathscr{E}^*(\gamma), \delta < \gamma+\omega^{\mathfrak{r}} \to \mathscr{E}(\delta).$$

From this, together with the previous sequent, we obtain by $\vee$–E and S:

$$\mathfrak{r} > 0, \mathfrak{C}, \mathscr{E}^*(\gamma), \delta = \gamma+\omega^{\mathfrak{r}} \vee \delta < \gamma+\omega^{\mathfrak{r}} \to \mathscr{E}(\delta)$$

further:

$$M, \vee\text{–}I, \vee\text{–}E, S \qquad \downarrow$$

$$\cfrac{\cfrac{\cfrac{\delta \leq \gamma+\omega^{\mathfrak{r}} \to \delta = \gamma+\omega^{\mathfrak{r}} \vee \delta < \gamma+\omega^{\mathfrak{r}}}{\mathfrak{r} > 0, \mathfrak{C}, \mathscr{E}^*(\gamma), \delta \leq \gamma+\omega^{\mathfrak{r}} \to \mathscr{E}(\delta)} \; S}{\mathfrak{r} > 0, \mathfrak{C}, \mathscr{E}^*(\gamma) \to \mathscr{E}^*(\gamma+\omega^{\mathfrak{r}})} \; \supset\text{–}I, \forall\text{–}I}{\mathfrak{r} > 0, \mathfrak{C} \to \forall\xi[\mathscr{E}^*(\xi) \supset \mathscr{E}^*(\xi+\omega^{\mathfrak{r}})]} \; \supset\text{–}I, \forall\text{–}I.$$

This completes the description of the transformation of a *TJ*-derivation up to ω_n into a *TJ*-derivation up to ω_{n+1}.

It should be noted, moreover, that our *TJ*-derivations are essentially 'intuitionist' *LJ*-derivations, i.e., *LK*-derivations with at most one succedent formula in each sequent. The formal exceptions occurring in the case of basic mathematical sequents could easily be eliminated by a different formulation of these concepts.

2.3. For the sake of comparison it should also be pointed out how a *TJ*-derivation up to ε_0 can be obtained by including techniques from *analysis*. (In this case ε_0 would, of course, also have to be included in the domain of numbers.)

(In the following, ϑ and ζ stand for bound, and ν and μ for free numerical variables.)

We write:

$$\mathfrak{G}(\mathfrak{v}) \quad \text{for} \quad \forall\xi\,[\mathscr{E}^*(\xi) \supset \mathscr{E}^*(\xi+\omega^{\mathfrak{v}})];$$

$$\mathfrak{J}_{\mathscr{E}} \quad \text{for} \quad \forall\zeta\,[\{\zeta > 0 \;\&\; \forall\vartheta[\vartheta < \zeta \supset \mathscr{E}(\vartheta)]\} \supset \mathscr{E}(\zeta)]$$

(compact formulation of the *TJ*-upper sequents).

From the section of the derivation given at 2.21, we easily obtain a derivation for

$$\nu < \omega, \mathfrak{G}(0) \supset \mathfrak{G}(\omega_\nu), \mathfrak{J}_{\mathscr{E}} \to \mathscr{E}(0) \supset \mathscr{E}(\omega_{\nu+1}).$$

From the section of the derivation given at 2.22, we furthermore obtain a derivation for $\mathfrak{J}_{\mathscr{E}} \to \mathfrak{J}_{\mathfrak{G}}$.

Both sequents together, with the help of the basic sequent

$$\mathfrak{J}_\mathfrak{G} \supset [\mathfrak{G}(0) \supset \mathfrak{G}(\omega_\nu)], \mathfrak{J}_\mathfrak{G} \to \mathfrak{G}(0) \supset \mathfrak{G}(\omega_\nu),$$

and structural inference figures, as well as a $\supset$–I, yield:

$$\nu < \omega, \mathfrak{J}_\mathfrak{G} \supset [\mathfrak{G}(0) \supset \mathfrak{G}(\omega_\nu)] \to \mathfrak{J}_\mathscr{E} \supset [\mathscr{E}(0) \supset \mathscr{E}(\omega_{\nu+1})].$$

We now leave the elementary number-theoretical formalism by allowing *bound predicate variables*. Thus we obtain, by $\forall$–E and $\forall$–I:

$$\nu < \omega, \forall \mathscr{E}\{\mathfrak{J}_\mathscr{E} \supset [\mathscr{E}(0) \supset \mathscr{E}(\omega_\nu)]\} \to \forall \mathscr{E}\{\mathfrak{J}_\mathscr{E} \supset [\mathscr{E}(0) \supset \mathscr{E}(\omega_{\nu+1})]\}$$

and now by a *CJ*-inference figure:

$$\mu < \omega, \forall \mathscr{E}\{\mathfrak{J}_\mathscr{E} \supset [\mathscr{E}(0) \supset \mathscr{E}(0)]\} \to \forall \mathscr{E}\{\mathfrak{J}_\mathscr{E} \supset [\mathscr{E}(0) \supset \mathscr{E}(\omega_\mu)]\}$$

and from this easily:

$$\mu < \omega \to \forall \mathscr{E}\{\mathfrak{J}_\mathscr{E} \supset [\mathscr{E}(0) \supset \mathscr{E}(\omega_\mu)]\}.$$

From this we obtain

$$\mathfrak{J}_\mathscr{E}, \mathscr{E}(0) \to \forall \zeta[\zeta < \varepsilon_0 \supset \mathscr{E}(\zeta)]$$

without essential difficulties and, finally, $\mathfrak{J}_\mathscr{E} \to \mathscr{E}(0) \supset \mathscr{E}(\varepsilon_0)$, i.e., the validity of transfinite induction up to ε_0.

§ 3. Demonstrations of nonprovability

We shall now prove that transfinite induction up to certain ordinal numbers is not provable in certain subsystems of elementary number theory; in particular, that transfinite induction up to the number ε_0 is not provable in the full system of elementary number theory.

The procedure resembles the consistency proof. This is natural since in both cases we are in fact dealing with the demonstration of a nonprovability, and since the statement of the nonprovability of transfinite induction up to ε_0 in elementary number theory entails the consistency of that theory; this is so since anything is provable from a contradiction, hence also **transfinite induction**. I shall, therefore, rely considerably on my consistency proof (*New*).

3.1. *The general schema for the demonstration of the nonprovability* of transfinite induction is this:

We define '*reduction steps*' for arbitrary *TJ*-derivations whose terminal number is not 0. (By the '*terminal number*' of a *TJ*-derivation we mean the number designated by the term ȝ.) A reduction step transforms a *TJ*-derivation into another *TJ*-derivation. In this process the terminal number of the derivation is, in general, preserved. But if the derivation is 'critical' (what this means will be explained later), then the terminal number diminishes; in this case the new terminal number can moreover be specified arbitrarily (smaller than the old number); the reduction step can always be carried out in such a way that the newly specified terminal number results. (Apart from this dependence on the specification of a a new terminal number, the reduction step can in each case be stated unequivocably.)

We furthermore correlate an *ordinal number* with each *TJ*-derivation, together with a rule for its calculation; we call it – in contrast with the terminal number – the '*value*' of the derivation. It is then proved that *with each reduction step the value diminishes*. As values only transfinite numbers below ε_0 will be used.

Once these concepts have been defined and the stated assertions proved, the following fundamental theorem follows:

The terminal number of a TJ-derivation cannot be larger than its value.

In order to see this we apply a transfinite induction on the value of the derivation:

If the value is 0, then the assertion is obviously true. (Since there is no smaller value, it is evident that no reduction step is applicable to such a derivation, i.e., it must have the terminal number 0.) Now suppose that the assertion has been recognized as true for all derivations whose value is smaller than an ordinal number α. (Where α is assumed to be larger than 0.) Suppose that an arbitrary derivation with the value α is given. If its terminal number is 0, then the assertion holds for that derivation. If this is not the case, and if the derivation is not critical, then a reduction step may be carried out which reduces a derivation to a derivation with the same terminal number and a smaller value. The assertion holds for the reduced derivation; hence also for the former derivation. If the derivation is critical, and if its terminal number were larger than its value, contrary to assumption, then the value could be taken as its new terminal number and the associated reduction step could be carried out: The result would be a derivation of smaller value, so that its value is therefore once again smaller than its terminal number, contrary to the induction hypothesis.

From this theorem the desired nonprovability theorems can be deduced immediately.

3.2. This deduction will now be carried out.

To begin with, the formalism of § 1 must be modified in several respects:

First, the connective $\supset$ will no longer be used; this does not, of course, represent a restriction, since $\mathfrak{A} \supset \mathfrak{B}$ can be replaced by its equivalent $(\neg \mathfrak{A}) \vee \mathfrak{B}$. (If this is done, the $\supset$-inference figures and the basic $\supset$-sequents become derivable from the inference figures for $\vee$ and $\neg$.) The concept of the '*TJ*-upper sequents' must be modified correspondingly at the same time; these therefore now have the form:

$$\mathfrak{r} > 0, \forall \mathfrak{x}[(\neg \mathfrak{x} < \mathfrak{r}) \vee \mathscr{E}(\mathfrak{x})] \to \mathscr{E}(\mathfrak{r}).$$

Second, we now admit further inference figures according to the following schemata ('*substitutions of terms*'):

$$\frac{\Gamma_1, \mathfrak{F}(\mathfrak{s}), \Gamma_2 \to \Theta}{\Gamma_1, \mathfrak{F}(\mathfrak{t}), \Gamma_2 \to \Theta} \quad \text{or} \quad \frac{\Gamma \to \Theta_1, \mathfrak{F}(\mathfrak{s}), \Theta_2}{\Gamma \to \Theta_1, \mathfrak{F}(\mathfrak{t}), \Theta_2},$$

where $\mathfrak{s}$ and $\mathfrak{t}$ may be replaced by terms without free variables, as long as they designate the *same* number. (It is easily seen that these inference figures could be replaced by applications of basic equality sequents and basic mathematical sequents of the form $\to \mathfrak{s} = \mathfrak{t}$, and are thus dispensable in principle, but convenient for the discussion that follows.)

Third, the concept of the *TJ*-derivation is finally modified as follows: The endsequent of a *TJ*-derivation shall now have the form:

$$\mathscr{E}(0) \to \mathscr{E}(\mathfrak{s}_1), \mathscr{E}(\mathfrak{s}_2), \ldots, \mathscr{E}(\mathfrak{s}_\nu),$$

the $\mathfrak{s}$'s must be numerical terms (ν a natural number); as '*terminal number*' we taken the smallest of the ordinal numbers represented by $\mathfrak{s}$'s.

The old form of the endsequent of a *TJ*-derivation is a special case of the new form; therefore, all nonprovability theorems that are proved for the new form also hold for the old form. We now develop one by one the definitions and proofs required for the demonstration of the nonprovability (according to 3.1).

3.3. *Definition of the reduction steps* for *TJ*-derivations.

3.31. Suppose that a *TJ*-derivation is given and that its endsequent is

$$\mathscr{E}(0) \to \mathscr{E}(\mathfrak{s}_1), \ldots, \mathscr{E}(\mathfrak{s}_\nu).$$

Suppose that none of the numerical terms $\mathfrak{s}$ are '0'.

The definition of a reduction step will, in essence, be taken over from *New*, § 3. Let us therefore first examine the essential differences between our present concept of a derivation and its counterpart in *New*:

The numerical terms now extend up to ε_0 (exclusively); in *New* they range only over the natural numbers. Functions are now admitted in arbitrary number; in *New* only the successor function was admitted. The predicate symbols are supplemented by the separate symbol $\mathscr{E}$, the basic sequents by the basic equality sequents, the inference figures by the 'substitutions of terms'; also noted must be the somewhat modified version of the *CJ*-inference figures. Finally the *TJ*-upper sequents are introduced, and so is the special form of the endsequent.

It will turn out that none of this necessitates any radical changes vis-à-vis the procedure employed in *New*, apart from the entirely new addition of *critical* reduction steps.

3.32. The reduction step begins with the same preparatory step as in *New*, 3.2: the replacement of 'redundant' free variables by 1.

The '*ending*' of the given derivation is defined somewhat differently from *New*, 3.2, as follows:

The ending includes all those derivational sequents which are encountered if we trace each individual path from the endsequent upwards and stop as soon as we reach the line of inference of an operational or *CJ-inference figure*.

The ending can therefore contain only structural inference figures and substitutions of terms and, after the preparatory step, no free variables. The uppermost sequents of the ending may be lower sequents of operational or *CJ*-inference figures as well as basic sequents of any one of these three kinds and *TJ*-upper sequents.

We now add to the preparatory step an additional step made necessary by the admission of function symbols:

All terms occurring in the ending are to be replaced by the numerical terms which result from the 'evaluation' of the occurring function symbols. (The function symbols occurring in the numerical terms themselves are, of course, retained.)

This replacement obviously turns all structural inference figures into correct inference figures of the same kind, as well as all basic sequents and *TJ*-upper sequents into other correct sequents of the same kind.

In the '*substitutions of terms*' upper and lower sequents become identical with one another; these inference figures are deleted.

Each *basic equality sequent* either assumes the form

$$\mathfrak{z} = \mathfrak{z}, \mathfrak{F}(\mathfrak{z}) \to \mathfrak{F}(\mathfrak{z}) \quad \text{or} \quad \mathfrak{z} = \mathfrak{t}, \mathfrak{F}(\mathfrak{z}) \to \mathfrak{F}(\mathfrak{t}),$$

where $\mathfrak{z}$ and $\mathfrak{t}$ are mutually *distinct* numerical terms. The first form is in each case replaced by

§ 3, DEMONSTRATIONS OF NONPROVABILITY   301

$$\frac{\mathfrak{F}(\mathfrak{z}) \to \mathfrak{F}(\mathfrak{z})}{\mathfrak{z} = \mathfrak{z}, \mathfrak{F}(\mathfrak{z}) \to \mathfrak{F}(\mathfrak{z})} \text{ (basic logical sequent)} \\ \text{(thinning),}$$

the second one by

$$\frac{\mathfrak{z} = \mathfrak{t} \to}{\mathfrak{z} = \mathfrak{t}, \mathfrak{F}(\mathfrak{z}) \to \mathfrak{F}(\mathfrak{t})} \text{ (basic mathematical sequent)} \\ \text{(thinnings and interchange).}$$

After this, basic equality sequents no longer occur in the ending.

Now, however, we must still amend those places of the derivation in which the uppermost sequent of the ending is the lower sequent of an inference figure. Here the evaluation of terms in general destroys the correctness of the inference figure, since the replacement is carried out only in the lower sequent. In such cases the change-over from the original lower sequent to the new form resulting from the evaluation of terms must be brought about by a number of 'substitutions of terms' according to the above schemata (to be inserted in the derivation). (Inference figures of this kind are therefore once again included in the ending. Their order is easily *unambiguously* determined.)

Finally, we require that in the case of a *CJ*-inference figure the term designated by $\mathfrak{t}$ in the schema of that figure must be retained in its *evaluated* form in the lower sequent of the *CJ*-inference figure. This clearly leaves the *CJ*-inference figure correct.

This completes the 'first preparatory step'.

3.33. We continue as in *New*:

If an uppermost sequent of the ending is the lower sequent of a *CJ*-inference figure, then a *CJ-reduction* is carried out. (In this case the entire reduction step has then been completed.)

Let us therefore consider such a *CJ*-inference figure (in order to make our choice unequivocal, we might select the figure standing on the extreme right). It has the form

$$\frac{\mathfrak{a} < \omega, \mathfrak{F}(\mathfrak{a}), \Gamma \to \Theta, \mathfrak{F}(\mathfrak{a}+1)}{\mathfrak{t} < \omega, \mathfrak{F}(0), \Gamma \to \Theta, \mathfrak{F}(\mathfrak{t})},$$

where $\mathfrak{t}$ is a *numerical term* by virtue of the preparatory step. Of this term it can be decided whether the number designated by it is smaller than ω or not. If $\mathfrak{t} < \omega$ is not true, we simply form

$$\frac{\mathfrak{t} < \omega \to}{\mathfrak{t} < \omega, \mathfrak{F}(0), \Gamma \to \Theta, \mathfrak{F}(\mathfrak{t})} \text{ (basic mathematical sequent)} \\ \text{(thinnings and interchanges).}$$

The derivation continues downwards as before; whatever had occurred above it is omitted; the reduction step has then been completed.

If $t < \omega$ is true, we replace the *CJ*-inference figure by structural inference figures as in *New*, 3.3, and above each figure we furthermore write (where $\mathfrak{m}$ equals $0, 1, 1+1, 1+1+1$, etc.):

(Basic mathematical sequent)

$$\frac{\to \mathfrak{m} < \omega \qquad \mathfrak{m} < \omega, \mathfrak{F}(\mathfrak{m}), \Gamma \to \Theta, \mathfrak{F}(\mathfrak{m}+1)}{\mathfrak{F}(\mathfrak{m}), \Gamma \to \Theta, \mathfrak{F}(\mathfrak{m}+1)} \text{ (cut)},$$

and below it:

$$\frac{\mathfrak{F}(0), \Gamma \to \Theta, \mathfrak{F}(t)}{t < \omega, \mathfrak{F}(0), \Gamma \to \Theta, \mathfrak{F}(t)} \text{ (thinning)}.$$

3.34. If the ending is nowhere bounded above by a *CJ*-inference figure, then the actual reduction step is preceded by a *second preparatory step* as in *New*, 3.4.

Here we can once again begin by introducing the concepts 'clustered' and 'cluster of formulae'. In the 'substitutions of terms', the formulae $\mathfrak{F}(\mathfrak{s})$ and $\mathfrak{F}(t)$ are considered to be clustered, so are the formulae of Γ_1, Γ_2, etc., occurring in the same places in the upper and lower sequents (as usual).

Now the formulae of a cluster therefore need not be entirely identical, but may differ formally in the values of their terms. The 'cut associated with a cluster' need now no longer exist in all cases, since the endsequent is not empty.

The second preparatory step is carried out precisely as in *New*, 3.42; it must be observed, however, that in the elimination of thinnings the omission of sequent formulae could eventually also affect the *endsequent*. If this should happen, we require that the original form of the endsequent is restored by carrying out thinnings at the end of the derivation. For the elimination of the basic logical sequents it is essential that the endsequent is not a basic logical sequent and that it is not derivable from such sequents by thinnings; this requirement is met, since the terminal number is not 0.

For the ending the following therefore holds: the only inference figures which it contains are 'substitutions of terms' and structural inference figures, with thinnings at most immediately above the endsequent. It contains no free variables. Its uppermost sequents are lower sequents of operational inference figures or basic mathematical sequents or *TJ*-upper sequents.

3.35. If the second preparatory step produces in the ending a 'cluster of formulae suitable for the application of an operational reduction' – i.e. (cf. *New*, 3.43), a cluster of formulae which possesses an associated cut and whose right and left sides contain at least one formula which is the principal formula of an operational inference figure (definitions of these concepts as in *New*, 3.41 and 1.32) – then an *operational reduction* as in *New*, 3.5, is carried out.

For this purpose we adopt the concept of *'level'* from *New*, 3.43. The schema of the reduction step is taken over without essential change; the only additional point that needs to be observed is that the operational inference figures may now be followed by 'substitutions of terms'. These are simply retained in the reduction step and any further 'substitutions of terms' that may be required for the 'evaluation' of the terms occurring in $\mathfrak{F}(\mathfrak{n})$ – or $\mathfrak{A}$, or $\mathfrak{B}$ – are adjoined, so that both cut formulae in the 'new cut' are really identical with one another.

With this operational reduction, the entire reduction step has once again been completed.

3.36. We shall now examine the question of what a derivation looks like to which none of the described reductions is applicable. In this case, we shall call the derivation 'critical' and shall state 'critical reduction steps' for it.

In *New* it was proved (3.43) that such cases could no longer arise; but now the former proof cannot be carried over completely. Here the essentially newly feature is that *TJ*-upper sequents can also occur as uppermost sequents of the ending and that the endsequent is not empty.

To begin with, it holds analogously to *New*: It is impossible for the derivation to contain neither an operational inference figure nor a *TJ*-upper sequent. For if this were the case, the endsequent $\mathscr{E}(0) \to \mathscr{E}(\mathfrak{z}_1), \ldots, \mathscr{E}(\mathfrak{z}_\nu)$ would have been derived from basic mathematical sequents alone without the use of logical connectives and variables. $\mathscr{E}(\mathfrak{t})$ could then be replaced by $\mathfrak{t} = 0$ throughout the entire derivation; this does not change the basic sequents, since $\mathscr{E}$ does not occur in basic mathematical sequents; the endsequent would thus become

$$0 = 0 \to \mathfrak{z}_1 = 0, \ldots, \mathfrak{z}_\nu = 0.$$

Since, by hypothesis, none of the numerical terms $\mathfrak{z}$ are '0', the endsequent would be 'false', whereas all uppermost sequents would be 'true'; this is impossible.

We can therefore be certain that at least one operational inference figure or *TJ*-upper sequent occurs in our derivation. We shall now try to adapt the proof from *New*, 3.43, in such a way that the *TJ*-upper sequents are given equal status with the lower sequents of operational inference figures, and we shall call their succedent formula the 'principal formula'. The concept of a prime formula will furthermore be restricted to formulae without $\mathscr{E}$ (as was done in § 1) so that the observations made at that point remain valid; although the formula belonging to the same cluster as a principal formula may now also have the form $\mathscr{E}(\mathfrak{t})$, and is thus of degree 0, it can nevertheless not be a 'prime formula'.

The 'substitutions of terms' cause no difficulties in the process if, as stipulated above, $\mathfrak{F}(\mathfrak{z})$ is considered as clustered with $\mathfrak{F}(\mathfrak{t})$.

Thus the considerations from *New* carry over; we must merely bear in mind that now the endsequent is not empty. As a result of this, there may not occur any cut which is suitable for an operational reduction; instead, *one* additional case can still arise: The endsequent contains a formula belonging to the same cluster as a principal formula. That formula can then only be a *succedent formula* of the endsequent since, by not containing logical connectives, a principal formula $\mathscr{E}(\mathfrak{t})$ is necessarily the succedent formula of a *TJ*-upper sequent and all formulae belonging to the same cluster as it are also succedent formulae (cf. *New*, 3.41), provided of course that the cluster extends up to the endsequent and therefore contains no 'cut formulae'. (From the same considerations it follows also that if both cut formulae of a cut belong to the same cluster as the principal formulae standing above them, then these principal formulae belong to *operational inference figures* and certainly not to *TJ*-upper sequents: For in the latter instance they would have to have the form $\mathscr{E}(\mathfrak{t})$, and on the 'right-hand side', where such a formula occurs only as an *antecedent* formula, it could not even be a principal formula.)

3.37. Thus, a 'critical derivation' necessarily has the following property:

One of the succedent formulae of the endsequent belongs to the same cluster of formulae as the succedent formula of one of the *TJ*-upper sequents belonging to the ending.

If there are several such formulae, we shall agree to choose that *TJ*-upper sequent with the above property which stands furthest to the right. Suppose that its succedent formula is $\mathscr{E}(\mathfrak{t})$. We now define the 'critical reduction steps':

We first specify an arbitrary ordinal number which is smaller than the terminal number of the derivation. Let it be designated by the numerical term $\mathfrak{z}^*$.

Then we replace the mentioned *TJ*-upper sequent – suppose that it runs: $\mathfrak{r} > 0, \forall \mathfrak{x}[(\neg \mathfrak{x} < \mathfrak{r}) \vee \mathscr{E}(\mathfrak{x})] \to \mathscr{E}(\mathfrak{r})$ – by the following derivational section:

Basic mathematical sequent

$$\frac{\rule{0pt}{0pt}\to \mathfrak{z}^* < \mathfrak{r}}{\neg\, \mathfrak{z}^* < \mathfrak{r} \to} \;\neg\text{-inference figure}$$

$$\frac{\neg\, \mathfrak{z}^* < \mathfrak{r} \to \mathscr{E}(\mathfrak{z}^*)}{} \text{ thinning}$$

$$\frac{\neg\, \mathfrak{z}^* < \mathfrak{r} \to \mathscr{E}(\mathfrak{z}^*) \qquad \mathscr{E}(\mathfrak{z}^*) \to \mathscr{E}(\mathfrak{z}^*)}{(\neg\, \mathfrak{z}^* < \mathfrak{r}) \vee \mathscr{E}(\mathfrak{z}^*) \to \mathscr{E}(\mathfrak{z}^*)} \;\vee\text{-inference figure}$$

$$\frac{\forall \mathfrak{x}\,[(\neg\, \mathfrak{x} < \mathfrak{r}) \vee \mathscr{E}(\mathfrak{x})] \to \mathscr{E}(\mathfrak{z}^*)}{\mathfrak{r} > 0,\, \forall \mathfrak{x}\,[(\neg\, \mathfrak{x} < \mathfrak{r}) \vee \mathscr{E}(\mathfrak{x})] \to \mathscr{E}(\mathfrak{z}^*),\, \mathscr{E}(\mathfrak{r})} \;\forall\text{-inference figure} \atop \text{thinnings.}$$

The new formula $\mathscr{E}(\mathfrak{z}^*)$ is 'carried forward' in the derivation (i.e., it is written down as the first succedent formula of each individual sequent in the derivational path leading from the above sequent to the endsequent). This leaves all inference figures correct (possibly by inserting interchanges) and the endsequent assumes the form:

$$\mathscr{E}(0) \to \mathscr{E}(\mathfrak{z}^*),\, \mathscr{E}(\mathfrak{z}_1),\, \ldots,\, \mathscr{E}(\mathfrak{z}_\nu).$$

We therefore obtain another correct *TJ*-derivation, in particular, a derivation with a lower terminal number, viz., the number designated by the term $\mathfrak{z}^*$.

This completes the definition of the reduction steps, and it meets the requirements laid down at 3.1.

3.4. Now follows the *correlation of 'values'* with the *TJ*-derivations, as well as the proof that with each reduction step the value of the derivation *diminishes*.

In both cases we can adopt the results from *New*, § 4, with minor modifications. The correlation of ordinal numbers proceeds as in *New*, 4.2, with the following additions:

Each *TJ*-upper sequent receives the ordinal number 5 (i.e., $\omega^0 + \omega^0 + \omega^0 + \omega^0 + \omega^0$), whereas all basic sequents receive the ordinal number 1, as in *New*. A 'substitution of terms' is given the same value as a structural inference figure (i.e., no value at all).

We must now verify that in a reduction step the value of the *TJ*-derivation diminishes (cf. *New*, 4.3).

In the 'first preparatory step' (3.32) the value of the derivation remains unchanged (since 'substitutions of terms', thinnings and interchanges do not 'count').

For the *CJ*-reduction (3.33) the following holds: In the simple case where $\mathfrak{t} < \omega$ is not true, the ordinal number of the (former) *CJ*-lower sequent after the reduction is 1 and was at least ω before (the same is true in the case where t is equal to '0'). The main case is dealt with as in *New*. The adjoining of $\to \mathfrak{m} < \omega$ may lead to several additions of 1 to the ordinal number for the lower sequent obtained after the reduction; but this does not prevent it from remaining smaller than the ordinal number of this sequent before the reduction.

The second preparatory step (3.34) is carried out as in *New*. The same holds for the operational reduction (3.35). In both cases the same minor modifications of the present version have obviously no influence on the calculation of value.

Still to be dealt with are the critical reduction steps (3.37): In the case of such reduction steps the place of a *TJ*-upper sequent with the ordinal number 5 is simply taken by a derivational section whose lowest sequent was correlated with the ordinal number 4. Thus the value of the total derivation is here also guaranteed to diminish.

3.5. *Conclusion*. Now all verifications required for the general proof schema (3.1) have been carried out and the theorem, stated in 3.1, follows:

The terminal number of a *TJ*-derivation cannot be larger than its 'value', as defined at 3.4.

This theorem, taken together with the relationship, stated in *New*, 4.4 (paragraph 2), between the degrees of the formulae occurring in a derivation and the ordinal number, i.e., the value of the derivation, yields the following individual results:

The terminal number of a *TJ*-derivation in which all formulae are at most of degree 0, or 1, generally: v, is always smaller than ω^ω, or ω_4, generally: ω_{v+3}.

These results can obviously still be considerably sharpened; I hope to be able to follow this up sometime in the future.

In the present context, however, the important theorem follows from the fact that the values are always smaller than ε_0:

Transfinite induction up to ε_0 and higher ordinal numbers is not provable in elementary number theory.

On the other hand it holds (cf. § 2 and the beginning of this paper): Transfinite induction up to any ordinal number *below* ε_0 is provable in elementary number theory.

These theorems call for a number of further observations. In § 2, we proved only the existence of *TJ*-derivations up to the numbers ω_v. Yet from a *TJ*-derivation another *TJ*-derivation with an *arbitrarily smaller* terminal number is easily obtained by replacing $\mathscr{E}(\mathfrak{v})$ throughout by $\mathscr{E}^*(\mathfrak{v})$ (in the sense of 2.2) and by making a number of simple additions to the endsequent and to the *TJ*-upper sequents. For every arbitrary terminal number below ε_0 a *TJ*-derivation with precisely this terminal number is therefore statable. (Every such number is, after all, smaller than a suitable ω_ν.)

A further remark is made necessary by the fact that the number ε_0 and higher ordinal numbers were not included in our domain of numbers so that in the system stated in § 1, transfinite induction up to ε_0 could not even be formulated. This is unimportant since the whole proof is obviously completely independent of this limitation of the domain of numbers. Arbitrarily many further ordinal numbers can indeed be introduced, as long as the following conditions are satisfied:

Precisely specified expressions (numerical terms) must be given for the unique designation of such numbers; all functions and predicates that are introduced must be decidably defined; finally, all basic mathematical sequents must be 'true' in the sense of these decidable definitions.

If these conditions are fulfilled, then our proof obviously remains fully applicable. From this it follows that if restricted transfinite induction is unprovable, according to our results, it cannot become provable by an extension of the *domain of numbers* as described above. It is in this sense that the following theorem is intended:

Transfinite induction up to ε_0 and higher ordinal numbers is not provable in 'elementary number theory'.

The mentioned conditions imposed on the domain of numbers – which are not required in their full strength for the validity of our proof, incidentally – correspond to the framework of 'elementary number theory'. The extent to which we can advance into the 'second number class' with 'elementary number-theoretical methods' is difficult to anticipate. We can, without doubt, go far beyond ε_0. Each 'segment of the second number class' obtained in this way becomes a domain of objects which must, by its very nature, be included in number theory or in theories logically equivalent to it (algebra etc.). This could be documented even more distinctly by mapping the ordinal numbers concerned onto the natural numbers[97]; the predicate < then becomes a more or less complicated predicate between the natural numbers, etc. This makes it completely clear that for such a segment of the second number class *transfinite induction* is a form of inference which, *in substance, belongs to elementary number theory*. The fact that transfinite induction even up to the number ε_0 is no longer derivable from the remaining number-theoretical forms of inference therefore reveals from a new angle the *incompleteness* of the number-theoretical formalism, which has already been exposed by Gödel[98] from a different point of view (viz., in relation to the nonprovability of number-theoretical *theorems*).

We might think that by including transfinite induction up to ε_0 in elementary number theory as a new form of inference, the incompleteness could be overcome. Yet an analogous incompleteness then no doubt arises in relation to a higher transfinite induction, etc. It seems likely that our result here is only a special case of a general affinity which exists between formally delimited techniques of proof, their possibilities of extension and newly arising incompletenesses, on the one hand, and the transfinite ordinal numbers of the second number class, transfinite induction and the constructive progression into the second number class on the other.

10. FUSION OF SEVERAL COMPLETE INDUCTIONS

In the following we intend to show that every mathematical proof in which the rule of complete induction is applied several times, can be so rearranged by means of certain simple fusions of inferences and concepts that only a single application of complete induction occurs in it.

For this purpose it is not necessary to specify a particular kind of formalization of mathematical proofs. It shall merely be presupposed that the concept of a 'mathematical proof' encompasses all forms of inference of 'predicate logic' and it must, of course, include the rule of complete induction. No further number-theoretical results need to be presupposed except for the admission of the primitive predicate '=' and its associated basic formulae (axiom formulae) $\mathfrak{n} = \mathfrak{n}$, as well as $\neg\, (\mathfrak{m} = \mathfrak{n})$, for arbitrary numerals $\mathfrak{n}$ and $\mathfrak{m}$, where $\mathfrak{n}$ and $\mathfrak{m}$ are distinct. Further mathematical concepts and their associated axioms may be admitted as desired. Note that the theorem to be proved is significant primarily for 'elementary number theory'. If elementary number theory is extended to 'analysis', then, as *Dedekind* has shown, complete induction becomes reducible to other forms of inference and the assertion of the theorem therefore loses its significance.

The proof runs as follows:

Let there be given a derivation (i.e., a formalized proof) with several occurrences of formalized complete induction. Each application of complete induction is logically equivalent to an application of the 'induction axiom' to a certain special proposition about the natural numbers, i.e., it can be expressed formally in such a way that a formula of the following form is asserted to hold:

$$\{\mathfrak{F}_v(1)\ \&\ \forall \mathfrak{x}\, [\mathfrak{F}_v(\mathfrak{x}) \supset \mathfrak{F}_v(\mathfrak{x}+1)]\} \supset \forall \mathfrak{y}\, \mathfrak{F}_v(\mathfrak{y}),$$

where $\mathfrak{F}_v$ stands for an arbitrary formula with one argument place for a numeral designating a natural number, and the formula is thus the formal counterpart of a proposition about natural numbers; the subscript v serves to distinguish the different complete inductions occurring in the derivation, i.e., it runs through the numbers 1, 2, ..., ρ, where ρ stands for the total number of complete inductions occurring in the derivation. ($\mathfrak{x}$ and $\mathfrak{y}$ designate arbitrary bound variables. The formulae $\mathfrak{F}_\nu$ can of course also contain free variables.)

We shall now deduce all of these ρ induction axiom formulae by applying a single formalized complete induction that fuses all of these formulae. This is done as follows: We construct the formula

$$[\mathfrak{b} = 1 \supset \mathfrak{F}_1(\mathfrak{a})] \,\&\, [\mathfrak{b} = 2 \supset \mathfrak{F}_2(\mathfrak{a})] \,\&\, \ldots \,\&\, [\mathfrak{b} = \rho \supset \mathfrak{F}_\rho(\mathfrak{a})],$$

briefly referred to as $\mathfrak{H}(\mathfrak{a})$. (Here $\mathfrak{a}$ and $\mathfrak{b}$ designate two free variables not yet occurring in the derivation.) By a *single* formal application of complete induction we obtain the formula:

$$\{\mathfrak{H}(1) \,\&\, \forall\mathfrak{x}[\mathfrak{H}(\mathfrak{x}) \supset \mathfrak{H}(\mathfrak{x}+1)]\} \supset \forall\mathfrak{y}\, \mathfrak{H}(\mathfrak{y}).$$

(Here it makes no difference whether complete induction is to be admitted in the form formalized in this axiom formula or whether any other version, possibly that of an inference figure, is chosen; by means of the latter we would then simply derive the above formula.)

From this formula we can now derive all of the ρ induction axiom formulae cited above along purely logical lines, i.e., without requiring another application of complete induction. Once this has been done, we have reached our goal. In order to prove this derivability, it should suffice to outline the main steps of the formal argument: In order to derive the induction axiom formula for $\mathfrak{F}_1$ from that of $\mathfrak{H}$, for example, we would have to begin by substituting 1 for $\mathfrak{b}$ in the latter formula (this is formally possible by means of successive ∀-introductions and ∀-eliminations in the case where no provision has been made for direct substitution as a permissible form of inference in the formalism). $\mathfrak{H}(1)$, for example, yields

$$[1 = 1 \supset \mathfrak{F}_1(1)] \,\&\, [1 = 2 \supset \mathfrak{F}_2(1)] \,\&\, \ldots \,\&\, [1 = \rho \supset \mathfrak{F}_\rho(1)],$$

a formula which can be proved to be equivalent to $\mathfrak{F}_1(1)$ by means of propositional logic alone, with the additional appeal to the truth of $1 = 1$ and the falsity of $1 = 2, \ldots, 1 = \rho$. The same holds for $\mathfrak{H}(\mathfrak{a})$ and $\mathfrak{F}_1(\mathfrak{a})$, for an arbitrary $\mathfrak{a}$. The entire induction axiom formula, therefore, may be rearranged in such a way, by applications of purely logical formalized forms of inference, that eventually $\mathfrak{F}_1$ replaces every occurrence of $\mathfrak{H}$, as was to be shown.

It might seem as though all that has been achieved is that one and the same formalized complete induction now occurs in ρ different places in the derivation, with the result that there are once again ρ individual, although identical, occurrences of complete induction. This observation, however, does not touch the heart of the matter, for we could fuse these duplicate induction inferences in a trivial way not only in thought, but also in form, so that only a single formalized induction actually occurs. In order to accomplish this we would have to prefix the induction axiom formula for $\mathfrak{H}$ with universal quantifiers quantifying all the free variables of $\mathfrak{H}$. We might denote the resulting formula by $\mathfrak{I}$, and the endformula of the derivation by $\mathfrak{E}$. This would yield a purely logical derivation with several formulae of the form $\mathfrak{I}$ as initial formulae (as well as possibly other mathematical axiom formulae as initial formulae; this does not affect our argument); the derivation can then be transformed in the familiar way into a (purely logical) derivation *without* such initial formulae and with $\mathfrak{I} \supset \mathfrak{E}$ for its endformula, ('deduction theorem' or, in the calculus of sequents, a trivial observation) and this, with the inclusion of *a single derivation for* $\mathfrak{I}$, once again yields a derivation for the original endformula $\mathfrak{E}$.

Our result shows that the *number* of complete inductions occurring in a number-theoretical proof is no measure of the 'complexity' of the proof in the context of metamathematics; although it does have some bearing on this point, it is not the number of inductions but their 'degree', i.e., the complexity of the induction proposition, that counts.

NOTES

[1] P. Hertz, Über Axiomensysteme für beliebige Satzsysteme, Math. Ann. **101** (1929); in connection with the above question cf., in particular, §§ 1 and 7. Other papers by P. Hertz on the same subject can be found in Math. Ann. **87** (1922) and **89** (1923) and in Ann. d. Philos. **7** (1928); in the following I shall refer to the last three papers as H.1, H.2, H.3, and to the first paper as H.4.

[2] Cf. note 1.

[3] Our definition of a 'proof' (and, correspondingly, of 'provability') differs from that of Hertz only with respect to Hertz's use of the 'syllogism' in place of the 'cut'. Cf. the next §.

[4] Hertz uses the expression (cf. H.3): 'q wird von $\mathfrak{p}_1, \ldots, \mathfrak{p}_\nu$ impliziert'.

[5] If the words 'sentence' ('theorem') and 'proof' are used informally as constituents of our language they are of course intended to mean something quite different from the purely formally introduced concepts of 'sentence' ('theorem') and 'proof' (and even under an intuitive interpretation the latter concepts are still considerably narrower than the former); the context should make it clear in each case how these concepts are intended.

[6] In H.3, Hertz proves the analogous theorem for his forms of inference. By virtue of § 3 we could appeal to Hertz's theorem, but we shall give a new proof since the 'normal form' of the 'proof' which emerges in the process will be needed later. Hertz's proof leads to an 'Aristotelian normal form' which is unsuitable for our purposes.

[7] Our normal proof corresponds roughly to the 'Goclenian normal proof' in Hertz. A corresponding analogue to the 'Aristotelian normal proof' could be formulated, but for it no theorem corresponding to theorem 2 would hold, as the following example shows:

$$\frac{e \to a \quad \dfrac{d \to b \quad ab \to c}{da \to c}}{ed \to c}$$

where $\mathfrak{p}_1 = d \to b$; $\mathfrak{p}_2 = ab \to c$; $\mathfrak{p}_3 = e \to a$; $\mathfrak{p}_4 = ed \to c$. It is easily seen that in this case no proof (by means of cuts and thinnings) is possible in which each upper sentence belongs to the $\mathfrak{p}$'s.

[8] Cf. note 6.

[9] Sentence systems consisting only of trivial sentences need not be excluded if an empty axiom system is admitted.

[10] Hertz's 'maximal nets' also include the associated net sentences; our propositions remain valid under this interpretation.

[11] An analogous procedure for finite systems is applied by Hertz in H.1, articles 36, 42, 43.

[12] Hilbert and Ackermann, Grundzüge der theoretischen Logik, Berlin, 1st ed. 1928, cited in the following as H.-A.

[13] Herr P. Bernays must be considered as the effective co-author of this theorem. When I communicated to him a proof of theorems VI and V, Herr Bernays noticed that the latter theorem could be brought into the considerably stronger form given in theorem IV. By making use of Bernays's argument, I was able to choose a way of proving all of these theorems jointly on the basis of theorem III, which is proved first.

[14] Cf. note 12.

[15] A. Heyting, Die formalen Regeln der intuitionistischen Logik und Mathematik, Sitzungsber. d. Preuss. Akad. d. Wiss., phys.-math. Kl. 1930.

[16] J. Herbrand, Sur la non-contradiction de l'arithmétique, J. Reine Angew. Math., **166** (1932) (§ 2).

[17] Cf. note 15. Heyting actually uses the additional rule: If $\mathfrak{A}$ and $\mathfrak{B}$ are true formulae, then $\mathfrak{A}$ & $\mathfrak{B}$ is a true formula. As Bernays observed, this rule can be derived by means of the axiom formulae and the other rules as follows: From $\mathfrak{A}$ and axiom formula 2.125 $\mathfrak{B} \supset \mathfrak{A}$ follows by rules 2.233 and 2.21; from axiom formula 2.123 $\mathfrak{B} \supset \mathfrak{A} \cdot \supset : \mathfrak{B} \& \mathfrak{B} \cdot \supset \cdot \mathfrak{A} \& \mathfrak{B}$ follows by replacement (2.233); both together yield $\mathfrak{B} \& \mathfrak{B} \cdot \supset \cdot \mathfrak{A} \& \mathfrak{B}$; from $\mathfrak{B}$ and 2.121 follows $\mathfrak{B} \& \mathfrak{B}$; and the last two results taken together therefore yield $\mathfrak{A} \& \mathfrak{B}$.

[18] D. Hilbert, Über das Unendliche, Math. Annalen **95** (1926), pp. 161–190.

[19] K. Gödel, Über formal unentscheidbare Sätze der Principia Mathematica und verwandter Systeme I, Monatsh. Math. u. Phys. **38** (1931), pp. 173–198.

[20] Cf. note 12.

[21] Cf. note 15, pp. 42–65.

[22] An important special case of the Hauptsatz has already been proved by Herbrand in a completely different way, cf. section IV, § 2.

[23] We take the symbols $\vee, \supset, \exists$ from Russell. Russell's symbols for 'and', 'equivalent', 'not', 'all' viz: $\cdot, \equiv, \sim, ()$, are already being used with different meanings in mathematics. We shall therefore take Hilbert's &, whereas Hilbert's symbols for equivalence, all, and not, viz.: $\sim, (), ^-$, again have already different meanings. Besides, the negation symbol represents a departure from the linear arrangement of symbols and is inconvenient for some purposes. We shall therefore use Heyting's symbols for equivalence and negation, and for 'all' we shall use a symbol (namely $\forall$) corresponding to $\exists$.

[24] Cf. note 21, p. 56.

[25] The following special case of theorem 2.1 has already been proved by Herbrand in a completely different way: If a formula $\mathfrak{P}$, in which the symbols $\forall$ and $\exists$ occur only at the beginning and span the whole formula (as we know, for every formula there exists a classically equivalent formula of this kind, cf. for example H.-A., p. 63), is classically derivable, then there is a sequent (the above midsequent) whose antecedent is empty and each one of whose succedent formulae results from $\mathfrak{P}$ by the elimination of the $\forall$ and $\exists$ symbols (including the variables next to them) and by the replacement of the bound object variables by free variables. Furthermore, this sequent is classically derivable without the use of $\forall$ and $\exists$ symbols, and from it we can derive $\rightarrow \mathfrak{P}$ simply by using those of our inference figures which we have designated by $\forall$–IS, $\exists$–IS, contraction in the succedent and interchange in the succedent. (In addition to theorem 2.1, we would still have to consider the case of a thinning in the succedent, but it is easy to see that such instances can always be avoided.) Cf. also: J. Herbrand, Sur le problème fondamental de la logique mathématique, Comptes rendus de la Société des sciences et des lettres de Varsovie, (Classe III), **24** (1931), p. 31, n.1.

[26] Earlier proofs may be found in the writings of J. von Neumann, On Hilbert's proof theory, Math. Zeitschrift **26** (1927), pp. 1–46; J. Herbrand, Sur la non-contradiction de l'arithmétique, J. Reine Angew. Math. **166** (1932), pp. 1–8.

[27] D. Hilbert, Die Grundlagen der Mathematik, Abh. math. Sem. Univ. Hamburg **6** (1928), pp. 65–85.

[28] V. Glivenko, Sur quelques points de la logique de M. Brouwer, Acad. Roy. Belg. Bull. Cl. Sciences, 5e. série, **15** (1929), pp. 183–188.

[29] Cf. note 15.

[30] A detailed and very readable discussion of these questions is contained in D. Hilbert's paper cited in note 18.

[31] In this connexion cf. also:
H. Weyl, Über die neue Grundlagenkrise der Mathematik, Math. Zeitschrift **10** (1921), pp. 39–79; and A. Fraenkel, Zehn Vorlesungen über die Grundlegung der Mengenlehre (or the relevant sections in Fraenkel's textbook on set theory).

[32] K. Gödel, cf. note 19.

[33] W. Ackermann, Begründung des 'tertium non datur' mittels der Hilbertschen Theorie der Widerspruchsfreiheit, Math. Ann. **93** (1925), pp. 1–36;
J. von Neumann, Zur Hilbertschen Beweistheorie, Math. Zeitschrift **26** (1927), pp. 1–46;
J. Herbrand, Sur la non-contradiction de l'arithmétique, J. Reine Angew. Math. **166** (1932), pp. 1–8;
G. Gentzen, #3 of this volume.

[34] There already exist several such formalizations and the present one follows more or less the established lines.

[35] Since the concept of a 'formula' is used quite generally for formalized propositions, the special case defined here should really be called a 'number-theoretical formula'. However, since no other 'formulae' occur in this paper, this modifier may be omitted. Corresponding remarks apply to the concepts of 'term', 'function symbol' etc.

[36] I shall not interpret such a formula as 'valid for arbitrary substitutions of numbers', as is usually customary in formal logic, since free variables are used in a more general sense in mathematical proofs; for example, cf. 4.53. Here, as in the case of bound ∃-variables, we should more appropriately speak of 'indeterminates' instead of 'variables', yet, for better or worse, 'variable' has become the generally accepted expression.

[37] For this we could write a single formula of the form

$$(\ldots ((\mathfrak{A}_1 \,\&\, \mathfrak{A}_2) \,\&\, \ldots) \,\&\, \mathfrak{A}_\mu) \supset \mathfrak{B}.$$

However, this would obscure the original structure of the mathematical proof; after all, in the proof the proposition 'if $\mathfrak{A}_1$ and $\mathfrak{A}_2 \ldots$ and $\mathfrak{A}_\mu$ hold, then $\mathfrak{B}$ holds' never occurred explicitly, the various propositions $\mathfrak{A}_1, \mathfrak{A}_2, \ldots, \mathfrak{A}_\mu$ occurred rather as assumptions and the proposition $\mathfrak{B}$ as a consequence of these assumptions.

[38] In #3, I am using the word 'sequent' in a more general sense than is necessary in the present context. The logical formalism developed here corresponds essentially to the 'NK-calculus' of the 'Investigations'. The 'LK-calculus' is also suitable for the consistency proof. In fact, the proof then becomes even simpler in parts, although less 'natural'.

[39] For 'propositional logic' (&, ∨, ⊃, ¬) cf. H.–A., p. 33; for 'predicate logic' (∀, ∃ included) cf. K. Gödel, note 19. The formalizations of the forms of inference used there can easily be shown to be equivalent with the formalization which I have chosen. (Cf. the proofs of equivalence in section V of my 'Investigations into logical deduction'.)

[40] Cf. W. Ackermann, Zum Hilbertschen Aufbau der reellen Zahlen, Math. Ann. **99** (1928), pp. 118–133.

[41] The 'Peano axioms' for the natural numbers are the result of such efforts (for example, cf. E. Landau, Grundlagen der Analysis, 1930). These axioms also contain complete induction, which I have included in the forms of inference. There is no fundamental

difference between forms of inference and axioms, since logical forms of inference can also be formulated as 'logical axioms' such as $\mathfrak{A} \mathbin{\&} \mathfrak{B} \to \mathfrak{A}$ for the &-elimination, etc.

[42] A proof for the 'redundancy of the ι' can be found in the book: Hilbert-Bernays, Grundlagen der Mathematik, I, Springer, 1st ed. 1934, pp. 422–457.

[43] Cf. the papers by Hilbert and Weyl cited in notes 30 and 31.

[44] Cf. D. Hilbert, Über das Unendliche, cited in note 18.

[45] Cf. A. Heyting, cited in note 15.

[46] K. Gödel, Zur intuitionistischen Arithmetik und Zahlentheorie, Ergebnisse eines math. Koll., Heft 4 (1933), pp. 34–38. – The result mentioned above was also discovered somewhat later by P. Bernays and myself independently of Gödel. Gödel also replaces $\mathfrak{A} \supset \mathfrak{B}$ by $\neg(\mathfrak{A} \mathbin{\&} \neg \mathfrak{B})$, this is unnecessary in my system of rules of inference since I am not using propositional variables.

[47] For example, cf. P. Bachmann, Die Elemente der Zahlentheorie, III, 10.

[48] I could here also use any other false minimal formula.

[49] Footnote added during the correction of the galley proof: Articles 14.1 to 16.11 were inserted in February 1936 in place of an earlier text. Ed.

[50] Readers acquainted with set theory should note: The system of 'ordinal numbers' here used is well-ordered by the $<$-relation, and the numbers with the characteristics 0, 1, 2, 3, 4, 5, etc. correspond, in that order, to the transfinite ordinal numbers $\omega+1; 2^{\omega+1} = \omega+\omega; 2^{\omega+\omega} = \omega \cdot \omega; 2^{\omega \cdot \omega} = \omega^\omega; 2(\omega^\omega) = \omega(\omega^\omega); 2[\omega^{(\omega^\omega)}] = \omega[\omega^{(\omega^\omega)}]$; etc.; the entire system corresponds to the 'first ε-number'. (In order to prove this the reader need merely consider the fact that the transition from the numbers with the characteristic ρ to the numbers with the characteristic $\rho+1$ described above correspond to the definition rule of the power of 2, and then apply the rules of transfinite arithmetic.) The 'theorem of transfinite induction' asserts nothing but the validity of transfinite induction for this segment of the second number class. The disputable aspects of general set theory do not, of course, enter into the consistency proof, since the corresponding concepts and theorems are here developed quite independently in a more elementary form than in set theory, where they are used in a much greater generality. – Similar connections between mathematical proofs or theorems and the theory of well-ordering, especially of the numbers of the second number class, are established in a paper by A. Church, A proof of freedom from contradiction, Proc. Nat. Acad. Sci. U.S.A. **21** (1935), pp. 275–281; and: E. Zermelo, Grundlagen einer allgemeinen Theorie der mathematischen Satzsysteme I, Fund. Math. **25** (1935), pp. 136–146.

[51] Also cf. K. Gödel, Über Vollständigkeit und Widerspruchsfreiheit, Ergebnisse eines math. Koll., Heft 3 (1932), pp. 12–13.

[52] Cf. P. Finsler, Formale Beweise und die Entscheidbarkeit, Math. Zeitschrift **25** (1926), pp. 676–682, and the paper by K. Gödel cited in note 19.

[53] Cf. the paper by P. Finsler cited in note 52.

[54] For example, cf.: L. E. J. Brouwer, Intuitionistische Betrachtungen über den Formalismus, Sitzungsber. d. Preuss. Akad. d. Wiss., phys.-math. Kl. (1928), pp. 48–52; and A. Heyting, Mathematische Grundlagenforschung – Intuitionismus – Beweistheorie, Erg. Math. Grenzgeb. **3** (1935), No. 4.

[55] Cf., e.g.: Hilbert-Ackermann, cited in note 12;
R. Carnap, Abriss der Logistik;
H. Behmann, Mathematik und Logik.

[56] H.-A., pp. 29–31.

[57] H.-A. call it the 'restricted predicate calculus'.

[58] H.-A., p. 65.

[59] Cf. note 55.

[60] Carnap, Abriss der Logistik, p. 21. – I am using the expressions 'Stufe' ((translated as 'type' in keeping with modern English usage, ed.)) and 'Typ' in the same sense as

Carnap and Behmann (in using 'Stufe' I am following Frege).
[61] Cf. B. Russell, Introduction to Mathematical Philosophy.
[62] H.-A., p. 93.
[63] Cf. K. Gödel, Über formal unentscheidbare Sätze der Principia Mathematica und verwandter Systeme I, cited in footnote 19, especially pp. 176-178, and R. Carnap, Logische Syntax der Sprache, Vienna, 1934, chapter III.
[64] Carnap, Abriss der Logistik, article 13.
[65] Cf. the paper cited in note 63, p. 176.
[66] Cf. p. 11 (H.-A.).
[67] Mathematische Annalen **112**.
[68] As far as number theory is concerned, a detailed discussion of these points can be found in section III of #4 of the present volume – Cf. also § 3 of #7.
[69] A. Heyting, Mathematische Grundlagenforschung – Intuitionismus – Beweistheorie. Erg. Math. Grenzgeb. **3** (1935), No. 4.
[70] G. Gentzen, #4 of the present volume. – It should be noted that in contrast with the rigour of proof found in the rest of the paper, section IV turned out to be rather sketchy due to a lack of space and time. A new version of the proof, together with a detailed exposition of the basic ideas involved, is presented in #8 of the present volume.
[71] K. Gödel, cited in note 19.
[72] Cf. note 70. I had to keep the presentation deliberately short, and believe that a detailed discussion of this point, which constitutes the crucial idea of the whole argument, would enhance the clarity of the exposition of the proof; I am hoping to be able, at some point, to publish such a discussion covering the case of the consistency proof for analysis at the same time.
[73] A. Church, An unsolvable problem of elementary number theory. Amer. J. Math. **58** (1936), pp. 345-363.
A. Church, A note on the Entscheidungsproblem, J. Symb. Logic **1** (1936), pp. 40-41.
A. Church, Correction to a note on the Entscheidungsproblem, J. Symb. Logic **1** (1936), pp. 101-102.
Cf. also: A. M. Turing, On computable numbers, with an application to the Entscheidungsproblem, Proc. London Math. Soc. 2, **42** (1937), pp. 230-265.
[74] W. Ackermann, Die Widerspruchsfreiheit der allgemeinen Mengenlehre, Math. Ann. **114** (1937), pp. 305-315.
[75] Th. Skolem, Einige Bemerkungen zur axiomatischen Begründung der Mengenlehre, Proc. 5th scand. math. congr. (1922), pp. 217-232.
Th. Skolem, Über einige Grundlagenfragen der Mathematik, Skr. Norske Vid.-Akad. Oslo, I, mat.-nat. Kl., **4** (1929).
[76] Th. Skolem, Über die Unmöglichkeit einer vollständigen Charakterisierung der Zahlenreihe mittels eines endlichen Axiomensystems, Norsk. mat. forenings skr., Ser. II, articles 1-22 (1933), pp. 73-82.
Th. Skolem, Über die Nicht-charakterisierbarkeit der Zahlenreihe mittels endlich oder abzählbar unendlich vieler Aussagen mit ausschliesslich Zahlenvariablen, Fund. Math. **23** (1934), pp. 150-161.
[77] Cf. H. Poincaré, Wissenschaft und Hypothese, German Edition, note by F. Lindemann, p. 246 of the first and second editions.
[78] Cf. H. Weyl, cited in note 31.
[79] Cf. note 73.
[80] L. E. J. Brouwer, Beweis, dass jede volle Funktion gleichmässig stetig ist, Proc. Akad. Wet. Amsterdam **27** (1924), pp. 189-193 and pp. 644-646.
[81] Cf. D. Hilbert, cited in note 18.
[82] Cf. J. Hjelmslev, Die natürliche Geometrie, Hamb. Math. Einzelschriften **1** (1923); also: Abhandl. Math. Sem. Hamburg **2** (1923), pp. 1-36.

[83] In the paper cited in note 69, p. 68.
[84] H. Weyl, Die Stufen des Unendlichen, Jena 1931, p. 17.
[85] Also cf. the Preface to the second edition of part I of van der Waerden's 'Moderne Algebra'.
[86] G. Gentzen, #3. In Gentzen #4 of the present volume, a formalism was introduced in section IV that differs somewhat from the formalism developed in section II. It was specifically designed for the proof in question and has no general significance.
[87] It should be mentioned, incidentally, that all basic logical sequents are also derivable in the new system and I therefore do not really have to admit such sequents any longer. Their retention has of course certain formal advantages.
[88] The proof of equivalence is to a large extent already given by the proof for the equivalence of the calculi *NK* and *LK* carried out in section V of my dissertation.
[89] In the earlier paper I have proved more generally the 'reducibility' of the endsequent of arbitrary derivations. Here I shall confine myself to consistency; this makes certain simplifications possible.
[90] The same reasoning, incidentally, underlies the proof of the 'Hauptsatz' of my dissertation.
[91] Cf. G. Hessenberg, Grundbegriffe der Mengenlehre, Sonderdruck a. d. Abh. d. Friesschen Schule, N.F. Vol. I, Book 4, pp. 479–706, Göttingen 1906.
[92] K. Gödel, cited in note 19.
[93] Cf. especially #4 article 16.2 of this volume.
[94] Cf. Hilbert-Bernays, Grundlagen der Mathematik, Vol. II, §§ 5 and 3c.
[95] Such calculation procedures for the above functions, even for general powers, can easily be taken from §§ 78 and 79 of Hessenburg's Grundbegriffe der Mengenlehre cited in note 91.
[96] Cf. #4 of the present volume, article 17.2.
[97] This has been done, for example, for the numbers below ε_0 in Hilbert-Bernays, Grundlagen der Mathematik, Vol. II, §§ 5 and 3c.
[98] In the paper cited in note 93.

GLOSSARY

Abbild — Counterpart
Abgrenzung — Delimitation
All-Zeichen — $\forall$-symbol
Anfangsfall der transfiniten Induktion — Restricted transfinite induction
Angebbarkeit — Statability
Annahme — Assumption
Anschaulich — Intuitive
Ansich — Actualist
Antezedens — Antecedent
Auffassung — Interpretation, point of view
Ausrechnung — Calculation
Aussage — Proposition
Äusserst — Terminal

Bedenklich — Disputable
Bedeutung — Meaning, significance
Begriff — Concept
Begriffsbildung — Derived concept
Beilegen (Sinn) — Ascribe (a sense)
Berechenbarkeit — Calculability
Beseitigung — Elimination
Bestimmt — Definite
Beweisfaden — Path
Beweismethode — Method of proof
Beweismittel — Technique of proof
Bezeichnung — Designation
Bund — Cluster

Durchlaufung — Running through

Eigenvariable — Eigenvariable
Einführung — Introduction
Endform — Reduced form
Endsequenz — Endsequent
Endstück — Ending
Endzahl — Terminal number
Ergebnis — Conclusion
Erreichbarkeit — Accessibility
Es-gibt-Zeichen — $\exists$-symbol

GLOSSARY

Faden	Path
Finit	Finitist
Folgerung	Consequence
Folgt-Zeichen	⊃-symbol
Formelbund	Cluster of formulae
Gipfelpunkt	Extremum
Gleichheits-Grundsequenz	Basic equality sequent
Grad	Degree
Grenzziehung	Delimitation
Grundsequenz	Basic sequent
Halbnetzsatz	Semi-net sentence
Hauptformel	Principal formula
Hauptprämisse	Major premiss
Herleitung	Derivation
Hilfsmittel	Technique
Hinterformel	Succedent formula
Höhe	Level
Inhaltlich	Informal
Kettenschluss	Chain-rule inference
Klammerzeichen	Quantifier
Logische Grundsequenz	Basic logical sequent
Logische-Zeichen-Schlussfigur	Operational inference figure
Mathematische Grundsequenz	Basic mathematical sequent
Misch-	Mix
Mischung	Mix
Mittelsequenz	Midsequent
Nebenformel	Side formula
Netz	Net
Nicht-Zeichen	¬-symbol
Numerus	Characteristic
Obersatz	Upper sentence
Obersequenz	Upper sequent
Oberste Sequenz	Uppermost sequent
Oder-Zeichen	∨-symbol
Operationsregel	Operational rule
Rang, Rangzahl	Rank
Reine Zahlentheorie	Elementary number theory
Richtig	True, valid, correct
Satz	Sentence, theorem, proposition, law
Schliessen	Deduction
Schluss	Inference
Schlussstrich	Line of inference

Schnitt	Cut
Schnittsatz	Conclusion of the cut
Sequenz	Sequent
Sinn	Sense
Sinnlos	Without sense
Sinnvoll	Significant
Spiegelbildlich-Symmetrisch	Dual
Stammbaumförmig	In tree form
Strukturänderung	Structural transformation
Struktur-Schlussfigur	Structural inference figure
Stufe	Type
Stufen des Unendlichen	Levels of infinity
Sukzedens	Succedent
Teilformel	Subformula
Termeinsetzung	Substitution of terms
Übereinandergeschachtelt	Nested
Unbedenklich	Indisputable
Unbestimmt	Indefinite
Untersatz	Lower sentence
Untersequenz	Lower sequent
Und-Zeichen	&-symbol
Verbunden	Clustered
Verdünnung	Thinning
Verknüpfungs-Schlussfigur	Operational inference figure
Verknüpfungszeichen	Connective
Vertauschung	Interchange
Vorderformel	Antecedent formula
Vorschrift	Rule, instruction
Wahlfreiheit	Option, choice
Widerlegung	Reductio ad absurdum
Widerspruchsfreiheit	Consistency
Widerspruchsherleitung	Contradictive derivation
Wirkungsbereich	Scope
Zahlterm	Numerical term
Zeichen	Symbol
Zusammenfassung	Fusion
Zusammenziehung	Contraction

INDEX OF SYMBOLS

$\forall$	4, 24, 70, 83, 140, 148, 161, 163, 169, 216, 313	$\vee\text{-}I$	77, 152, 293
		$\vee\text{-}E$	77, 78, 152, 181, 293 (cf. also: distinction of cases)
&	54, 70, 83, 140, 148, 153, 163, 164, 169, 181, 253, 313		
		$\forall\text{-}I$	77, 78, 152, 183, 188, 190, 203
$\vee$	54, 61, 70, 78, 83, 140, 148, 149, 152, 163, 165, 169, 171, 253, 254, 313		
		$\forall\text{-}E$	77, 152, 203
		$\exists\text{-}I$	77, 153
$\supset$	54, 70, 83, 140, 148, 149, 153, 167, 169, 171, 206, 253, 290, 313	$\exists\text{-}E$	77, 78, 153
		$\supset\text{-}I$	77, 78, 153
		$\supset\text{-}E$	77, 153
$\neg$	54, 70, 77, 83, 140, 148, 149, 153, 168, 181, 206, 253, 313	$\neg\text{-}I$	77, 79, 183, 188, 190
		$\neg\text{-}E$	77, 78
		&-IS	83, 84
$(\mathfrak{x})$	54	&-IA	83, 84
E	54, 61	$\vee$-IS	84
$\mathfrak{S}^{\mathfrak{b}}(\mathfrak{S}^{x_1 \ldots x_\nu}_{\mathfrak{y}_1 \ldots \mathfrak{y}_\nu})$	58	$\vee$-IA	84
$\vee$	70, 218	$\forall$-IS	84
$\wedge$	70, 72, 79, 218	$\exists$-IA	84
$\supset\subset$	70, 313	Subst $\mathfrak{A} \binom{\mathfrak{b}}{\mathfrak{c}}$	214
$\exists$	70, 83, 140, 148, 153, 159, 161, 163, 164, 169, 170, 171, 216, 253, 313	ε_0	11, 12, 16, 17, 233, 284, 285, 287 seq.
$\rightarrow$	70, 71, 201	fu_1	289
actualist meaning of –	206	fu_2	289
finitist meaning of –	206	ω	230 seq., 278 seq.
&-I	77, 152, 203, 293	ω_α	288
&-E	77, 152, 181, 203, 293	#	279

INDEX OF AUTHORS

Ackermann, W. 12, 14–16, 24, 27, 28, 53, 54, 56, 57, 61, 65, 68, 74, 86, 103, 104, 114, 117, 156, 214, 216, 236, 240, 313–316
Anderson, J. 25

Bachmann, P. 14, 27, 315
Bar-Hillel, Y. 28
Behmann, H. 315, 316
Benacerraf, P. 28
Bernays, P. vi, 1, 2, 6, 8–10, 20, 24, 26–28, 201, 313, 315, 317
Beth, E. W. 2, 7, 24–26, 28
Black, M. 10, 26
Bolzano, B. 3, 24
Brouwer, L. E. J. 3, 10, 11, 18, 19, 21, 135, 224, 227, 234, 235, 245–247, 249, 314–316

Cantor, G. 3, 10–12, 16, 26, 230, 242
Carnap, R. 315, 316
Church, A. 24, 27, 239, 240, 245, 315, 316
Craig, W. 7, 25
Curry, H. B. 6, 18, 21, 24, 25, 28

Davis, M. 24
Dedekind, R. 224, 309
Denton, J. 6
Dreben, B. 6

Euclid 5, 144, 145, 148–150, 155, 170, 248

Fermat, P. 161, 162, 166, 176, 239
Finsler, P. 315
Fitch, F. B. 12, 13, 27
Fraenkel, A. 28, 240, 314
Frege, G. 4, 24, 68

Gauss, C. F. 225

Gentzen, G. vi, vii, viii, 1–28, 201, 314, 316, 317
Glivenko, V. x, 116, 314
Gödel, K. xi, 3, 4, 7, 8, 11, 15–18, 24, 25, 67, 138, 169, 193, 197, 199, 212, 213, 215, 229, 232, 233, 236, 238–240, 242, 284, 287, 308, 313–317
Goldbach, C. 140, 161
Goodstein, R. L. 26

Herbrand, J. 6–8, 13, 25, 56, 236, 313, 314
Hertz, P. ix, 1, 2, 29, 31, 32, 40, 312
Hessenberg, G. 317
Heyting, A. 3, 56, 58, 61, 65, 68, 69, 75, 81, 106, 117, 236, 249, 313, 315, 316
Hilbert, D. vii, x, 3, 4, 6, 8, 9, 13, 14, 16, 18–22, 24, 26–28, 53, 54, 56, 57, 61, 65, 67, 68, 69, 73–75, 81, 86, 103, 104, 114–117, 135, 165, 214, 216, 222.
Hilbert, D. and Ackermann, W. 14, 24, 27, 53, 54, 56, 57, 61, 65, 68, 74, 86, 103, 104, 114, 117, 214, 313–315
Hilbert, D. and Bernays, P. 6, 9, 26, 27, 315, 317
Hjelmslev, J. 18, 28, 316

Jaśkowski, S. 4, 24
Johnstone, H. 25

Kalmár, L. 27
Kanger, S. 7, 25
Kant, I. 19, 22, 28
Kemp Smith, N. 28
Ketonen, O. 7, 8, 25, 26
Kleene, S. C. 7, 9, 25–27
Kneale, W. 5, 25
Kneebone, G. T. 24
Kneser, H. vii
Kolmogorov, A. N. 3

INDEX OF AUTHORS

Kronecker, L. 224, 244

Lambek, J. 2, 24
Landau, E. 314
Langford, C. H. 25
Leblanc, H. 5, 25
Legendre, A.-M. 156
Lewis, C. I. 6, 25
Lindemann, F. 316
Lorenzen, P. 12, 13, 19, 25, 27, 28
Löwenheim, L. 7
Lukasiewicz, J. 4

MacCall, S. 24
Mostowski, A. 28
Matsumoto, K. 7, 25

Ohnishi, M. 7, 25

Peano, G. 53, 136, 314
Poincaré, H. 3, 224, 316
Prawitz, D. 5, 24, 25
Putnam, H. 28

Quine, W. V. O. 20, 28

Rasiowa, H. 7, 25
Rosser, J. B. 18, 19, 28
Russell, B. 3, 4, 12, 13, 24, 68, 69, 75, 133, 135, 162, 214, 313, 316
Russell, B. and Whitehead, A. N. 4, 12, 13, 24, 214, 235

Schmidt, H. A. 14, 27
Scholz, H. vi, viii, 24
Schröter, K. 2, 24
Schütte, K. 12, 14, 15, 27, 28
Sikorski, R. 7, 25
Skolem, Th. xi, 7, 15, 238, 241–243, 247, 316

Takeuti, G. 12–15, 27, 28
Tarski, A. 2, 3, 9, 14, 24, 27
Thomason, R. 25
Turing, A. M. 245, 316
Tymieniecka, A. 24

Vaihinger, H. 19, 20, 28
van der Waerden, B. L. 317
van Heijenoort, J. 24–28
von Neumann, J. 236, 313, 314

Waisman, F. 20, 28
Wang, H. 13, 27
Weierstrass, K. 17
Weyl, H. vii, 3, 21, 24, 165, 224, 234, 235, 244, 245, 249, 314–317
Whitehead, A. N. 12, 13, 24, 214, 235
Woodger, J. H. 27

Zeller, Rev. 170
Zermelo, E. 240, 315

INDEX OF SUBJECTS

𝔄, 42
Absolute consistency 9, 14, 138, 228, 237
Accessible ordinal number 192, 195
Ackermann function 156
Actualist (cf. also: *an sich*) 19
  – concept of a function 245
  – concept of a set 134
  – interpretation and intuitionism 235
  – interpretation and Skolem's theorem 241
  – interpretation and the continuum 242
  – interpretation and the stability of a reduction rule 173
  – interpretation of classical analysis 243 seq.
  – interpretation of Fermat's last theorem 162
  – interpretation of infinity 18, 224 seq., 235, 247
  – interpretation of transfinite propositions 162
  – mathematics and constructivism 248–250
  – mathematics and existence 248
  consistency of the – interpretation 228
  intuitionist view of – propositions 19, 21, 229, 247, 249
  sense of - propositions 21, 195, 201, 226, 229, 247, 250
Adjunction (cf. also: thinning; assumption) 151, 152
Algebra 2, 200, 223, 236
Als ob (cf. also: *as if*) 19,28
Analysis 3, 65, 134, 137, 143, 197, 199, 200, 223 seq., 235, 242 seq., 250, 296, 309, 316
  classical – 18, 246, 247
  consistency of – 12, 15, 16, 136, 227 seq., 232, 236

constructivist (intuitionist) – 18, 244 seq.
  ramified – 12, 14, 15, 27
An sich (cf. also: actualist) 18, 19, 28, 134
Antecedent 2, 30, 71
  – formula 151, 254
  empty – 72
  interchange in the – 84, 151
  omission of – formulae 151
Antinomies of set theory 3, 23, 132–136, 158, 162, 224, 227, 228, 234–237
  – and indirect existence proofs 235
  – and infinity 234
  – and logicism 237
  – and the natural numbers 133, 136
Applications of the *Hauptsatz* 103 seq.
Applied mathematics 249
Argument 71
Arithmetic (cf. also: elementary number theory) 3, 8, 53, 55, 110, 136
  axiom formulae of – 56, 57, 111, 112, 114, 115
  classical – 3, 4, 53, 56
  consistency of – 8, 66, 67, 69, 112, 136, 197
  consistency of – and Gödel's theorem 67
  formalized – 55
  informal – 55
  intuitionist – 4, 53, 55
As if (cf. also: *als ob*) 18–21, 23, 28, 201, 250
Assumption (cf. also: adjunction) 150, 164, 165, 254
  – calculus 5
  – formula, 74–76, 82
  –s in proofs 150
  transfinite –s 165
Auxiliary
  – calculus 128 seq.
  – concept of 'satisfaction' 33
  – symbol 70, 71, 217

INDEX OF SUBJECTS

Axiom 55, 68, 69, 136, 144, 151, 194
- formula 55, 56, 111, 112, 114, 309
- of choice 214, 217, 221
- of comprehension 216, 220
- of infinity 12, 13, 214, 222, 240, 241
- of reducibility 13
- of set formation 216
-s of elementary number theory 56, 57, 144, 151, 155–157, 164
-s of geometry 68
- system 2, 3, 39, 238
- system for analysis and Skolem's theorem 242
- system for set theory 240
consistency and the – of infinity 222
independent – system 2, 39
logical –s 315
Peano –s 136, 314
Axiomatic
- method 3, 17
weakness of the – method 240
Axiomatic set theory 18, 235

Basic
- equality sequent 290, 300
- formula 74, 75, 81, 116, 216
- logical sequent 151, 154, 177, 179, 257, 290
- mathematical sequent 151, 177, 257, 260, 290
- sequent 13, 83, 123, 151, 257, 290
reduction rule for – sequents 177
Bound
- predicate variable 297
- variable 55, 141, 144, 215, 288
replacement of – variables 57, 113, 152
Brackets 54, 71, 141, 215

Calculability 245
Calculation procedure 158, 159, 161, 175, 239, 244, 245
Calculus 2, 5, 7, 68, 74, 75, 81, 116, 128 seq., 252
Categorical algebra 2
Chain rule 180 seq.
  purpose of the – 181
Change of levels 281
Characteristic
- of an ordinal number 187
- significance of the – for the reduction step 191
Choice 175–177, 184, 196–198
- set 217

axiom of – 214, 217, 222
free – sequence 245, 246
Circulus vitiosus 134
CJ 293
CJ-inference figure 256, 262, 265, 290
  degree of a – 257
CJ-reduction 262, 264
Classical
- analysis 18, 246, 247
- arithmetic 3, 4, 53, 56
- logic 4–6, 53, 66, 68, 103
consistency of – predicate logic 103
consistency of – arithmetic and Gödel's theorem 67
constructivist objections to – analysis 226, 227
equivalence of the – calculi 128 seq.
the – calculus *NK* 75, 81
Classification
- of forms of inference 144, 148 seq.
- of mathematics 223
Closed sentence system 29, 38, 39
Cluster
- associated with a formula 267
- of formulae 267, 269, 302
cut associated with a – 267, 302
left side of a – 267
right side of a – 267
uppermost formula of a – 267
Clustered formulae 266, 302
Combinatorial topology 223
- and the consistency proof 200
Complete induction 12, 68, 145, 153, 154, 180, 183, 190, 194, 197, 205, 207, 231, 256, 309, 314
- and completeness 154
- and consistency proofs 8, 139, 194, 197, 198, 212, 262
- and constructivism 225
- and the complexity of proofs 232, 284, 309, 311
- and the cut 262
- and the correlation of ordinal numbers 188
- and the finitist interpretation 166
alternative forms of – 155
arithmetic without – 69, 115
normal form of – 145
Completed infinity 225, 230, 245
- and mathematical existence 248
Completeness (cf. also: incompleteness)
- and complete induction 154

– and Gödel's theorem 240
– of axiom systems 238
– of formalisms 7, 143, 198, 233, 307
– of the rules of inference 7, 154
Complex
– functions 224
– of elements 30
Complexity
– extremum 262
– of a derivation 259, 261, 275
– of a proof and complete induction 232, 284, 309, 311
– of a proof and the ordinal numbers 11, 12, 186, 261
maximal – 262
measure of – 11, 12, 186, 270, 311
Conclusion
– of a 'chain-rule' inference 181
– of a cut 31
– of an inference 30, 148
– of a syllogism 32
Conjunction (cf. also: &) 70
Connective (cf. also: logical –; logical symbol)
terminal – 254
Consequence (cf. also: logical consequence) 2, 3, 33
Consistency 4, 53, 217, 222, 227, 236, 260, 261
– and derived concepts 157, 193
– and constructive forms of inference 261
– and the empty endsequent 261
– and Gödel's theorem 197, 229, 232, 233, 236, 238–240, 284, 287
– and non-denumerability 247
– and the axiom of infinity 222
– and the statability of a reduction rule 11, 177, 211, 213
– of analysis 12, 15, 16, 136, 227 seq., 232, 236
– of arithmetic 8, 66, 67, 69, 112, 136, 197
– of elementary number theory 4, 132, 136, 228, 287
– of geometries 136
– of intuitionist arithmetic 67
– of predicate logic 14, 103, 214
– of propositional logic 214
– of ramified analysis 14
– of set theory 136, 232, 240, 241
– of the actualist interpretation 228
– of the ramified theory of types 13
– of the simple theory of types 13, 14, 214
– proof 8, 229, 239, 250

– proof and algebra 200
– proof and complete induction 8, 139, 194, 197, 198, 212, 262
– proof and existential propositions 201
– proof and Gödel's theorem (cf. Gödel's theorem)
– proof and Hilbert's programme (cf. Hilbert's programme)
– proof and indisputable forms of inference 138, 171, 193, 228
– proof and the natural calculus 252
– proof and the statability of a reduction rule 11, 177, 211, 213
– proof and transfinite induction 8, 231, 232, 261, 286, 297
– proof for analysis 227, 228, 236, 246, 247, 316
– proof for elementary number theory 21, 27, 252
– proofs and incompleteness 17, 240
absolute – 9, 14, 138, 228, 237
difficulties involved in – proofs 229 seq.
intuitionist objections to – theory 200, 201
sense of the results established in – proofs 200
significance of denumerability for – proofs 247
transferability of the – proof 198, 199
value of – proofs 200, 201
Constant symbols 70, 111
Construction rule 160, 194, 211, 244
Constructive
– analysis and natural geometry 18
– concept of a set 134
–, finitist, and intuitionist methods 18, 227
– forms of inference in consistency proofs 261
Constructivism (cf. also: idealism)
– and complete induction 225
actualist mathematics and – 248, 250
fundamental principle of – 225, 235, 247
Constructivist 227, 239, 250
– and indirect existence proofs 250
– concept of a number 244
– interpretation and Skolem's theorem 241
– interpretation of analysis 243
– interpretation of infinity 224 seq., 237, 286
– objections to classical analysis 226, 227
– point of view and indisputable forms of inference 228

– principle of set formation 225
– proof of the theorem of transfinite induction 253, 261, 285, 286
– techniques and Gödel's theorem 239
– techniques in proof theory 228, 237, 239
need for a – consistency proof for classical analysis 246
significance of the – point of view 250
Continuum 20, 243 seq., 250, 251
actualist interpretation and the – 242
Contraction 84, 256
– in the antecedent (cf. also: omission) 84
– in the succedent 84, 129
Contradiction (cf. also: empty sequent)
– in mathematics 114, 133, 138, 241, 255
derivability of a – 112
law of – 79, 86, 154
proof by – (cf. reductio (ad absurdum))
Contradictive derivation 261
reduction step on a – 264
Correctness of the rules of inference (cf. also: informal –) 159
Correlation
– of ordinal numbers with derivations 11, 16, 187, 188, 261, 279
– of values 305
Counterexamples 7
Critical
– derivation 298, 303
– reduction step 300, 303, 304
property of a – derivation 304
Cut 2, 5, 15, 31, 32, 84, 256, 312
– and complete induction 262
– and the syllogism 32
– and the *Hauptsatz* 5
– associated with a cluster 267, 302
– element 31
– formula 256
degree of a – 256
eliminability of the – 15, 28

Decidable
– formula in a proof 269
– functions and predicates 160, 174, 194, 199, 211, 288
– propositions 160
Decidability of the predicate calculus and Fermat's last theorem 239
Decision
– problem 6, 66, 69, 238, 239
– procedure 7, 69, 160, 161 seq.
– rule 161, 198

Dedekind cut 224
Deduction (cf. also: natural –) theorem 311
Definability and the Herbrand-Gentzen theorem 7
Definite
– function 54, 141
– number 141
– object 54, 111
– predicate 54, 141
– proposition 53, 65, 142, 158
Definition 14, 156, 160
– of the objects of proof theory 194, 211
– table 158
– tables and decision rules 161
Degree
– of a *CJ*-inference figure 257
– of a cut 256
– of a derivation 89
– of a formula 15, 71, 254
– of an induction proposition 311
– of a sentence 43
level and – 276
Denumerability 29, 200, 223, 229, 241–243, 246, 247
– and algebra 223
– and Skolem's theorem 241
significance of – for consistency proofs 247
Dependent formulae and propositions 76, 150, 152
Derivable formula in arithmetic 55, 112
Derivability and logical consequence 2, 3, 24, 33
Derivation (cf. also: proof) 55, 72, 74, 151, 179, 216, 257, 258, 287, 309
– in the ordinary sense 291
– in tree form 73, 258
comparison of different concepts of a – 181, 260
complete induction in a – 309
complexity of a – 259, 275
contradictive – 261
critical – 298, 303
degree of a – 89
equivalence of –s 116
LJ- – 83
LK- – 83
modified concept of a – 179, 287
NJ- – 75
NK- – 81
order of a – 107, 108
ordinal number of a – 11, 187–189, 279

rank of a – 89
reduction step on a – 179 seq.
simplification of a – 11, 186, 191, 262, 263, 274, 275, 284
TJ- – 287, 291, 293
value of a – 298
Derived concept 57, 139, 144, 155–157, 160, 193
Description operator (cf. 'the ... such that')
D-inference figure 73
D-formula 73
Disjunction (cf. also: ∨) 70, 77
Disputable forms of inference 158, 170, 197, 261
Distinction of cases (cf. also: ∨-E) 78, 153, 254
Dots 54
Double negation
 elimination of the – 168–170
 law of – 53, 55, 57
D-S-formula 73
Dual(ity) 86, 137, 138, 259

E 77
E 54
$\mathscr{E}$ 289
Eigenvariable 77, 84, 255
Element
 cut – 31
 ideal – 18, 247, 250
 net – 41
 sentence – 29, 30
Elementary
 – formula (cf. also: minimal formula; prime formula) 54, 71
 – inference 145
 – number theory 3, 8, 68, 132, 136–138, 154, 197, 198, 223 seq., 239, 250, 287, 292, 309
 – number theory and bound predicate variables 297
 – number theory and transfinite induction 292, 307
 axioms of – number theory 155–157
 consistency of – number theory 4, 8, 132, 136, 228, 240, 287
 derived concepts in – number theory 155–157
 formalization of – number theory 138, 139, 252
 forms of inference in – number theory 148, 149

functions in – number theory 143, 156
incompleteness of – number theory 307, 308
indisputable forms of inference of – number theory 158
predicates in – number theory 143, 156
Eliminability
 – of negation 66, 67, 81
 – of the cut 15
Elimination
 – of double negation 81, 153, 168–170, 181, 204, 259
 – of logical connectives 5, 80, 82, 148, 150, 168, 258, 259, 262, 263, 269, 293
 – theorem (cf. *Hauptsatz*)
Empty
 – antecedent 72
 – endsequent and consistency 261
 – sequent 16, 72, 103, 255, 261
 – succedent 72, 82, 112
 nonderivability of the – sequent 16, 103
Endformula 55, 73, 151
Ending 264, 300
Endsequent 6, 257
 – after the reduction step 181
 – and the subformula property 6
 empty – 261
 ordinal number of the – 279
 reducibility of the – 317
Equality 110, 111, 216, 309
 axiom formulae for – 309
 basic – sequent 290, 300
Equivalence 70
 – between formulae and sequents 115
 – of classical calculi 128 seq.
 – of derivations 116, 260
 – of formulae 115
 – of intuitionist calculi 116 seq.
 – of logical calculi 69, 115, 116, 128, 131
 – of net elements 49
 – of sequents 115
Examples
 arithmetic axiom formulae 56, 57, 111, 112, 114, 115
 axioms of elementary number theory 157
 basic logical sequents 180
 basic mathematical sequent 260
 basic sequents 86, 257
 circulus vitiosus in analysis 134
 contradiction 262
 degrees of formulae 254

derivations 258, 260
forms of inference 148
formulae 54, 141, 215, 253
inference figure 257
LJ-derivation 85
LK-derivation 85
mix 89
natural deductions 74, 75
natural sum 296, 297
NJ-derivations 79, 80
ordinal numbers 187
prime formula 253
proof 144
reduction rule 176
sentence system without an independent axiom system 40
sequent 255
subformulae 71
subformula property 88
terms 54, 141
theorem in proof theory 137
true formulae 74
true numerical propositions 114
Existence 246, 248–250
Existential
– proposition 201, 226
– quantifier (cf. also: E-symbol; ∃-symbol) 4, 54, 70, 77
Excluded middle (cf. law of the –)
Expression 53, 54, 70, 211
Extended *Hauptsatz* (cf. sharpened *Hauptsatz*)
Extension of a formalism 17, 114
Extremum
  complexity – 262
  relative – 263

False
  – formula 79, 114
  – sequent 254, 269
Falsity (cf. truth)
Fan theorem 8, 26
Fermat's last theorem 161
– and provability 166
– and the decidability of the predicate calculus 239
– and the existence of a decision rule 161
– and the reduction rule 176
actualist interpretation of – 162
finitist interpretation of – 166
Figure 70, 72
  inference – 72, 73, 82, 83, 249, 289

inference – schema 77, 83, 84, 117, 125, 129, 255, 256, 290, 291
proof – 66, 72
Finite
– mathematics 158 seq.
– sentence system 39
– sets of natural numbers 142, 143
Finiteness
– of the reduction procedure 186, 191, 193
proof of – 191 seq., 195, 197
Finitist 10, 18, 169, 227, 250
– forms of inference 135, 139
– interpretation of $\forall$, &, $\exists$, $\vee$ 163–165, 169
– interpretation of $\supset$, $\neg$ 167–170
– interpretation and the statability of a reduction rule 173, 201, 206
– interpretation of complete induction 166
– interpretation of Fermat's last theorem 166
– interpretation of number-theoretical axioms 164
– nature of transfinite ordinal numbers 277, 278, 285
– sense of actualist propositions 201
– sense of transfinite propositions 162
– techniques of proof 171, 193, 194, 210, 212
Hilbert's – point of view 4, 9, 10, 22, 23, 67, 222, 237
Follows 212
Formal
– arithmetic in *LK* 110 seq.
– system 68, 233
Formalization
– of elementary number theory 138, 139, 252
– of logical deductions 68
– of proofs 137, 138, 228, 258
– of propositions 69, 70, 139, 140
– of techniques of proof 139, 143
– of the forms of inference in elementary number theory 149 seq.
– of transfinite induction 291
Formalized arithmetic 55
Forms of inference 30, 68, 75, 138, 139, 144, 228
– and axiom systems 238
– and decision procedures 161 seq.
– and restricted transfinite induction 17, 308

330 INDEX OF SUBJECTS

– for sequents 150, 151, 255, 256
classification of the – 144, 148 seq.
delimitation of the – in proof theory 138, 228
disputable – 158, 170, 197, 261
finitist – 135
formalization of the – 149 seq.
indisputable – 158, 171, 193, 197, 200, 228, 229
informal completeness of the - 34
informal correctness of the – 33
intuitionist delimitation of the – 135, 163
Formula 54, 70, 140, 141, 214, 253, 289, 314
 adequacy of the definition of a – 142
 antecedent – 151, 254
 assumption – 74–76
 basic – 74, 75, 81, 116, 216
 cut – 256
 decidable – 269
 degree of a – 15, 71, 254
 D- – 73
 dependent – 76
 derivable – 55, 112
 D-S- – 73
 elementary – 54, 71
 end– 73
 false – 79, 114
 initial – 73–75
 lower – 72
 minimal – 142, 159
 mix – 88, 207
 prime – 253, 289, 304
 principal – 87, 256, 304
 purely logical – 66, 70
 reducible – 262
 S- – 72
 side – 87
 succedent – 151, 254
 terminal symbol (connective) of a – 71, 254
 transfinite – 142
 true – 55, 56, 74, 114
 upper – 72
 uppermost – of a cluster 262
Formulae
 – and propositions 140, 141
 – and sequents 72
 axiom – of arithmetic 56, 57, 111, 112, 114, 115
 axiom – of intuitionist logic 56
 clustered – 266, 302
 cluster of – 267, 269, 302

equivalence of – 73, 115
formally identical S- – 72, 73
Four-colour problem 223
Fractions 142
Free
 – choice sequence 245, 246
 – variable 55, 141, 144, 215, 288
 replacement of a – variable 57, 113
Function 69, 110, 156, 158, 160, 174, 194, 198, 199, 211, 245, 246, 260, 287, 288
 – symbol 140, 141, 288
 –s in elementary number theory 143, 156
 –s in *New* 253
Fundamental
 – conjecture 15
 – principle of constructivism 225, 227, 235, 247
 – principle of constructivism and classical analysis 247, 250
 – principle of intuitionism and the actualist interpretation 235
 – principle of proof theory 162
 – theorem of algebra 236
Gentzen's – principle 162

G 293
General set theory (cf. also Gentzen's different use of this concept on p. 240) 3, 224 seq.
Geometry
 – and infinity 224
 axioms of – 68
 natural – 18, 248
 pure (Euclidean) – 18, 248, 250
GLC 13, 14
Gödel numbers (correlation of natural numbers) 197
Gödel's equivalence theorem 169
Gödel's theorem 4, 8, 11, 16, 17, 25, 67, 138, 193, 197, 238 seq., 242, 287
 – and consistency proofs 197, 229, 232, 233, 236, 238–240, 284, 287
 – and constructivist techniques of proof 239
 – and incompleteness 233, 238, 240, 308
 – and restricted transfinite induction 16, 232, 284
 – and Skolem's theorem 242
 – and the consistency of arithmetic 67
 – and the reduction rule 213
 – on completeness 240
 – on decidability 239

Goldbach's conjecture 140, 161
– and decidability 161, 168

Hauptsatz (cf. also: sharpened *Hauptsatz*)
 5–7, 15, 68, 69, 83, 85, 87, 88, 317
– and intuitionist propositional logic 103
– and the cut 5
– and the decision rule 69
 applications of the – 103 seq.
 proof of the – 88 seq., 101 seq.
 proof of the – for *LJ* 101 seq.
 proof of the – for *LK* 88 seq.
Herbrand theorem 6, 313
Herbrand-Gentzen theorem (cf. also: sharpened *Hauptsatz*) 7, 313
Hilbert's programme 3, 4, 8, 18, 135, 214, 222, 227, 236, 238
– and philosophy 237

I 77
IA 83
Ideal
– element 18, 247, 250
– point 247, 248, 250
– proposition 247
 intuitionist concepts as – elements 250
 real numbers as – elements 250
 sense of – propositions 247, 248
Idealization of reality 23, 248
Idealism (cf. also: constructivism) 19
ℑ 95
Immediate inference (cf. also: thinning) 31
Implication (cf. also: ⊃) 10, 70, 77
Impredicative definition 14
Incompleteness (cf. also: completeness)
– and calculation procedures 245
– and consistency proofs 17, 198 seq.
– and Gödel's theorem 233, 238, 240, 308
– and restricted transfinite induction 16, 17
– and the axiomatic method 17
– and the reduction procedure 17
– of elementary number theory 307, 308
– of formal systems 7, 143, 198, 233, 307
Indefinite
– number (cf. term)
– proposition 142
Independent
– axiom system 2, 39
– sentence system 39
Indeterminate 314
Indirect
– existence proofs 226, 235

– existence proofs and the antinomies of set theory 235
– proof (cf. also: reductio (ad absurdum)) 169, 250
 sense of –ly proved existential propositions 201, 226
Indisputable forms of inference 158, 229
– and consistency proofs 200
– and the constructivist point of view 228, 250
– and the proof of finiteness 197
– and the reduction procedure 197
Induction (cf. also: complete induction; transfinite induction)
– axiom 309
– proposition 145
– step 145
 degree of an – proposition 311
Inference 69
– figure 72, 76, 82, 83, 255, 289
– figure schema 77, 83, 84, 117, 125, 129, 255, 256, 290, 291
 D- – figure 73
 elementary – 145
 immediate – (cf. also: thinning) 31
 line of – 255
 operational – figure 82
 operational – figure schema 84, 256
 predicate – figure 107
 propositional – figure 107
 rule of – 5, 151–155, 159, 216
 structural – figure 82, 85, 86
 structural – figure schema 83, 84, 256
 TJ- – figure 291
Inferences in Euclid's proof 145 seq.
Infinite
– domain of objects 222
– sentence system 29, 40
 analysis and – sets of objects 142, 224
 reduction rule for sequents over – domains 176
Infinity 3, 223 seq., 234, 237
– and geometry 224
– and the antinomies of set theory 234
 actualist interpretation of – 18, 224 seq., 235, 247
 axiom of – 12, 13, 214, 222, 240, 241
 axiom of – and consistency proofs 222, 240
 completed – 225, 230, 245
 completed – and mathematical existence 248

constructivist interpretation of – 224 seq., 237, 286
intuitionist and finitist uses of – 11
intuitionist interpretation of – 235
levels of – 3, 223
potential – 160, 162, 195, 196
Informal
– arithmetic 55
– completeness of the forms of inference 34
– correctness of the forms of inference 33
– sense of a formula 141
Initial
– formula 73, 74, 76
– sentence 35
– sequent 83
Integer 142
Interchange 84, 129, 151, 256
– in the antecedent 84, 151
– in the succedent 84, 129
Interdependence of the logical connectives 171
Introduction of logical connectives 5, 80, 82, 148, 150, 168, 258, 262, 263, 269, 293
Intuitionism
fundamental principle of – 225, 235
Intuitionist 3, 10, 135, 169, 227, 235, 250
– analysis 18, 244 seq.
– arithmetic 4, 52, 54
– concepts as ideal elements 250
– consistency of classical arithmetic 4
– delimitation of the forms of inference 135, 163
– interpretation of infinity 235
– logic 4–6, 53, 58, 66, 68, 167, 168
– methods in proof theory 10
– objections to consistency proofs 198, 200
– propositional logic and the decision problem 6, 69, 103
– propositional logic and the *Hauptsatz* 103
– use of infinity 11
– view of the connectives $\supset$ and $\neg$ 167
– view of the elimination of double negation 168, 169
– view of the law of the excluded middle 3, 169
consistency of – arithmetic and Gödel's theorem 67
consistency of – predicate logic 103
equivalence of the – calculi 116 seq.

Irrational number 243, 244
IS 83

J 293

L 293
Law
– of contradiction 79, 86, 154
– of double negation (cf. also: elimination of double negation) 53, 55, 57
– of the excluded middle 3, 6, 14, 16, 53, 68, 69, 75, 81, 82, 85, 86, 105, 154, 169, 170, 226, 259
Left
– rank of a derivation 89
– side of a cluster 267
Level
– and degree 276
– of a sequent 15, 270, 303
–s of infinity 3, 223
–s of mathematics 3, 223, 233
LHJ 69, 116
LHK 69, 116, 117
Line of inference 255
Linear
– order of proofs 231
– sentence 30
– sentence system 2, 39
LJ 69, 81, 82, 86, 105, 106, 115, 116, 123
– – derivation 83
proof of the *Hauptsatz* for – 101 seq.
LK 4, 8, 13, 16, 69, 81, 86, 106, 110, 112, 115, 116, 128, 252, 314
duality of – 86, 259
proof of the *Hauptsatz* for – 88 seq., 106
Logic 3, 68, 314
classical – 4–6, 53, 66, 68, 103
intuitionist – 4–6, 53, 58, 66, 68, 167, 168
modal – 5–7
predicate – 53, 66, 68, 216, 309, 314
propositional – 30, 33, 216, 314
Logical
– axioms 315
– connective 140, 141, 143, 148
– consequence and derivability 2, 3, 24, 33
– inferences and the antinomies 133
– symbol 54, 70, 313
basic – sequent 151, 154, 177, 179, 257, 290
elimination of – connectives 5, 80, 82, 148, 150, 168, 258, 259, 262, 263, 269, 293
equivalence of – calculi 69, 115, 116, 128, 131

formalization of – deduction 68
interdependence of the – connectives 171
introduction of – connectives 5, 80, 82, 148, 150, 168, 258, 262, 263, 269, 293
scope of a – symbol 54
Logicism 135, 237
Logistic calculus 5, 75, 82
Lower
 – formula 72
 – sentence 31, 32
 – sequent 72, 255
L-system 4–7, 25

𝔐 90
M 293
Major premiss 184
Mantissa of an ordinal number 187
Mathematical
 – axiom (cf. also: axiom of elementary number theory) 151, 287, 290
 basic – sequent 151, 177, 257, 260, 290
Mathematics
 – over large finite domains 160
 applied – 249
 levels of – 3, 223, 233
Maximal
 – complexity 262
 – net 41, 312
Measure
 – of complexity 11, 12, 270, 311
 ordinal numbers as –s of complexity 186
Meta
 – logic 238 seq.
 –mathematics (cf. also: proof theory) 5, 11, 15, 18, 21, 27, 137, 238 seq.
 –theorem (cf. theorem in proof theory)
Method (cf. also: axiomatic method)
 – of infinite descent 155
 –s of proof 238
Midsequent 6, 106, 108
 properties of the derivation of the – 114
Minimal
 – formula 142, 159
 – term 142
Mix 88
 – formula 88, 207
 – sequent 207
Modal logic 5–7
Model 7, 136, 224, 241, 243

Natural
 – calculus and consistency proofs 252

 – deduction 2, 4, 7, 25, 74, 75
 – geometry and constructive analysis 18
 – geometry and physics 248
 – intuitionist derivations 75
 – numbers 53, 136, 142, 223, 230
 – numbers and Skolem's theorem 243
 – numbers and the antinomies of set theory 133, 136
 – numbers as ordinal numbers 232, 284
 – proof 263
 – succession of inferences 293
 – sum of ordinal numbers 278
 abandonment of the – concept of a sequent 255
 calculus of – deduction 2, 7, 68, 81
 insufficiency of the – numbers as ordinal numbers in proof theory 232
Negation (cf. also: ¬) 66, 67, 70, 155
Net
 – element 41
 – sentence 41
 maximal – 41, 312
Nest of intervals 245
New 287 seq.
 functions in – 253, 288
NJ 68, 69, 75, 82, 115, 116,
 – -derivation 75
NK 4, 68, 69, 81, 82, 116, 128, 314
 – -derivation 81
Nonconstructive mathematics 18
Nondenumerability 229
 – and consistency proofs 241
Nonderivability of the law of the excluded middle in LJ 105
Nonequality and negation 67
Nonexistence of a largest prime number 144
Nonstandard model (cf. discussion of Skolem's theorem on p. 243)
Normal form 312
 – of a proof 5, 35, 69, 312
 – of complete induction 145
 – theorem (cf. Hauptsatz; sharpened Hauptsatz)
N-system 4, 5
Number
 – theory (cf. arithmetic; elementary – theory)
 constructivist concept of a – 244
 finitist interpretation of – -theoretical axioms 164
 terminal – 297, 299
Numeral 54, 141, 200, 253

Numerical
- term 253, 262, 288, 291
- value 181
- variable 288

Object 69
- variable 70, 111, 140, 314
Objects
- of analysis 223, 224
- of finite mathematics 158
- of proof theory 138, 194, 211, 228
infinite sets of – 142, 224
Omission (cf. also: contraction in the antecedent)
- of an antecedent formula 151
Operational
- inference figure 82, 255
- inference figure schemata 84, 256
- reduction 262 seq., 303
- rules of inference 5, 55, 57
Option (cf. choice)
Order
- of a derivation 107, 108
- of a sequent (cf. ordinal number) 15
- type 285
Ordinal number 3, 27, 107, 187, 230, 261, 287, 288, 298, 315
- of a derivation 11, 187–189, 279
- of the endsequent 279
- as measure of complexity 186
–s and *TJ*-derivations 298
accessible – 192, 195
characteristic of an – 187
correlation of –s with derivations 11, 16, 187, 188, 261, 279
mantissa of an – 187
natural numbers as –s 232, 284
natural sum of –s 278
transfinite – 11, 26, 187, 229, 230, 261, 277 seq., 288

Partialaussage 164
Path 73, 258
Peano axioms 53, 136, 314
Philosophy
- and Hilbert's programme 21, 237
- and the mathematician 234
Physics 21, 227, 249, 250
- and classical analysis 18, 247
- and natural geometry 248
Potential infinity 160, 162, 195, 196
Predecessor 27, 110, 111

Predicate 69, 156, 158, 194, 198, 211, 287, 288
- in elementary number theory 143, 156
- inference figure 107
- logic 53, 66, 68, 216, 309, 314
–s in finite mathematics 158
- symbol 111, 140, 141, 253, 288
- variable 289, 292
classical – logic 4–6, 66
consistency of – logic 14, 103, 214
decidability of the – calculus and Fermat's last theorem 239
decidable – 160, 174, 194, 199, 211, 288
decision procedure for – logic 239
decision rule for –s 160, 198
intuitionist – logic 4–6, 66
introduction of – symbols in logical calculi 111
Premiss 30, 181
major – 184
Prime 140
- formula (cf. also: minimal formula) 253, 289, 304
- number 144
Principal formula 87, 256, 304
Principle of duality 137
Principia Mathematica (cf. Russell-Whitehead)
Procedure (cf. calculation –; decision –)
Projective geometry 137, 247
- and ideal points 247, 250
Proof (cf. also: derivation) 31, 35, 69, 136, 138, 144, 216, 228, 309, 312
- figure 66, 72
- of a sentence 30
- of consistency (cf. consistency –)
- of finiteness and the indisputable forms of inference 193, 195, 197
- of the *Hauptsatz* 88 seq., 101 seq., 317
- of transfinite induction (cf. also: *TJ*-derivation) 195, 232, 285–287
complexity of a – and complete induction 311
complexity of a – and the ordinal numbers 11, 261
direct – 165
formalization of –s 137, 138, 228, 258
indirect existence – 226, 235
indirect – (cf. also: reductio (ad absurdum)) 169, 250
methods of – 238
natural – 263

INDEX OF SUBJECTS

normal form of a – 5, 35, 69, 312
purely logical – 69
simplification of a – 11, 186, 191, 262, 274, 275
well-ordering of –s 12, 231
Proof theory 137, 138, 144, 200, 211, 222, 227, 228
  constructivist techniques in – 228, 237, 239
  fundamental principle of – 162
  intuitionist methods in – 10
  objects of – 194, 211, 228
Proposition 69
  –s and formulae 140, 141
  decidable – 159
  definite – 142
  existential – 201, 226
  formalization of –s 69, 70, 139, 140
  ideal – 247
  indefinite – 140, 142
  sentence as a – 30
  transfinite – 21, 161, 162, 195
  true – 54, 217
Propositional
  – inference figure 107
  – logic 30, 33, 216, 314
  – variable 54, 70
  consistency of – logic 214
Provability 32, 33, 53, 65, 312
  – and Fermat's last theorem 166
  – and Gödel's incompleteness theorem 240
  – and logical consequence 2, 3, 24, 33
  necessary and sufficient conditions for – 8

Quantifier (cf. also: E-symbol; $\exists$-symbol; ($\mathfrak{X}$)-symbol; $\forall$-symbol) 4, 54, 70, 77
  existential – 4, 54, 70, 77
  universal – 4, 24, 54, 70, 313

Ramified
  – analysis (cf. also: analysis) 12, 14, 15
  – theory of types (cf. also: simple theory of types) 13, 135, 214 seq.
  consistency of the – theory of types 13
Rank of a derivation 90
  left – 89
  right – 90
Real
  – function 243, 245, 246
  – number 14, 136, 224, 241, 243–246
Realism (cf. also: actualist interpretation) 19

Recursive definition 27, 53
Reduced
  – form of a sequent 11, 26, 174 (13.4) 207 seq.
  – sequent 104
Reducibility
  – of an endsequent 317
  – of a derived sequent 285
  – of a formula 262
  axiom of – 13
Reductio (ad absurdum) (cf. also: indirect proof) 79, 153, 168, 169, 181, 204, 206
Reduction
  – of derived sequents 179 seq.
  – of true sequents 175
  – procedure 17
  – procedure and incompleteness 17
  – procedure and the indisputable forms of inference 197
  – rule 11, 17, 173, 175 seq., 199, 202–213
  – step 173 seq., 179, 181 seq., 189, 261, 263, 264, 272, 274, 297, 299
  – step and complete induction 180, 183
  critical – step 300, 303, 304
  extension of the – procedure 17, 198
  finiteness of the – procedure 186, 191, 193
  operational – 262–264, 266, 268 seq., 303
  stability of a – rule 173, 175 seq., 196, 204, 206
  upper bound for the number of – steps 197
Reinterpretation of theorems and proofs 223
Relative extremum 263
Relativity
  – of the concept of a set 241
  Skolem's theorem of – 241
Replacement
  – in a schema 76
  – of bound variables 57, 152
  – of free variables 57, 113
  – of propositional variables 57
Restricted
  – transfinite induction (cf. also: transfinite induction) 8, 10, 12, 15–18, 26, 27, 287 seq.
  – transfinite induction and incompleteness 16, 17
  – transfinite induction and reduction rules 17
Restriction
  – on LJ-inference figures 6, 83

–s on variables 58, 77, 84, 117, 153, 216, 255
Right
 – rank of a derivation 90
 – side of a cluster 267
Rules
 – for $\forall$ and $\exists$ in the simple theory of types 216
 – of inference 5, 151–155, 159, 216
 – of inference and the 'chain rule' 180 seq.
 – of replacement 57, 290
 completeness of the – of inference 154
 correctness of the – of inference 159
 operational – 5, 55, 57
Russell's antinomy 3, 133, 162, 214

$\mathfrak{S}$ 46
$\overline{\mathfrak{S}}$ 46, 48
S 293
Satisfaction 33
Schema
 – for basic formulae of LHJ 116, 117
 – for CJ-inference figures 256
 – for TJ-inference figures 291, 298
 inference figure – 77, 83, 84, 117, 125, 129, 255, 256, 290, 291
 replacement in a – 76
Scope of a logical symbol 54
Second number class 10, 16, 26, 230, 233, 286, 307, 308
Semantic tableaux 7
Semi-net sentence 42
Sense
 – of actualist propositions 21, 195, 201, 226, 229, 247, 250
 – of directly proved propositions 165
 – of ideal propositions 247, 248
 – of indirectly proved existential propositions 201, 226
 – of the results established in consistency proofs 200
 – of transfinite propositions 161, 162
 informal – of a formula 141
Sentence 2, 29, 30, 33, 312
 degree of a – 43
 initial – 35
 lower – 31
 net – 41
 semi-net – 42
 tautologous – 30
 trivial – 30
 upper – 31, 32

Sentence system 1, 2, 29, 38, 39, 40
 closed – 38, 39
 finite – 39
 infinite – 29, 40
 linear – 2, 39
Sequent 2, 71, 82, 150, 151, 252, 254, 255, 289, 314
 – calculus 5
 –s and formulae 72
 basic equality – 290, 300
 basic logical – 151, 154, 177, 179, 257, 290
 basic mathematical – 151, 177, 257, 260, 290
 basic – 13, 83, 123, 151, 257, 290
 empty – 16, 72, 103, 255, 261
 equivalence of –s 115
 false – 254, 269
 forms of inference for –s 150, 151, 255, 256
 initial – 83
 level of a – 15, 270, 303
 lower – 72, 255
 mix – 207
 order of a – (cf. also: ordinal number) 15
 reduced – 104
 reduced form of a – 11, 26, 174 (13.4), 207 seq.
 reducibility of an end– 317
 symmetrization of –s 259
 true – 174, 175, 254, 269, 290
 upper – 72, 255, 291
 uppermost – 257
Set 133, 215, 216
 – of all sets 224, 225
 – theory (cf. also: general – theory; axiomatic – theory) 137
 – theory and Skolem's theorem 241
 actualist concept of a – 134
 antinomies of – theory (cf. antinomies)
 axiomatic – theory 18, 235
 axiom of – formation 216
 choice – 217, 222
 consistency of – theory 136, 232, 240, 241
 constructivist-concept of a – 134
 constructivist principle of – formation 225
 general – theory 3, 224 seq.
 model for – theory 241
 relativity of the concept of a – 241
S-formula 72
 formally identical –e 72

Sharpened *Hauptsatz* (cf. also: Herbrand-Gentzen theorem) 6–8, 15, 25, 106, 110, 112, 313
– and modal logic 6
Side formula 87
Simple theory of types 13, 14, 214
Simplification of a proof 11, 186, 191, 262, 263, 274, 275, 284
Skolem's theorem 15, 241–243
– and constructivism 241
– and denumerability 241, 247
– and elementary number theory 243
– and Gödel's theorem 242
– and the actualist interpretation 241
𝔖q 108
Statability of a reduction rule 173, 175 seq., 196, 204, 206
– and consistency 11, 177, 211, 213
– and the actualist interpretation 173
– and the finitist interpretation 173, 201, 206
– and the truth of a sequent 11, 173, 175–177, 199, 206, 207
Strengthened *Hauptsatz* (cf. sharpened *Hauptsatz*)
Structural
– inference figure 82, 85, 86
– inference figure schemata 83, 84
– rules of inference 5
– transformation 151, 180
– transformations and the 'chain rule' 180
Subcomplex and the assignment of truth-values 33
Subformula 71, 159
– property 6, 87
– property and the endsequent 6
Substitution of terms 299
Succedent 2, 30, 72
– formula 151, 254
empty – 72, 82, 112
interchange in the – 84, 129
Successor function 253
Syllogism (cf. also: cut) 2, 32, 312
Symbol 53, 54, 69, 70, 140, 141, 313
–s as objects of proof theory 138, 194, 211, 228
–s for definite functions 70
–s for definite predicates 70
–s for definite propositions 70
auxiliary – 70, 71, 217
constant – 70, 111

function – 140, 141, 288
logical – (cf. also: logical connective) 54, 70, 313
predicate – 111, 140, 141, 253, 288
terminal – 71
variable – 70
Symmetrization of sequents 259
Symmetry
– between the logical connectives 259
– of a formalism and the operation reduction 269
– of the schemata 86
Syntactic variable 54, 70, 90, 142, 152, 214, 255

𝔗 46, 48, 50
Tautologous sentence 30
Techniques of proof 139, 143, 237, 242
– and incompleteness 308
finitist – 171, 193, 194, 210, 212, 238, 239
Term (cf. also: indefinite number) 54, 141, 159, 253, 289, 314
evaluation of –s 159
minimal – 142
numerical – 253, 262, 288, 291
substitution of –s 299
variable – 253
value of a – 159
Terminal
– connective 254
– number 297, 299
– symbol of a formula 71
Tertium non datur (cf. law of the excluded middle)
The . . . such that 157
Theorem 69
– in a formal system 69, 70
– in proof theory 137, 238 seq.
– of transfinite induction 192, 193, 195, 253, 285, 315
constructivist proof of the – of transfinite induction 285, 286
Thinning (cf. also: immediate inference) 30, 83, 255
– and the syllogism 32
– in the antecedent 83
– in the succedent 83
TJ-derivation 287, 291–293
– and the ordinal numbers 298
reduction step on –s 299
TJ-inference figure 291

TJ-upper sequent 291
Transfinite
— assumptions 165
— formula 142
— induction (cf. also: restricted — induction) 231, 232, 284 seq.
— induction and Brouwer's proof of the uniform continuity of functions 246
— induction and elementary number theory 292, 397
— induction and Gödel's theorem 16, 239
— induction and the consistency proof 8, 231, 232, 261, 286
— induction theorem (cf. theorem of — induction)
— ordinal numbers 11, 26, 187, 229, 230, 261
— propositions 21, 161, 162, 195
— propositions and decision rules 161
— propositions and the elimination of the double negation 169
actualist sense of — propositions 162
finitist sense of — propositions 162
formalization of — induction 291
proof of — induction 285
sense of — propositions 161, 162
Transformation
— of a proof (derivation) 60, 117 seq., 217 seq., 293
structural — 151, 180
structural —s and the 'chain rule' 180
Tree form
derivations in — 73, 258
Trivial sentence 30
True
— formula 55, 56, 74, 114
— proposition 54, 217
— sequent 174, 175, 254, 269, 290
Truth
— of a formula 11, 56, 82, 114, 158, 159, 254
— of a minimal formula 159
— of a proposition in finite mathematics 158
— of a sequent 159, 176, 293

— of a sequent and the statability of a reduction rule 11, 173, 175–177, 199, 206, 207
conditional — 82
Truth-values 33, 159, 161, 218
Type 13, 215, 315

Universal quantifier (cf. also: ($\mathfrak{x}$)-symbol; $\forall$-symbol) 4, 24, 313
Upper
— formula 72
— sentence 31, 32
— sequent 72, 255, 291
Uppermost
— formula of a cluster 267
— sequent 257, 291
Urteilsabstrakt 165

Value
— of a derivation 298
— of a term 159
correlation of —s 305
numerical — 181
Variable
—s in elementary number theory 141
—s of different type 215
— term 253
bound — 55, 141, 144, 215, 288
bound predicate — 297
free — 55, 141, 144, 215, 288
numerical — 288
object — 70, 111, 140, 314
predicate — 289, 292
propositional — 54, 70
replacement of —s 57, 76, 113, 152
restrictions on —s 58, 77, 84, 117, 153, 216 255
syntactic — 54, 70, 90, 142, 152, 214, 255
Verschärfter *Hauptsatz* (cf. sharpened *Hauptsatz*)

Well-ordering of proofs 12, 231
Weyl's concept of a number 244

Zermelo-Fraenkel axioms for set theory 240